Smart Sensors, Measurement and Instrumentation

Volume 40

Series Editor

Subhas Chandra Mukhopadhyay, School of Engineering, Macquarie University, Sydney, NSW, Australia

More information about this series at http://www.springer.com/series/10617

Nabil Derbel · Olfa Kanoun
Editors

Advanced Methods for Human Biometrics

Springer

Editors
Nabil Derbel
Department of Electrical Engineering
Engineering High School of Sfax
University of Sfax
Sfax, Tunisia

Olfa Kanoun
Chair for Measurement and Sensor Technology
Department of Electrical Engineering and Information Technology
Chemnitz University of Technology
Chemnitz, Germany

ISSN 2194-8402 ISSN 2194-8410 (electronic)
Smart Sensors, Measurement and Instrumentation
ISBN 978-3-030-81981-1 ISBN 978-3-030-81982-8 (eBook)
https://doi.org/10.1007/978-3-030-81982-8

This Springer imprint is published by the registered company Springer Nature Switzerland AG
The registered company address is: Gewerbestrasse 11, 6330 Cham, Switzerland

Preface

This book entitled *Advances Methods for Human Biometrics* reports on recent developments in the field of biomedical authentication by manifold methods based on biological signals, image processing as well as measurements of human characteristics such as fingerprints and iris. With its content, the book covers a wide range of technologies and recognition techniques including advanced biomedical approaches and algorithm based methods. The intent is to provide insight in corresponding theories, applications, and perspectives of the state of the art and future developments relevant to the field of biomedical engineering for the field of human biometrics, as well as the paradigms and methodologies behind them. This book is a good reference for graduate students, researchers, educators, engineers and scientists and includes thirteen chapters structured into four parts as follows.

Part I of this book focuses on recognition methods based on measurements of human characteristics and comprises three chapters treating advances in analysis methods of fingerprints and iris-based biometry.

- Chapter 1 entitled "Efficient Fingerprint Analysis Based on Sweat Pore Map" investigates analysis methods of fingerprint recognition.
- Chapter 2 entitled "Fingerprint Recognition Based on Level Three Features" contains advanced studies in the recognition of person by fingerprints.
- Chapter 3 is entitled "Fractal Analysis for Iris Multimodal Biometry" which is a novel technology that ensures a high level of security for the automatic recognition of individuals based on their iris characteristics.

Part II is dedicated to the authentication by biological Signals and comprises three chapters reporting about advances in ECG and Surface EMG based biometric authentication.

- Chapter 4 entitled "Security with ECG Biometrics" describes the advantages of ECG as biometrics and provides a general overview of ECG as a biometrics technology by addressing algorithms, filters and classifiers.
- Chapter 5 entitled "ECG Biometric System for Human Recognition Based on the Possibility Theory" uses the possibility theory for the recognition by an ECG biometric system.

- Chapter 6 entitled "Surface EMG Based Biometric Person Authentication by a Grasshopper Optimized SVM Algorithm" presents a biometric person authentication based on the use of Electromyography (EMG) signals measured at the surface of a human body.

Part III presents algorithm-based methods of multimodal authentication and comprises four chapters.

- Chapter 7 entitled "Tracklet and Signature Representation Using Part Appearance Mixture Approach in the Context of Multi-shot Person Re-Identification" is dedicated to the recognition of persons in a video surveillance scene, showing an increasing interest
- Chapter 8 is entitled "A Novel Approach for Speaker Recognition in Degraded Conditions". It concerns the human voice as an ideal data source for identifying people in several applications, caused by the increasing needs for security in public locals.
- Chapter 9 entitled "Visual Methods for Sign Language Recognition: A Modality-Based Review" is based on recent advances in human actions recognition, exploiting the ascension of GPU-based learning from massive data, and getting closer to human-like performances.
- Chapter 10 entitled "A Software Architecture for Developing Disease Registries" proposes a novel software architecture for developing disease registries taking into account several aspects of disease registries.

The fourth and last part concerns biomedical characteristics for person authentication and comprises four chapters.

- Chapter 11 entitled "3D Textures Analysis Based on Features Extraction" uses the Decimal Descriptor Patterns operator for analyzing textures. In fact, it presents an excellent description of the spatial structure of image texture, based on structural analysis of pattern distributions.
- Chapter 12 is entitled "Image Processing and Analysis for Decision Making Applied to Melanoma". In fact, malignant melanoma is a terrible disorders causing death. The object of the modern dermatology is the early screening of skin cancer, in order to reduce the mortality rate.
- Chapter 13 is entitled "Biomedical Computer Aided Design Systems: Application to Alzheimer Disease". In fact, Alzheimer's disease is very frequent among elderly people. In the literature, there are extensive researches toward the early detection and diagnosis of this problem. This chapter proposes a new Computer-Aided Design system to diagnose Alzheimer's by analyzing Magnetic Resonance Images of the brain.

Sfax, Tunisia — Nabil Derbel
Chemnitz, Germany — Olfa Kanoun
Mai 2021

Contents

Part I
Authentication Based on Measurements of Human Characteristics

Chapter 1
Efficient Fingerprint Analysis Based on Sweat Pore Map

Samia Barhoumi, Houda Khmila, and Imene Khanfir Kallel

Abstract Sweat pores in fingerprints are more accurate for person's discrimination than the general ridge flows and pattern configurations or minutiae details (associated with local ridge and furrow structure). However, to use them efficiently in biometrics for Fingerprint Identification Systems (AFIS), they need to be correctly extracted and characterized. The approach proposed in this chapter follows a conventional sequence of steps, namely, a pre-processing phase, a characterization phase, a matching phase and, finally, a decision phase. The proposed contribution originality rather lies in the details and techniques used to achieve each phase. Sweat pores are tracked on ridges after their isolation based only on their gray levels. We characterize them by two different signatures related to their local features: positions and orientations. Based on these features, three signatures other than position and orientation are extracted. In fact, each pore has a set of neighbors that have to be not changed in number, position and orientation with respect to the concerned pore. For the matching process, we employ a strategy that brings similar fingers and repels different fingers preparing, then, scores for the decision phase. The global signature, created by concatenation of only efficient vectors, is then assessed. Identification results of the global signature, show high performances even reaching the highest level.

Keywords Sweat pore detection · Pore registration · Fingerprint level 3 features · Features extraction · Identification

---

S. Barhoumi (✉)
University of Sfax, Cybernics Team—ISBS & CEM Lab, ENIS, Sfax, Tunisia
e-mail: semya.barhoumi@gmail.com

H. Khmila
University of Sfax, Cybernics Team—ISBS & CEM Lab, ENIS & University of Gabes, ENIG, Sfax, Tunisia
e-mail: houdakhmila@gmail.com

I. K. Kallel
University of Sfax, Cybernics Team—ISBS, & CEM Lab, ENIS, Sfax, Tunisia
e-mail: imen.khanfir.kallel@gmail.com

iTi, IMT Atlantique, Technopole Brest France, Plouzané, France

N. Derbel and O. Kanoun (eds.), *Advanced Methods for Human Biometrics*, Smart Sensors, Measurement and Instrumentation 40,
https://doi.org/10.1007/978-3-030-81982-8_1

1.1 Introduction

Identification via fingerprints is a widely studied subject. Roughly speaking, details located in fingerprints can be generally categorized under three levels (Jain et al., 2004); namely (i) the first one includes general ridge flow and pattern configuration, (ii) the second one consists of minutiae details associated with the local ridge and furrow structure, including formations, defined as a ridge ending, bifurcation, dot, or combinations thereof and (iii) the third one takes into account all dimensional attributes of a ridge, such as ridge path deviation, shape, width, edge contour, pores, incipient ridges, creases, breaks, scars and other permanent details. Level 1s details, are not sufficient for individualization, but can be used for exclusion. The relationship between level 2 details enables individualization. Level 3 involves microscopic details such as width, shape and position of sweat pores, con-tours, folds and other details. Information of this level is unique and can be used adequately but it requires high-resolution processed images (at least the resolution of 1000 dpi) (Bennet and Perumal, 2011; Zhao et al., 2009).

Therefore, the use of this level remained restricted for a long time. Nevertheless, level 3 becomes increasingly used for individual's identification and the most used detail is pores. This is due to the ad-vantages of using pores on AFIS. In fact, pores are one of the most reliable features in fingerprint images. They are distinguished by their stability over time, as well as their uniqueness (Pyo et al., 2016; Zhao et al., 2010c). In addition, pores are ample on fingerprints: 1400 pores can be found in a fingerprint image while from 20 to 40 pores are sufficient for identification (Chen and Jain, 2007; Malathi and Meena, 2010).

In this context, this chapter fits. Pores are first extracted from the fingerprint images by using an approach based on the concept of regional maxima. After pore detection, a novel method is proposed to characterize pores based on their local features. Once all these features are extracted, the matching process is achieved and the results are evaluated. The proposed contribution consists of two points: the first one concerns the information relative to the extracted pores: positions, orientations, number of neighbors, distance between the extracted pore and each one of his neighbors and finally the relative orientation of each neighbor to the concerned pore; and the second one concerns the matching process where only reliable pores are used. Moreover, the approach allowing to measure this reliability is also described.

This chapter is organized as follows: in Sect. 1.2, some surveys related to sweat pore extraction are evoked. In Sect. 1.3, the proposed strategy is detailed. Section 1.4 is dedicated to the performance evaluation and to results discussion. The last section provides a conclusion and some perspectives for this work.

1.2 Related Works

Researchers have evinced the power of fingerprint sweat pores in identifying individuals and even in detecting aliveness or alteration. Then, there have been a number of published researches of pore detection and corresponding feature extraction, mainly in order to increase the recognition accuracy. Jain et al. (2006) proposed a hierarchical matching system using features at all the three levels extracted from 1,000 ppi fingerprint scans. The used level-3 features (which consist of pores and ridge contours) were automatically extracted using Gabor filters and wavelets transform. The matching process was locally achieved using the Iterative Closest Point (ICP) algorithm (Kryszczuk et al., 2004). Their experiments show that Level-3 features carry significant discriminatory information. The authors proved that there is a relative reduction of 20% in the Equal Error Rate (EER) of the matching system when Level-3 features are used in combination with Level-1 and Level-2 features.

Kryszczuk et al. (2004) combined Level-2 and Level-3 to build their own recognition system. The authors used Gabor wavelets filtering for the ridge structure enhancement, before extracting minutiae, by the use of a skeletal representation of the ridge structure. To detect the sweat pores, they binarized original grayscale images and then focused on the areas of white pixels entirely surround-ed by black pixels and being in between an upper and a lower threshold previously defined. The authors considered ridge skeleton and minutiae coordinates as Level-2 features while sweat pore coordinates were considered Level-3 features. Before comparing a small portion of a test fingerprint with the reference fingerprint, they employed an alignment procedure provided at the global level using a method correlating the ridge line structure. The match score of the ridge structure was computed by normalized the resulting correlation, while the minutiae and pore features were compared based on a geometric distance criterion. It is worth noticing that their correlation based approach is often time consuming and its accuracy is highly dependent on the way the parameter space is quantized.

Due to the development of neural networks and high-powered GPUs, some researches have used a deep network approach for the extraction of pores. Jang et al. (2017) proposed a technique based on convolutional neural networks to extract pores. They used the deep networks to detect pores in detail in a large area of high resolution fingerprint image. The refinement of the pore information is made by finding the local maxima to identify pores with different intensities in the fingerprint image. Labati et al. (2018) also utilized a neural network to extract pores in heterogeneous fingerprint images. Genovese et al. (2016) used a similar idea for pore extraction from live-scanned fingerprints. Zhang et al. (2013) have patented in 2013 their partial fingerprints recognition approach that consists in four steps. First, they extracted features including ridge orientations, valley images, minutiae, and pores from at least two fingerprint fragments. For pore extraction, an adaptive method was developed based on a dynamic anisotropic pore model that they proposed. It was built to accurately describe pores by using orientation and scale parameters (Zhao et al., 2010b). Pore Valley Descriptors (PVD) were used to represent the pore features based on their locations on ridges, their orientations and the ridge orientation field and valley structures in their neighborhoods. The second step consists on aligning the fingerprint fragments and then matching the pores and minutiae on the fingerprint fragments after applying the estimated alignment transformation. Finally, they calculated a final matching score based on a pore matching score and minutiae matching score, identifying a person based on the result of the final matching score. Bennet and Perumal (2011) proposed a hierarchical matching system using features at all the four levels (the Level-4 considers oscillated patterns) extracted from 1.000 ppi fingerprint scans. The authors automatically extracted Level-3 features, pores and ridge contours, using Gabor filters and wavelets trans-form. Feature's matching was then locally achieved using the Iterative Closest Point (ICP) algorithm and Level-4 features, being oscillated patterns including curve scanned DCT. Matching step led to measure the recognition rate using the k-nn classifier. It is worthwhile to notice that in many studies, researchers join specific features of Level-3 with those of Level-2 as described in Jain et al. (2006; Zhao et al., 2009, 2010a, 2010b). Despite methods variety and the researcher efforts to build a robust and sufficient AFIS, there still remain shortcomings in already developed systems. Indeed, the two types of features are either separately matched and then fused at the score level or consecutively matched, as minutiae is firstly considered and then the match of pores is based on the minutiae matching results to provide a unique score. In the two cases, dependency between matching minutiae and matching pores remains. Thus, accuracy and performance of pores matching is restricted by those of minutiae matching and this will probably reduce the efficiency of the discriminatory power of pores.

The approach adopted by Zhao et al. (2009) joined minutiae and pores too, but with relatively more efficient way. They proposed a direct matching pore method independently of the minutiae matching. Nevertheless, the complexity of the pore descriptor implies its dimension reducing. Nguyen and Jain (2019) proposed an end-to-end automatic pore extraction and matching system to analyze the utility of pores in latent fingerprint identification.

In order to avoid these weaknesses, this chapter proposes a novel strategy that employs only Level-3 features, specifically those related to sweat pores.

1.3 Proposed Approach

A novel strategy for directly matching fingerprint pores is proposed. It consists, first, on detecting pores based on their intensity level in the fingerprint image with consideration to a priori information such as brightness of these pores in the image and their location on ridges. Then, five types of features related to pores are extracted, namely their centroid positions and their orientations. Based on those local features of the pore and those of his neighbors three signatures, namely the number of neighbors of each pore, the distance between each extracted pore and each one of his neighbors and the relative orientation of each neighbor to the concerned pore, are thus obtained for each fingerprint image. 5 signatures are thus obtained for each fingerprint image.

The proposed method integrate the alignment process in the matching one to avoid the complicated transformations of classic methods. In the beginning of the matching step, an alignment process is performed so that only pores having acceptable correspondences with fingerprint image references ones are considered. Such pores are called "winner pores". Final matching score is only based on that of the "winner pore" allowing then person's identification. In the following, each of the proposed approach steps is detailed.

1.3.1 Step 1: Pores Detection

Sweat pores locate on ridges of the fingerprints. Pores have two types of spatial appearance: they can be closed or opened. As shown in Fig. 1.1a a closed pore appears like an isolated luminous blob on the dark ridge and an opened pore intersects, either totally or partially, with its adjacent luminous valley as seen as in 1.1b.

Based on these facts, we decided to take advantage of both spatial and colorimetric pore appearance as well as their location, in order to detect them on the fingerprint images. Pores exhibit intensity peaks in ridge dark areas. But the contrast is not enough to efficiently extract them. Therefore, pore brightness is first intensified using a Laplacian of Gaussian filter (LoG) given by:

$$LoG\,(x, y) = -\frac{1}{\pi\sigma^4}\left[1 - \frac{x^2 + y^2}{2\sigma^2}\right]\exp\left(-\frac{x^2 + y^2}{2\sigma^2}\right) \tag{1.1}$$

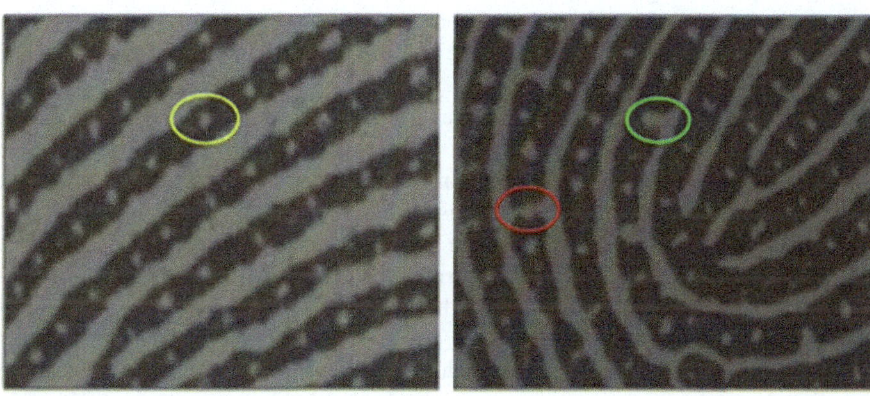

(a) (b)

Fig. 1.1 Two fingerprint images: **a** A closed pore is circled with yellow, **b** two opened pores are circled with red and green

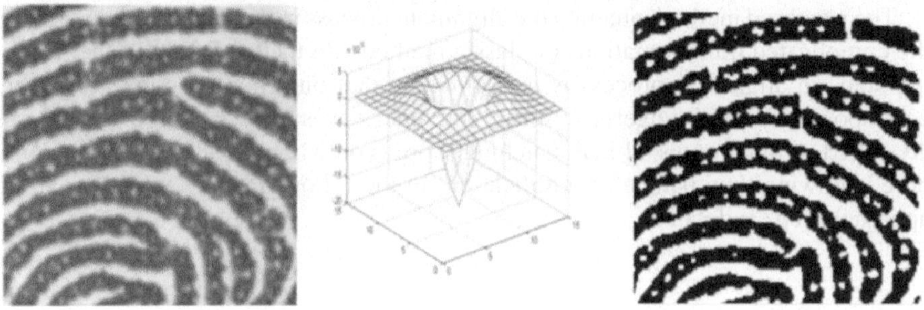

(a) (b) (c)

Fig. 1.2 *LoG* filtering : **a** Original image, **b** form of filter, **c** filtered image

where x and y are the coordinates of one pore. The filter kernel standard deviation σ is empirically adjusted depending on the image resolution (in our case it was set to 2).

The structure of this filter, shown in Fig. 1.2b is well adapted to the shape of pores and allows to enhance their brightness in the fingerprint image. Following such a filtering process, we get a fingerprint image where pore level is set to 255 as seen in Fig. 1.2c.

The only information about pores appearance is not enough to make them distinguishable from all other parts of the image. Indeed, pore intensities are similar to those of valleys. Then, to isolate pores we decide to take into account their positions in the image. A morphologic mask is built to hide valleys allowing then restricting looking for pores only on ridges. We construct such a mask by first binarizing the LoG-filtered image, then filling holes corresponding to pores and finally inversing the binary image after cleaning it by the use of some morphologic operators. The result of this operation is seen in Fig. 1.3.

Then, we look for the regional maxima on the enhanced fingerprint image. These regions are the blobs with high intensity values inside and low intensity values on their outer limits. We call pore map the resulting image. However, these blobs are not the only areas extracted. Other regions appear on the pore map: Very small areas resulting from various kinds of noise are also extracted as well as other templates, with relatively larger surface, that may correspond to injuries (see Fig. 1.4). Two size-thresholds, minimum and maximum are then used to eliminate all extracted artifacts as conducted in Zhao et al. (2010a).

To refine the extracted areas, a procedure of filling holes completes the cleaning operation to get the final pore map. An example of filled holes is shown in Fig. 1.5.

Figure 1.6 shows final pore map extracted from a fingerprint fragment.

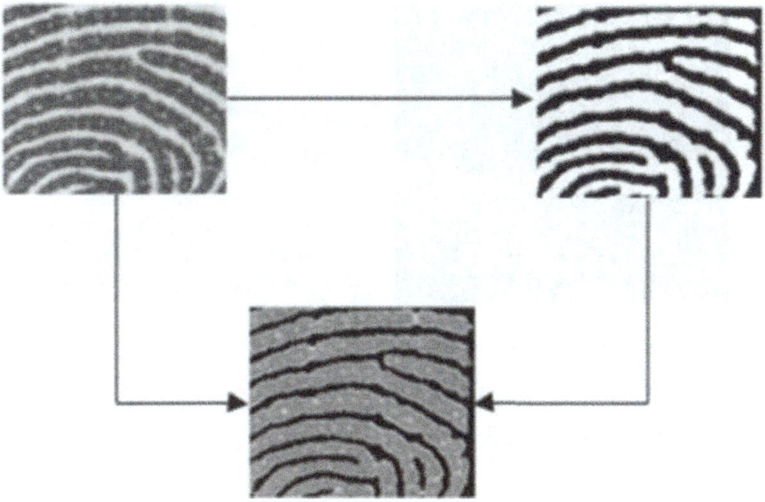

Fig. 1.3 Application of morphologic mask

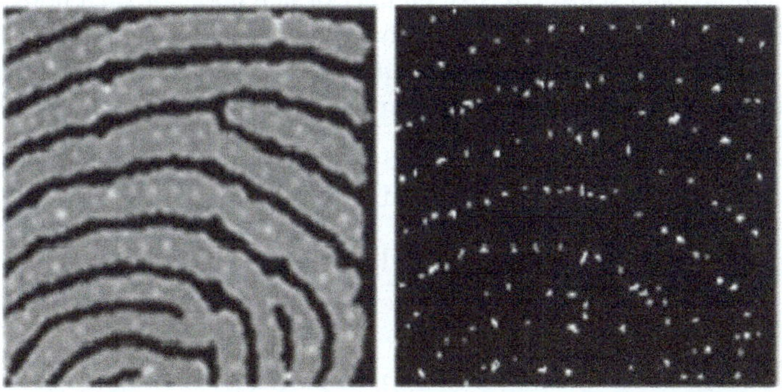

Fig. 1.4 Pore map before cleaning operation

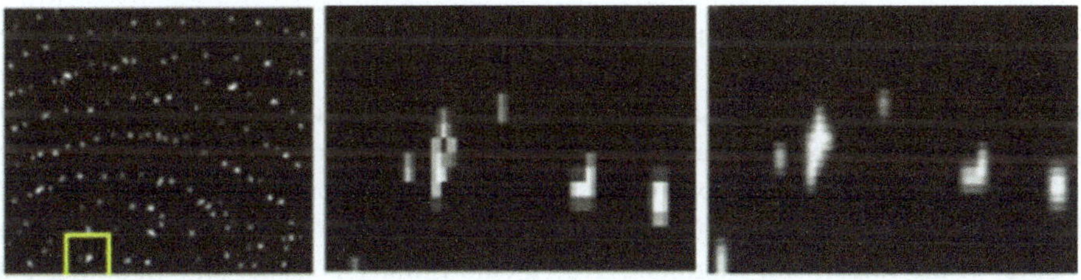

Fig. 1.5 Filling holes

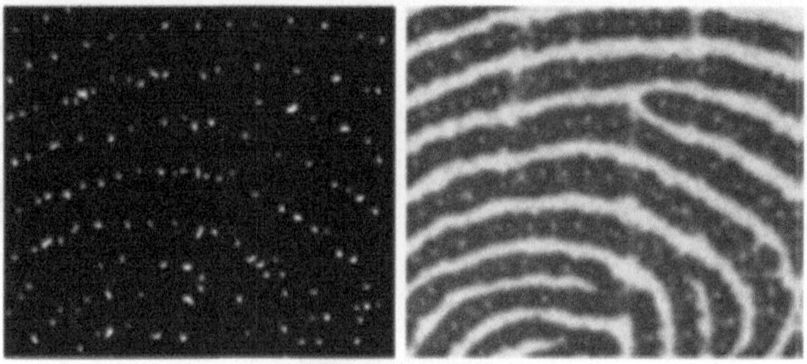

Fig. 1.6 Marking pores detected on the original image

1.3.2 Step 2: Features Extraction

Based on the constructed pore map, different types of information are used in the construction of the feature vectors: geometric (such as location of pores on ridges, their orientations) and statistical (like number of neighbors of each pore, distances between the concerned pore and his neighbors and relative orientations of the neighbors based on the concerned pore). For each pore, we associate a descriptor that contains local features of this pore. The process is as follows:

- The first characteristic is related to pore-positions. For each blob of the map, the coordinates of its centroid are computed as follows:

$$x_{blob} = \frac{1}{\sum p \in blob} \sum_{p \in blob} x_p \tag{1.2}$$

$$y_{blob} = \frac{1}{\sum p \in blob} \sum_{p \in blob} y_p \tag{1.3}$$

  Therefore, to each pore-map corresponds a position-signature that consists of two vectors of as many lines as there are pores in the map.
- The second characteristic is related to pore orientations. Considering the fingerprint image as an oriented texture, several methods have been developed to estimate the local orientation at each pixel in this image. For this work, we adopt the method of Zhao developed in Zhao et al. (2010b) to estimate the local orientation at each pore on the fingerprint fragment. This method firstly consists on smoothing the fingerprint image with a Gaussian filter. Then, we compute, on each pore centroid (x, y) of the pore map, the gradients $G_x(x, y)$ and $G_y(x, y)$ using Sobel operator respectively along x and y directions. The orientation of the pore centroid (x, y) is estimated by the following equation:

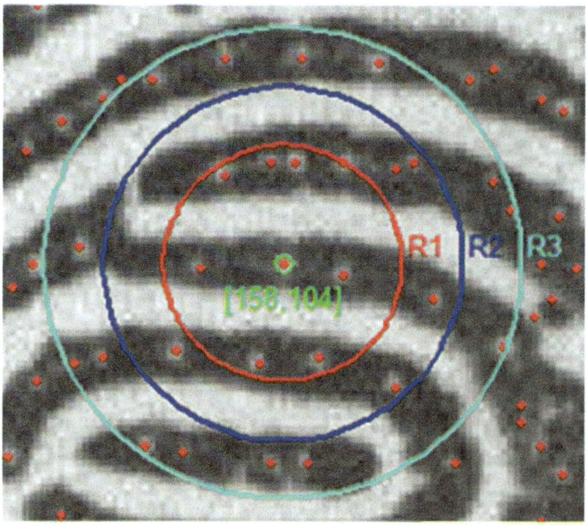

Fig. 1.7 Circles used in extraction of neighbors of pore in position [185, 104]

$$\theta(x, y) = \frac{\pi}{2} + \frac{1}{2} \arctan\left(\frac{\overline{G}_{xx} - \overline{G}_{yy}}{2\overline{G}_{xy}}\right). \tag{1.4}$$

where G_{xx}, G_{yy} and G_{xy} are obtained by the Gaussian smoothing of square gradients performed as follow:

$$\overline{G}_{xx}(x, y) = G_x(x, y) \times G_x(x, y). \tag{1.5}$$

$$\overline{G}_{yy}(x, y) = G_y(x, y) \times G_y(x, y). \tag{1.6}$$

$$\overline{G}_{xy}(x, y) = G_x(x, y) \times G_y(x, y). \tag{1.7}$$

The obtained orientation values lie in the interval $[0, \pi]$.

To extract the next features, we firstly set up three circular areas whose centers are the concerned pore and their radius are respectively $R_1 = 20pixels$, $R_2 = 30pixels$ and $R_3 = 40pixels$. Figure 1.7 shows the three circular areas used to extract the neighbors of one pore located in position [184; 104].

- In this step, we firstly extract the number of neighbors for the concerned pore in the circle having radius R_1.
- In addition, noting that the arrangements of pore vis-a-vis each other have specific organization into each fingerprint fragment, we decide to characterize them using the distance between the concerned pore and each one of their neighbors. The result affects for each pore a set of values that represent the distances between the pore and his neighbors. The Euclidean distance between two points (Fig. 1.8) is used here to calculate the distances between pores with the formulation:

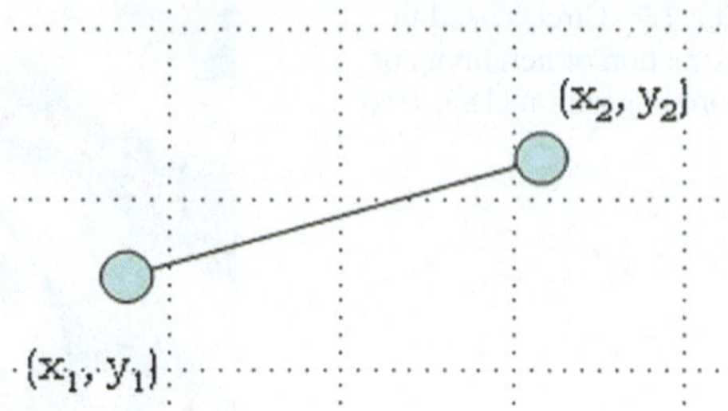

Fig. 1.8 Euclidean distance between two pores with position respectively (x_1, y_1) and (x_2, y_2)

$$d_{ij} = \sqrt{(x_i - x_j)^2 + (y_i - y_j)^2}. \tag{1.8}$$

where (x_i, y_i) and (x_j, y_j) are the coordinates of each pore. This is the third characteristic of the fingerprint image presented in one vector.

- Then, we calculate, on one vector, the relative orientation of the neighbors based on the features of the concerned pore. We use the method presented in Joyeux and Penczek (2002) to calculate the new coordinates of the neighbors of the pore serving at computing their new orientations relatives to the concerned pore. This method consist on 2D alignment transformation: translation and rotation by using the following matrix:
  - Translation matrix

$$T = \begin{bmatrix} 1 & 0 & \Delta x \\ 0 & 1 & \Delta y \\ 0 & 0 & 1 \end{bmatrix} \tag{1.9}$$

  - Rotation matrix

$$R = \begin{bmatrix} \cos\theta & -\sin\theta & 0 \\ \sin\theta & \cos\theta & 0 \\ 0 & 0 & 1 \end{bmatrix} \tag{1.10}$$

The coordinate of the pore after transformations are:

$$P' = T \times R \times \begin{bmatrix} x \\ y \\ 1 \end{bmatrix} \tag{1.11}$$

We, then, obtain a global signature that consists structures of as many lines as there are pores in the map. The structure contains the position of the pore, his orientation and the information related to the neighbors of the interested pore.

1.3.3 Step 3: Pores Alignment

During the capture step, it is practically impossible to maintain the same pose of fingerprints as well as we can be sure to guaranty the same positions of pores on the images. Many types of deformations could occur between images taken for the same finger, such as translation, rotation and nonrigid deformation, bringing to improper matching.

To remedy such alteration in fingerprints, numerous fingerprint alignment methods appeared in the literature. Some of them seek to re-move the rigid and non-rigid transformations, between two different poses, by modeling them. Such a process is often very complex and costly in time processing. Therefore, and since it has been shown that considering only translation and rotation already works well in most cases, most of research works merely compensate non rigid transformations. Hence, in the literature, the fingerprints alignment step consists on retrieving the transformation parameters between fingerprint images and removing the transformation between them.

In this chapter, we propose to do otherwise. Indeed, convinced that any change made to the image may degrade its quality and subsequently impair accuracy of the extracted parameters, we propose to seek the pairs of pores that can correspond to each other between two images, according to their characteristics, without transforming either of them. We call such pairs of pores "winner pores". In the next step, matching will only deal with these "winner pores". To find them, we proceed as follows: Let I and I' be two finger print images (reference image and candidate image). Each of them possesses a pore map: $M = p_1, p_2, \ldots, p_n$ for I and $M' = p'_1, p'_2, \ldots, p'_n$ for I'. To find accurately the correspondent couple of pores we proceed as follows, according to each of the computed features, based on appropriate metrics, detailed in the next sub-section (Table 1.1):

- For each pore p_i from M' we look for the closest pore p'_j from M',
- And for each pore p'_j from M' we look for the closest pore p_i from M,

Table 1.1 Distance matrix between two pore maps. $dist'_i = \min[d(p_i, p'_j)]_{i=1..n}$ and $dist_j = \min[d(p_i, p'_j)]_{j=1...m}$

Candidate pore-map M'

		p'_1	...	p'_j	...	p'_m	Formed pairs
Reference pore-map	p_1	$d(p_1, p'_1)$	...	$d(p_1, p'_j)$	...	$d(p_1, p'_m)$	$dist'_1$
	...	...	...	...	...	...	...
	p_i	$d(p_i, p'_1)$	...	$d(p_i, p'_j)$	...	$d(p_i, p'_m)$	$dist'_i$
	...	...	...	...	...	...	...
	p_n	$d(p_n, p'_1)$	...	$d(p_n, p'_j)$	...	$d(p_n, p'_m)$	$dist'_n$
Formed pairs		$dist_1$	...	$dist_j$	...	$dist_m$	

The pores that will be matched are those who have a number of neighbors equal to ±4 in first circular area with radius R_1. It happens that a pore p_i from the M pore-map appears to be the closest pore to more than only one pore from the M' pore-map. To avoid this problem, a set of thresholds is used. The difference between the number of neighbors of each pore must not exceed 6 for the second circular area having as radius R_2 and 10 for the third circular area having as radius R_3. The other pores that did not be selected for the matching, are considered as impostor pores two the candidate one. Finding closest pore is not sufficient to ensure that pores of a formed pair correspond to each other, they have to be sufficiently close. So, we merely retain pairs for which pore distance is less than a threshold empirically fixed in advance. Retained pores are called "winner pores". This process discards many pores, but it does not bother the matching step since it has been proven by Locard's studies in LOCARD (1912) that 20–40 pores are sufficient to generate patterns distinctive enough to identify a fingerprint. Moreover, it alleviates the matching step.

1.3.4 Step 4: Pores Matching

The matching step aims to provide a distance score between two fingerprint fragment images I and I' (reference and candidate). Let us recall that the bio-metric signature that we propose in this work consists of structures that contains sub-signatures as we have pores in the map. This sub-signatures are formed by: a position-signature (two vectors), an orientation-signature (one vector), distance of the neighbors (one vector) orientations of the neighbors respected to the concerned pore (one vector). For each pore correspondent a couple. The couple contains two vectors: one is the distances of neighbors and the other is their related orientations. To calculate the distance between the couples, one metric is adequately used: the Euclidean distance given according to the Eq. 1.8. For each sub-signature, the alignment process is first performed to generate the set of "winner pores". Then, a distance score, between the reference image I and the candidate image I' is calculated to judge whether fingers corresponding to I and I', belong (or not) to the same person.

1.4 Experiments and Performance Evaluation

1.4.1 Data Base

To assess our approach, we use the PolyU HRF database (Liu et al., 2010; Zhao et al., 2010a, 2009) established by a research team in the Biometric Research Center of the Polytechnic University of Hong Kong (Zhao et al., 2010b). Only the partial fingerprint image database DBI consisting of 1080, has been adopted. In fact, given what was asserted in 2004 in Kryszczuk et al. (2004), Morier (2004), the recognition of fingerprints takes advantage from pores and their own features significantly in small fingerprints area. The full signature file is extracted from 50 fingers with each finger having 10 samples taken in two sessions, so in totally 500 signatures from 500 fingerprint images.

1.4.2 Training and Test Process

The issue of a biometric identification system is reduced to a classic classification problem. The population to classify gathers all distances calculated between signatures extracted from each pair of fingerprint images. Two classes are considered: Genuine A and impostors I. To carry out the evaluation of the above described fingerprint identification system, the 1080 images of the considered data base, are divided into 2 data sets, namely a training set and a test set. Each one includes 540 images. During the training phase, also called off-line phase, the training images are used to:

- Set thresholds used in the alignment process to generate the set of "winner pores";
- Draw the distance distributions, according to each feature, to define appropriate thresholds, later used as decision criterion during the test phase;
- Generate signatures and store them in the system database to serve as references when matching files during the test phase.

The test set is a set of entry fingerprint images employed as candidate images.

1.4.3 Feature Matching

Each extracted feature is used for matching two fingerprints. Recall that the matching phase typically defines the similarity metric between two fingerprint images. It determines whether a given pair of images belongs to the same finger or to two different fingers, depending on the fact that the calculated similarity is superior or not to a predetermined threshold. The similarity metric is based on a correspondence concept between two fingerprint images. The distance between two feature vectors x and y, one stored and the other presented as a candidate is computed; using the Euclidean distance. We have carried out 2950 different matches between the 500 fingerprint images. Recall that in the adopted database, we have for each finger 10 samples in 2 sessions. Each session contains 5 images. The set of matches is divided into 2 classes:

- Genuine matches: they are computed between feature vectors from the same finger. The feature signature of each image is matched with all the other feature signatures of the same fingerprint images in the same session leading in all to 500 matches;
- Imposter matches: they are computed between feature vectors extracted from 2 different fingers. The feature vector of the first image of each finger on the second

Table 1.2 Distance matrix between two individuals of pores

Individual 1					
Individual 2	$d(p_1, p'_1)$	...	$d(p_1, p'_j)$	...	$d(p_1, p'_m)$
	...	...	...	...	...
	$d(p_i, p'_1)$	...	$d(p_i, p'_j)$	...	$d(p_i, p'_m)$
	...	...	...	...	...
	$d(p_n, p'_1)$	...	$d(p_n, p'_j)$	...	$d(p_n, p'_m)$

Table 1.3 Vector of the minimum values of the matrix of distances between the pores $(p'_1 \dots p'_m)$ and pores $p_i, i = 1 \dots n$

Individual 1				
$\min_{i=1 \dots n} [d(p_i, p'_1)]$	...	$\min_{i=1 \dots n} [d(p_i, p'_j)]$	...	$\min_{i=1 \dots n} [d(p_i, p'_m)]$

Table 1.4 Vector of the minimum values of the transpose of the matrix of distances between the pores $(p_1 \dots p_m)$ and pores $p_j, j = 1 \dots m$

Individual 2	$d(p_1, p'_1)$
	...
	$d(p_i, p'_1)$
	...
	$d(p_n, p'_1)$

session is matched with the feature vector of the first image of all the other fingers in the first session leading to 2450 matches;

The matching of different attributes adopted is done at the judged pores according to their positions. To achieve such judgment, we followed an original strategy to calculate the distances between the pores using common metrics. The first step of this strategy consists on calculating the Euclidean distances be-tween each pair of pores, coming from two fingers and giving the matrix "d" shown in Table 1.2. Pores $(p'_1 \dots p'_m)$ are those of the individual 1. The pores $(p_1 \dots p_n)$ are those of the individual 2.

The vector calculating the minimum of each column of the matrix "d" provides, for each pore of the finger of individul 1, the pore which is closest in the finger of the individul 2, such as shown in Tables 1.3 and 1.4.

Once the pairs of pores closest to each other are found, a threshold is imposed on the distances between them. Indeed, the fact of reporting that p_x of finger2 is the closest pore to p'_y of the finger1 is not enough to confirm that p_x is actually p'_y. It must be proved that the distance between them is very close to zero. If this is simply not zero, that is due to slight differences inevitably produced during the acquisition of the fingerprint image. We then proceeded to calculate the similarity between two individuals, as a ratio between the number of pores judged correspondents, according to the experimentally chosen threshold, and the total number of pairs of pores evaluated. The following equation illustrates this relationship:

$$similarity = \sum \frac{number\,(mindistance < threshold)}{number\,(evaluated\ pairs)} \quad (1.12)$$

The difference between the two individuals is simply the complement to 1 of the similarity calculated by the following equation:

$$distance = 1 - similarity \quad (1.13)$$

1.4.4 Performance Evaluation

In order to evaluate the performance of the proposed approach, a set of 500 images acquired from 50 fingers is used. For each one of the 50 fingers of the adopted database, two files are given. Each file represents a session containing 5 images. We used the images in the file of the second session for training and those of the file of the first session for the test. So, it has 250 images for training and 250 images for testing. The distances calculated for the two classes Genuine and Impostors are normalized between 0 and 1. To represent their distributions, we calculate their frequency of occurrence in an interval of width equals to 0.01, over the entire interval [0, 1]. Thus, a distribution of distances is actually a representation of the frequencies of occurrence, based on the distance normalized to the interval [0,1]. For best viewing, we propose to smooth each of the two classes. The distribution of the generated distances is shown in Fig. 1.9 It clearly shows the existence of two separated classes with a slight overlap.

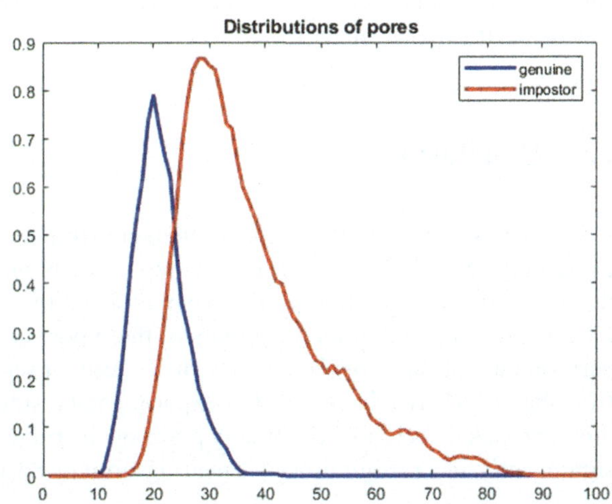

Fig. 1.9 Distributions of distances of pore signatures

Table 1.5 Range of scores obtained from the proposed approach

	EER	CAR	CRR	FAR	FRR
Score range	0.1392	0.8615	0.8608	0.1392	0.1385

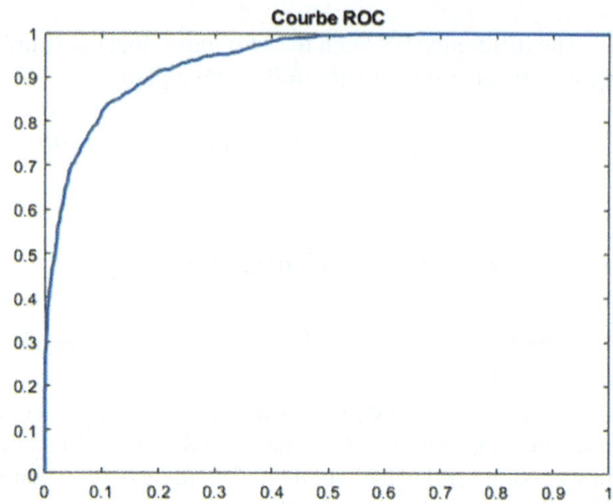

Fig. 1.10 ROC curve of the proposed approach

The Receiver Operating Characteristic (ROC) curve summarizes the matching performing by plotting the False Rejection Rate (FRR) against the False Acceptation Rate (FAR) (Fawcett, 2006).

Table 1.5 shows clearly the distinguished and satisfying scores obtained by applying the proposed approach. Figure 1.10 illustrates the ROC curve for an Equal Error Rates (EER) equal to 0.4139.

As it can be seen in the Fig. 1.10, the ROC curve rises from (0,0) to (1,1) with a surface area equal to 0.9410. This indicates a good classification performance on the test set of fingerprint images.

1.5 Conclusion

A new approach was proposed in this chapter to improve the fingerprint identification using only level 3 features. The first step consists on pores extraction. Then with the detected corresponding pores, orientations at these pores are detected and next local features related of the neighbors of these pore are determined. To evaluate the performance of the proposed approach, we used a set of partial fingerprint images from the PolyU HRF database. The experimental results demonstrated above all that this approach detect more accurately pores on the ridges of a finger. Then it improves significantly the recognition by ensuring around 86.15% for the recognition rate and a value of the area under the ROC curve close to 0.9410. In the present work, system identification of individuals by fingerprint recognition has proved especially significant efficiency and high performance. The proposed method in this chapter is designed to high resolution fingerprint images. It is not exclusive to the partial fingerprint images but rather it can be applied to the full images.

References

Bennet, D., & Perumal, S. A. (2011). Fingerprint matching using hierarchical level features. *International Journal of Computer Science Issues (IJCSI), 8*(1), 284.

Chen, Y., & Jain, A. K. (2007). Dots and incipients: Extended features for partial fingerprint matching. In *2007 Biometrics Symposium* (pp. 1–6). IEEE.

Fawcett, T. (2006). An introduction to ROC analysis. *Pattern Recognition Letters, 27*(8), 861–874.

Genovese, A., Munoz, E., Piuri, V., Scotti, F., & Sforza, G. (2016). Towards touchless pore fingerprint biometrics: A neural approach. *2016 IEEE Congress on Evolutionary Computation (CEC)* (pp. 4265–4272). IEEE.

Jain, A., Chen, Y., & Demirkus, M. (2006). Pores and ridges: Fingerprint matching using level 3 features. In *18th International Conference on Pattern Recognition (ICPR'06)* (Vol. 4, pp. 477–480). IEEE.

Jain, A. K., Ross, A., & Prabhakar, S., et al. (2004). An introduction to biometric recognition. *IEEE Transactions on Circuits and Systems for Video Technology, 14*(1).

Jang, H.-U., Kim, D., Mun, S.-M., Choi, S., & Lee, H.-K. (2017). Deeppore: Fingerprint pore extraction using deep convolutional neural networks. *IEEE Signal Processing Letters, 24*(12), 1808–1812.

Joyeux, L., & Penczek, P. A. (2002). Efficiency of 2D alignment methods. *Ultramicroscopy, 92*(2), 33–46.

Kryszczuk, K. M., Morier, P., & Drygajlo, A. (2004). Study of the distinctiveness of level 2 and level 3 features in fragmentary fingerprint comparison. In *International Workshop on Biometric Authentication* (pp. 124–133). Berlin: Springer.

Labati, R. D., Genovese, A., Muñoz, E., Piuri, V., & Scotti, F. (2018). A novel pore extraction method for heterogeneous fingerprint images using convolutional neural networks. *Pattern Recognition Letters, 113*, 58–66.

Liu, F., Zhao, Q., Zhang, L., & Zhang, D. (2010). Fingerprint pore matching based on sparse representation. In *2010 20th International Conference on Pattern Recognition* (pp. 1630–1633). IEEE.

Locard, E. (1912). Les pores et la decouverte des criminels biológica. *Lyon.*

Malathi, S., & Meena, C. (2010). A novel approach for fingerprint recognition based on pores. *International Journal of Computer Science and Research, 1*, 10–14.

Morier, P. (2004). Automatic extraction of level 2 and level 3 features for fragmentary fingerprint comparison.

Nguyen, D.-L., & Jain, A. K. (2019). End-to-end pore extraction and matching in latent fingerprints: Going beyond minutiae. arXiv:1905.11472.

Pyo, M., Lee, J., Baek, W., Lee, C. W., Park, B. J., & Kim, J.-M. (2016). Sweat pore mapping using hydrophilic polymer films. *Journal of Nanoscience and Nanotechnology, 16*(12), 12263–12267.

Zhang, D., Zhao, Q., Nan, L., & Lu, G. (2013). Partial fingerprint recognition. US Patent 8,411,913.

Zhao, Q., Liu, F., Zhang, L., & Zhang, D. (2010a). Parallel versus hierarchical fusion of extended fingerprint features. In *2010 20th International Conference on Pattern Recognition* (pp. 1132–1135). IEEE.

Zhao, Q., Zhang, D., Zhang, L., & Luo, N. (2010b). Adaptive fingerprint pore modeling and extraction. *Pattern Recognition*, *43*(8), 2833–2844.

Zhao, Q., Zhang, D., Zhang, L., & Luo, N. (2010c). High resolution partial fingerprint alignment using pore-valley descriptors. *Pattern Recognition*, *43*(3), 1050–1061.

Zhao, Q., Zhang, L., Zhang, D., & Luo, N. (2009). Direct pore matching for fingerprint recognition. In *International Conference on Biometrics* (pp. 597–606). Berlin: Springer.

Chapter 2
Fingerprint Recognition Based on Level Three Features

Houda Khmila, Samia Barhoumi, Imene Khanfir Kallel, Nadia Smaoui, and Houda Derbel

Abstract Nowadays one of the most popular topics in fingerprint recognition academic research is the high-resolution fingerprint identification way. This technique has been prominently attractive to the worldwide scientific community thanks to the possibility of using level 3 features like pores which cannot be detected in lower resolution images. In this context, this chapter proposes two contributions: First, a pore detection method for high-resolution fingerprint image based on the morphological operation (skeletonization) and the labeled connect component method. Second, a new method to match pores without an alignment. Our matching approach is based on the contextual characteristics of the pore. It consists of positions and orientations of pore neighbors, which are defined as polar coordinates given in the polar system centered on the considered pore. The proposed algorithms are tested on a high-resolution fingerprint database. Experimental results show that our methods outperform the existing algorithms.

H. Khmila (✉)
University of Sfax, Cybernics Team—ISBS & CEM Lab, ENIS, Tunisia & University of Gabes, ENIG, Sfax, Tunisia
e-mail: houdakhmila@gmail.com

S. Barhoumi
University of Sfax, Cybernics Team—ISBS & CEM Lab, ENIS, Sfax, Tunisia
e-mail: semya.barhoumi@gmail.com

I. K. Kallel
University of Sfax, Cybernics Team—ISBS, & CEM Lab, ENIS, Tunisia, Sfax, France
e-mail: imen.khanfir.kallel@gmail.com

iTi, IMT Atlantique, Technopole Brest France, Sfax, France

N. Smaoui
University of Sfax, Cybernics Team—ISBS, & CEM Lab, ENIS, & University of Gabes, ISIMG, Sfax, Tunisia
e-mail: nadia.smaoui@isimg.tn

H. Derbel
University of Sfax, CEM Lab, ENIS and Faculty of Sciences of Sfax, Sfax, Tunisia
e-mail: h.derbel@fss.rnu.tn

N. Derbel and O. Kanoun (eds.), *Advanced Methods for Human Biometrics*,
Smart Sensors, Measurement and Instrumentation 40,
https://doi.org/10.1007/978-3-030-81982-8_2

Keywords Fingerprint recognition · Pore extraction · Pores matching · High resolution · Level 3

2.1 Introduction

Over the past few years, many popular Automatic Fingerprint Identification Systems (AFIS) focused on Level 1 features (global characteristics) and Level 2 features (local characteristics) have been broadly used to classify authentic individuals from sharper (Jain et al., 2006). However, they can't achieve this with high accuracy, their approach still depends on the quality of the image to do. Hence, the use of Level 3 characteristics (e.g. sweat pores and incipient ridges) as additional matching data has attracted the attention of many researchers. As one of level 3 features, pores have been proved to be very distinctive. Thus, it has gained tremendous interest in the development of Automatic Fingerprint Identification Systems (AFIS). The main process of introducing pores into fingerprint identification includes two steps: pore extraction and matching. Accordingly; we propose in this study a novel fingerprint pore detection approach based on a skeleton image and the labeled connect component object in the image. Then, as our second contribution, we present a pore matching approach based on the number of neighbor pores, the distance between each one of them, and the candidate pore as well as their orientations. This chapter is organized as follows.

First, we will present a background of biometric systems and some properties that make them useful in the classification of authentic individuals. Second, we will extend potential and recent related works of pore detection. Then we will illustrate the proposed approach. Third, we will expose a rival state-of-the-art method for pore matching. Moreover; we will describe our proposed approach. After that, we will discuss the experimental results and analysis. Finally, a brief conclusion is presented.

2.2 Biometry Background

Biometric is a global technique aimed at establishing the identity of a person by analyzing one of its behavioral, physiological characteristics.

- Behavioral biometric: we can cite one of the most recent modalities such as eye movements, also typing on the keyboard by analyzing the speed of the pressure force on the keyboard or the movement of the arms. The signature is also a behavioral biometric seen that one can analyze pen pressure and its speed of movement. But these biometrics usually depend on the physical or moral state of the person which can change the information provided by these modalities.

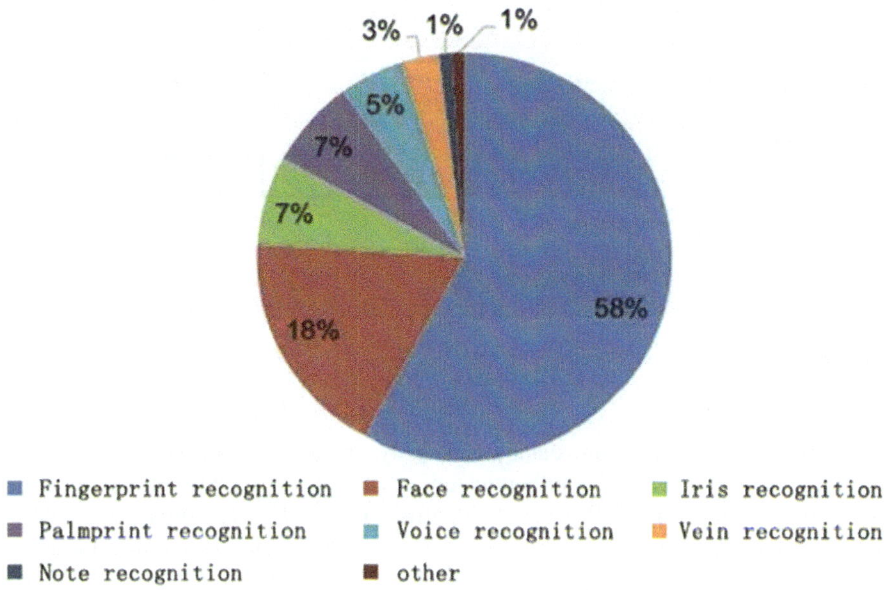

Fig. 2.1 2015 Global biometric market structure

- Morphological biometrics: these are the physiological modalities that can be found at all individuals, such as fingerprints, the shape of the hand, facial features, the shape of the iris or the hand, the venous pattern of the retina.

Biometric systems grant a high level of security and comfort. The identification systems based on biometric, check identity rather than knowledge, for example, a password or pin number or ownership of a key card (Prabhakar et al., 2007). Human biometric characteristics have been shown to be different and cannot be transferred and the information about these characteristics cannot be stolen, forgotten or lost. In the last few years, fingerprint technology has contributed to the generation of higher profits, according to the report published by the International Biometric Group, as shown in Fig. 2.1.

2.2.1 Biometric Systems

A biometric system is essentially a pattern recognition system. It is mainly composed of two essential phases: enrollment and recognition. In the first stage, the user is prompted to provide several samples (generally between 3 and 5) of the biometric modality by means of an acquisition device. This step is usually followed by a preprocessing step which aims to improve the quality of the acquired modality. Then comes the most credible step of extracting the discriminant features from the

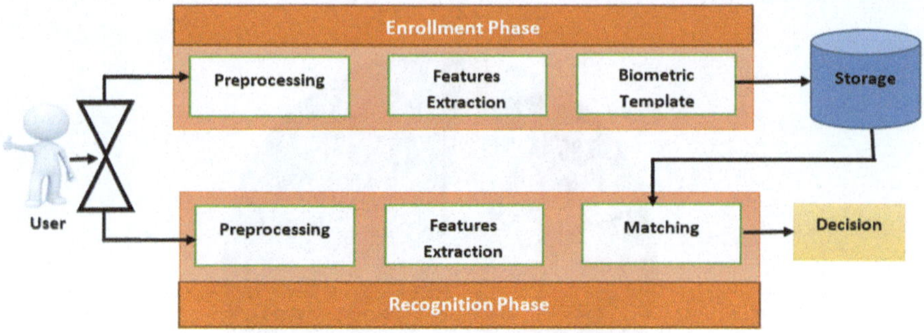

Fig. 2.2 Typical biometric system architecture

modality that are stored in a database. Finally, in the recognition step, the same modality of the person to be authenticated, or identified, is captured and the same procedures are performed. If we are in identification mode, we compare the features to all those contained in the database. In case of verification, the extracted features are compared only to those of the person declared and the decision to accept or refuse authentication is announced by the system. Figure 2.2 shows the block diagram of a biometric system.

2.2.2 *Biology of the Fingerprint*

A fingerprint is the image of the epidermis of one fingertip, usually acquired on a flat surface such as a sheet of paper or a scanner, but also on any rougher surfaces, especially on crime scenes, or acquired directly in space by photo or camera (Stosz and Alyea, 1994). The epidermis of the fingertips is characterized by a complex pattern of relief, consisting of ridges and valleys. This epidermis pattern is formed in the fetus around the seventh month of pregnancy. Compared to other biometrics such as facial features, the fingerprint has the advantage of very little variation during the life of an individual, and even after death, with the exception of major skin breakdowns due to scarring or diseases. Based on the ridges design and their appearance fingerprint features are arranged into three levels (Maltoni et al., 2009).

- *Level-1 features or global features*

Enclose the fingerprint from a global perspective with which are attached to all global characteristics related to the oriented texture pattern exhibited by the ridge/valley structure. These include the orientation field, ridge frequency, and singularities. Although these features don't carry any information about the individuality of a fingerprint, they can be used as exclusive tools to decrease the search space at the matching step, by ensuring fingerprint classification, as well as to make a difference between a fingerprint and palm print patterns (Figs. 2.3 and 2.4).

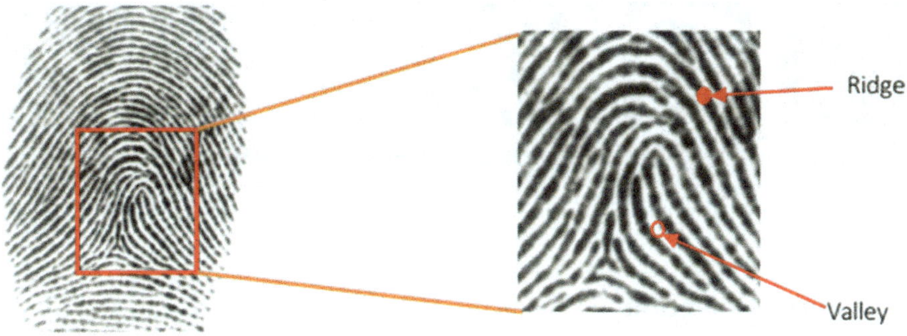

Fig. 2.3 Fingerprint images with marked ridge and valley

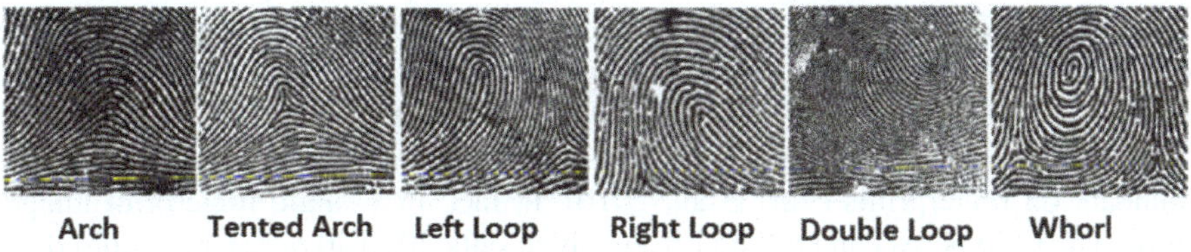

Fig. 2.4 Global features of fingerprint images

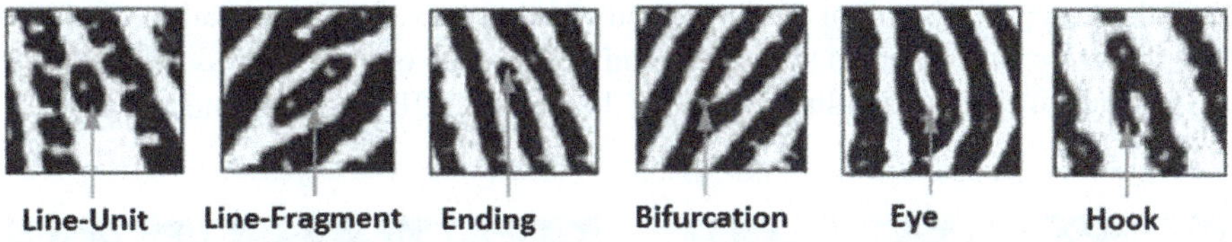

Fig. 2.5 Different types of minutiae

- *Level-2 features*
  Refer to all local details related to the paths of the ridges. The most noticeable level-2 characteristics are local discontinuities exhibited by the ridges as shown in the Fig. 2.5. In fact, ridges often run in continuous curves and suddenly terminate at specific points locations called ridge endings. Others split at some points called bifurcations to yield other ridges. Both ridge endings and bifurcations points are called minutiae.
- *Level-3 features*
  Fingerprints once acquired with high resolution, typically 1000 dpi, exhibit some useful information that is not visible to the naked human eye in the standard resolution, given by major sensors in the market working typically in 500 dpi. This information is attached to the pores and shapes of the ridges. Figure 2.6 shows some characteristics of level 3.

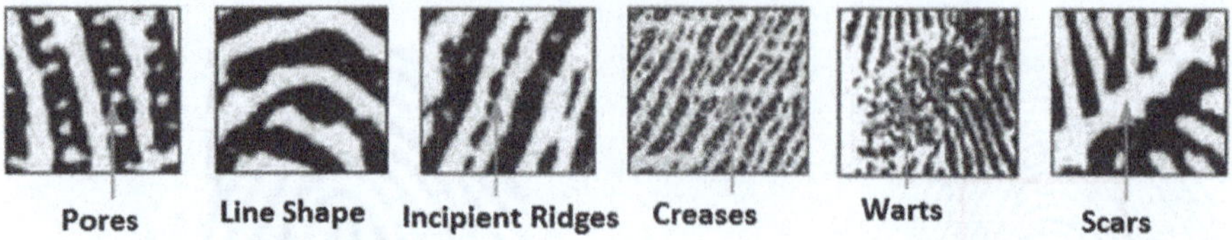

Fig. 2.6 Different types of level 3 features

2.3 Pores Detection

- *The notion of fingerprint pores*

Pores are the most used level 3 features due to the high accuracy they bring to the identification. They are the microscopic holes that form on fingerprints ridges since the sixth month of gestation, to duct the sweat-gland to the surface of the epidermis. Due to their anatomy, the pores are resistant to be damaged or mimic, also they are abundant on fingerprints (Puri et al., 2010). In fact, in a single fingerprint image, more than 1400 pores can be found from which only 20 to 40 pores are required for fingerprint identification. There are 9 to 18 pores along a centimeter of a ridge. A pore can be seen as open on one print, yet as shut on another print of the same finger depending on the pressure of the finger and whether it exudes transpiration. A closed pore shows up as a secluded spot on the ridge, while an open pore is associated with one or both of the surrounding valleys (Chen et al., 2013) as presented in Fig. 2.7.

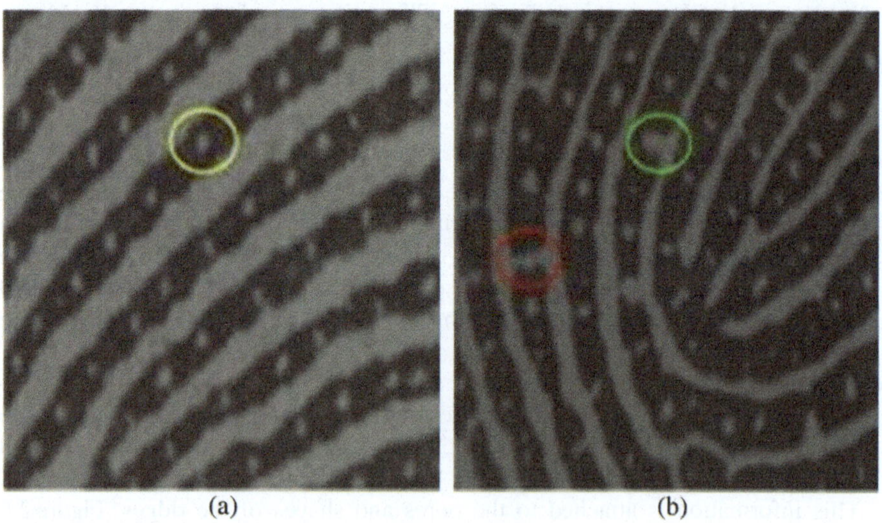

Fig. 2.7 Two fingerprint images where in **a** a closed pore is circled with yellow and in **b** two opened pores are circled with red and green

Thus, the shape and size of pores can change from one impression to another, so the essential thought of the proposed pore extraction strategy is to extract open pores and closed pores independently. We precise that in the green circle, the pore is connected to one valley, and in the red circle, the pore is connected to two valleys.

2.3.1 Related Works

Numerous studies have been proposed in the literature aiming at detecting pores in the fingerprint image. The difference of Gaussian filters is used by Zhao et al. (2010b), the fingerprint image is divided into blocks to fit the Gaussian scales to the ridge frequencies on the blocks. The small-scale output is subtracted from the large-scale output after binarization. Consequently, a post-processing stage starts to remove the spurious pores from the initial result. The drawback of this approach is that the circular assumption of pores does not hold well on real fingerprint images. Jain et al. (2006), defined a pore model using a Mexican hat wavelet transform. In such a model, ridges are detected using Gabor filters representing different orientations. Finally, the two responses are combined and the pores are extracted by empirical thresholding. But, this method is not adapted to the variations of pores. In de Paula Lemes et al. (2014), the authors suggested an accurate approach called Dynamic Pore Filtering (DPF) that does not use an expensive processing technique and adapts itself to different sizes and shapes of pores. This method is divided into five steps: (1) extraction the region of interest and equalization of images; (2) estimation of the global threshold for binarization; (3) estimation of the pore size; (4) classification of pores; and (5) determinate the location of pores. The authors in Zhao et al. (2010a), introduced adaptive pore extraction methods that first, divided the fingerprint image into background and foreground blocks. Then A local instantiation of the suitable pore model is attained depending on the dominant ridge orientation and frequency on each foreground block. The pores are finally extracted by filtering the block with the adaptive model. This technique gives an accepted result but it uses an expensive technique like the ridge orientation and the frequency estimation. Labati et al. proposed a method based on Convolutional Neural Networks (CNN) (Labati et al., 2018). First, they highlighted the position of the pores then they estimated the coordinates of those pores after that, the feature is extracted from the area around the estimated pores. Finally, they discard the incorrectly detected pores. Recently, (Khmila et al., 2019) presented an approach based on the crossing number method to detect open pores and the size of objects in the thinned image for the closed pores detection.

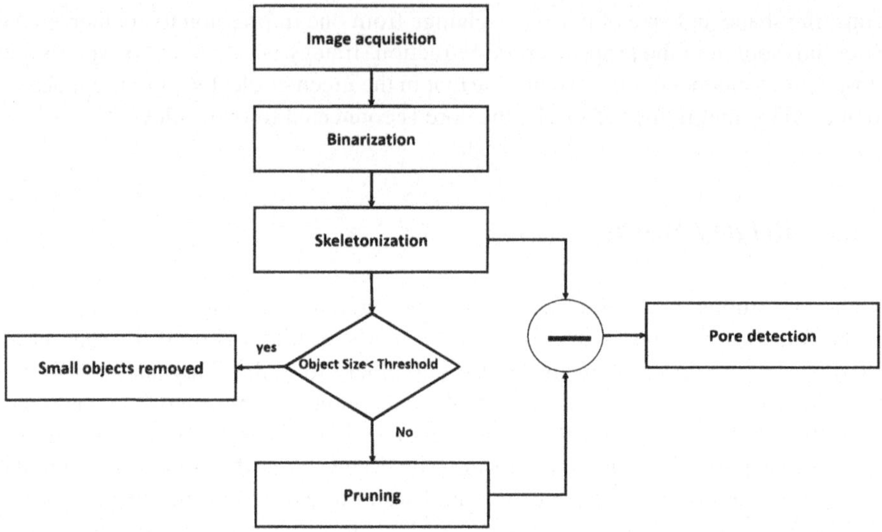

Fig. 2.8 Flowchart of the proposed pore detection method

2.3.2 Proposed Method

This chapter proposes a new method to detect pores in the fingerprint image that summarized in the flowchart presented in Fig. 2.8.

- *Binarization step*: Binarization is the second phase after the acquisition step of our approach. It consists of transforming an image from 256 levels (grayscale image) into 2 levels (black and white image). The binary image is also named image thresholding that's why the binarization process needs a threshold value to classify all pixels with values above this threshold as white and the other pixels as black (Fig. 2.9).
- *Skeletonization step*: After the binarization step, a thinning process is started. Image thinning is a fundamental pre-processing step that plays a vital role in many applications such as pattern recognition, fingerprint classification, medical applications, etc. It is a morphological operation, used to eliminate selected foreground pixels from binary images. The algorithm recursively removes any point with more than one neighbor in the foreground, provided that this does not disconnect the region containing that pixel locally and takes the pixels on the borders of the foreground regions with at least one neighbor in the background (Jain et al., 2006) (Fig. 2.10).
- *Remove small objects step*: To remove the small object from the obtained image we need a method that computes the size of each region, for that we chose to use the connected component labeling (CCL). It is a central algorithm between low-level image processing (filtering) and high-level image processing (recognition, decision). CCL consists of providing a unique number to each connected component

of a binary image. The objective of the CCL is to mark all the pixels belonging to an object, also called a connected component, with the same label in the source image. In order to do this, the algorithm scans the input binary image in raster order and decides how to label each foreground non-zero pixel on the basis of the labels already assigned to the pixels within a chosen neighborhood. Typically, neighborhoods consisting of four or eight pixels are used. The Fig. 2.11 shows a binary image and the result of the CCL algorithm.
Afterward, the component in which their size above certain threshold values (determined empirically) are kept in the skeleton image and the other are removed (Fig. 2.12).

- *Pruning step:* Subsequently, an operation of pruning applied on the remaining ridge to remove the spur and the endpoints of each branch or ridge i.e. Remove the pixel that has only one neighbor. For example:

$$\begin{bmatrix} 0\,0\,0\,0 \\ 0\,0\,0\,0 \\ 0\,0\,1\,0 \\ 0\,1\,0\,0 \\ 1\,1\,0\,0 \end{bmatrix} \textit{becomes} \begin{bmatrix} 0\,0\,0\,0 \\ 0\,0\,0\,0 \\ 0\,0\,0\,0 \\ 0\,1\,0\,0 \\ 1\,1\,0\,0 \end{bmatrix}$$

- *Subtracting phase:* The pixel subtraction operator takes two images as input and produces as output a third image whose pixel values are simply those of the first image minus the corresponding pixel values from the second image. The purpose of this operation is to find changes between two images.

$$g\,(x, y) = f_1\,(x, y) - f_2\,(x, y)\,. \tag{2.1}$$

where f_1 and f_2 are the two input images and g is the output third image. In this stage, we subtract the obtained image of the last step from the first thinned image to obtain a cart of detected pores (Fig. 2.13).

2.4 Pores Matching

2.4.1 Related Works

The current methods of matching pores are generally classified into two principal categories (Xu et al., 2019): alignment-based methods and direct pore matching methods.

- The alignment-based methods align the fingerprint images using different features other than pores. Then the pores, included in the aligned area, are matched using their geometric distances.

Fig. 2.9 Fingerprint binary image

Fig. 2.10 Fingerprint thinned image

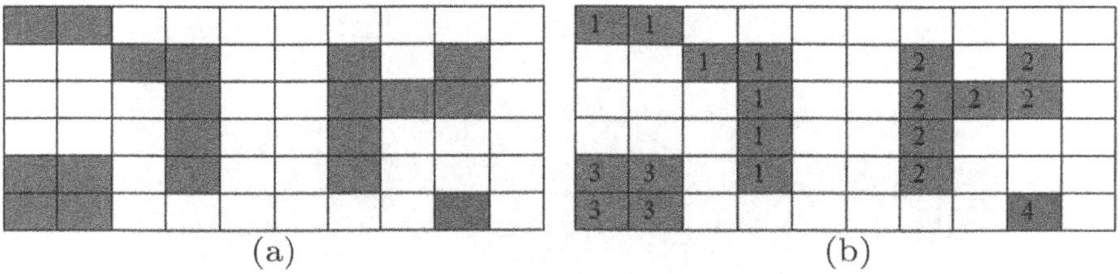

Fig. 2.11 **a** Input image, **b** connect component labeled image

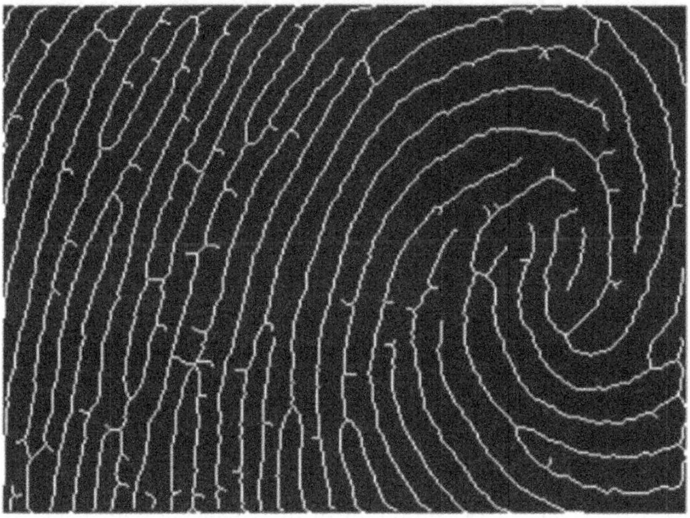

Fig. 2.12 Result of small object removing from the fingerprint image

- The direct pore matching methods match directly the pores of the fingerprint images independently of the other features.

Typical methods are introduced in the following subsections. For instance, (Jain et al., 2006) proposed a method that they denoted MICPP. This method extracts the minutiae in fingerprint images, then it matches them. After that, pores, located in a rectangular neighborhood to each aligned pair of minutiae, are matched using a modified iterative closest point (ICP) algorithm.

Authors in Lu and Xu (2017) proposed another method for matching pores in two fingerprint images. Their method is based on the deterministic annealing algorithm. They, first, used singular points (cores and delta) to align the compared fingerprint images. Then they matched minutiae using the alignment result. Afterward, a convex hull is established for each fingerprint image to minimize the effect of deformation. And the pores located in these convex hulls are matched. On the other hand, authors in Labati et al. (2018), Liu et al. (2010) and Zhao et al. (2009) proposed a direct pore matching method for the identification system. This method includes two principal

Fig. 2.13 Pore extraction

steps: correspondence establishment and refinement. The first one consists of calculating the score similarity between pores, based on their local features, and using it to build one to one correspondences. The second one refines the result of correspondences using the RANSAC (RANdom SAmple Consensus) or WRANSAC (Weighted RANSAC) algorithm.

Recently, (Cui et al., 2014) proposed in their paper a pore matching method independent from the minutiae-and-ridge and based on the polar histogram. They generated, firstly, a descriptor, formed by relative locations of each pore, based on the polar histogram. Secondly, they establish the correspondences between pores from the reference and the candidate images. Finally, they use RANSAC algorithm to refine the result and then the final matching score is determined. More recently, (Kim et al., 2018) proposed a stable pore matching algorithm that excludes dependencies on both minutiae-ridges and fingerprint-device to enhance the performance and the utilization of pores for the identification process. These methods suffer from several defects. For example, the result of the based matching minutiae method heavily depends on the accuracy of the detection algorithm of minutiae in the high-resolution fingerprint images. Also, the method based on direct pore matching suffers from a high consuming computing time because of the huge number of matches done in the matching step.

2.4.2 Proposed Method

Each extracted pore is characterized by its cartesian coordinates as well as the orientation of the ridge to which it belongs. As we know that the matching based on the alignment approach depends on the reliability of the algorithm of the alignment and the method of the detection of the minutia or the singular points that are used. In this chapter, a new method of matching pores without an alignment is proposed. Our approach is based essentially on the contextual information of pore. It consists of four steps: (i) local feature extraction, (ii) local similarity estimation for pore couple, (iii) determination of the threshold of the global similarities using possibility models, (iv) score computing and (v) decision making. A flowchart of the proposed method is presented in Fig. 2.14.

2.4.2.1 Local Feature Extraction

For every extracted pore p from the fingerprint image, 3 neighboring zones $Z1$, $Z2$ and $Z3$ with different radius respectively $R1 = 20\ pixels$, $R2 = 30\ pixels$, and $R3 = 40\ pixels$ are used to determine the pore relative features. In each of the three zones, we count the number of neighbor pores of $p(nb_1, nb_2, nb_3)$. The pore p is also characterized by its Euclidean distances Δd to each of its neighbors of the largest neighboring region $Z3$, as well as its deviations $\Delta\theta$ to the same neighbors, that correspond to the differences between their respective orientations.

2.4.2.2 Local Similarity Estimation for Pore Couple

After the extraction of local features, a local similarity ℓ_s between pore pair $p_{1,i}$ and p_{2j} of image 1 and image 2 is calculated using the following equation:

$$\ell_s = \sqrt{\frac{n^2}{(nb_{1,i,3} \times nb_{1,j,3})}} \tag{2.2}$$

where $nb_{1,i,3}$ and $nb_{1,j,3}$ are the numbers of neighbor pores of $p_{1,i}$ and $p_{2,j}$ respectively, and n is the number of corresponding neighbor pores.

If $p_{1,i,k1}$ and $p_{2,j,k2}$ are neighbors of respectively $p_{1,i}$ of image 1 and $p_{2,j}$ of image 2, they are judged as corresponding neighbor pores, if their distances to $p_{1,i}$ and $p_{2,j}$ respectively, are sufficiently close, and the same criterion is respected for orientation deviations. It means that the following conditions are verified:

$$(\Delta d_{p1,i,k1} - \Delta d_{p2,j,k2})^2 < \xi$$

$$(\Delta\theta p_{1,i,k1} - \Delta\theta p_{2,j,k2})^2 < \beta$$

where ξ and β are two threshold values determined empirically.

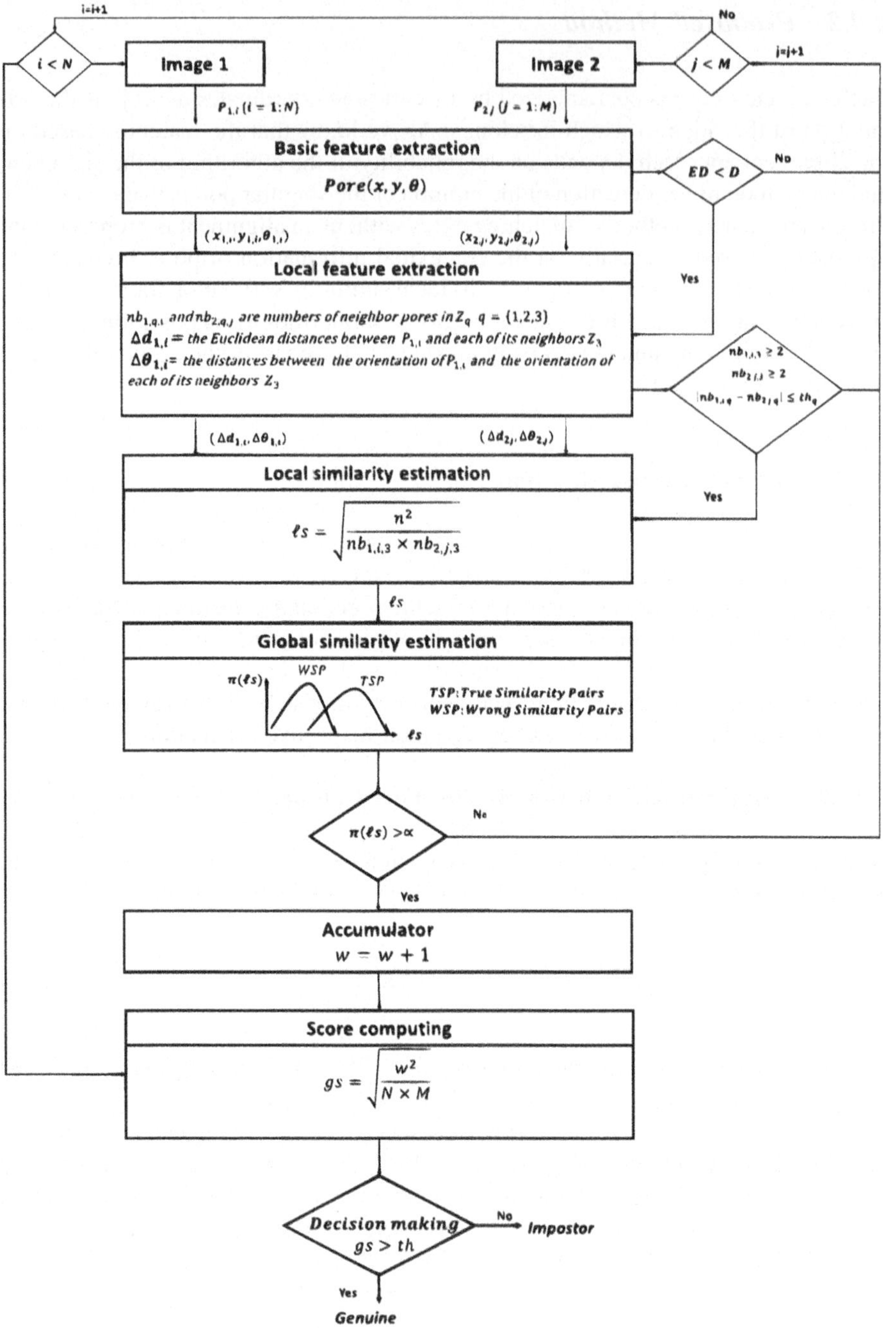

Fig. 2.14 Flowchart of the proposed matching method

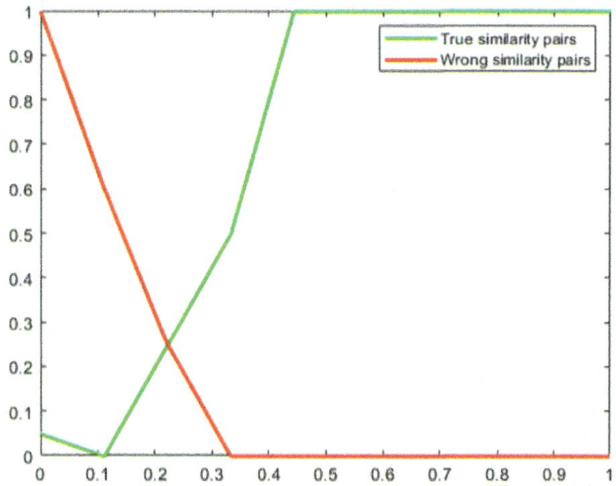

Fig. 2.15 Possibility distribution of the true and the wrong similarity pairs

2.4.2.3 Global Similarity Estimation by Possibility Model Thresholding

ℓ_s is a similarity estimation between two examined pores, namely $p_{1,i}$ and $p_{2,j}$, belonging respectively to image 1 and image 2. ℓ_s is ranged inside [0, 1]. A low value means that the examined matched pores are different and a value near to 1 means that the pores $p_{1,i}$ and $p_{2,i}$ can be the same. So, in order to decide about the similarity or the difference between two pores according to ℓ_s measure, we use a threshold denoted α. In this section, we present an experimental methodology to efficiently determinate a suitable threshold α. In the first step, we manually select 21 true neighbor pores, then the ℓ_s similarities of these pairs are calculated. The same operation is run on 21 other wrong neighbor pores. In the second step, two possibilistic models of respectively true neighbor pore class and wrong neighbor pore class, are built using ℓ_s. Possibility distributions are performed using Dubois and Prade asymmetric approach (Dubois et al., 1993). Figure 2.15 illustrates the two built possibilistic models.

It can be seen from this figure that the pairs of pores which have a similarity value lower than 0.22 have a possibility value to be wrong pairs bigger than to be true pairs and vice versa. To be certain that these pairs can be correctly matched, we set a threshold $\alpha = 0.5$ on the possibility values of the true pairs.

2.4.2.4 Score Computing

In this step, we detail the computing process of the global similarity score g_s between image 1 and image 2. We call it the global similarity process because it uses the previous steps to produce a final score. Firstly, for each pore of image 1 we look for a possible corresponding pore among the pores of image 2. The couple of pores, which verifies the following conditions, contributes to global score computing process.

- Condition 1: the two pores $p_{1,i}$ and $p_{2,j}$ have at least two neighbors in the first zone.

$$nb_{1,i,3} \geq 2$$

and

$$nb_{2,j,3} \geq 2$$

- Condition 2: the Euclidean distance ED between pores $p_{1,i}$ and $p_{2,j}$ lower than $D\ pixels$.
- Condition 3: in each zone, the difference of number neighbors must follow this condition;

$$|nb_{1,i,q} - nb_{2,j,q}| \leq th_q$$

Local similarity ℓ_s of each contributing pore couples is estimated. After that its possibility value $\pi(\ell_s)$ is determined and compared to the threshold α. The mated pores are classified as winner pores when $\pi(\ell_s) \geq \alpha$. Finally, the global similarity score g_s is computed using the equation:

$$gs = \sqrt{\frac{w^2}{N \times M}} \tag{2.3}$$

where N and M are the numbers of pores in I_T and I_M respectively. And w is the number of winner pores.

2.4.2.5 Decision Making

Construction of gs distribution for genuine and for impostors. Deduction of the corresponding threshold to the equal error rate (th). This threshold is used for decision making. If gs value superior to this threshold, the two fingerprints are for the same person and If gs value inferior to this threshold, the two fingerprints are for two different persons.

2.5 Experimental Results

2.5.1 Database

The Hong Kong Polytechnic University High-Resolution Fingerprint (PolyU-HRF) database (Liu et al., 2010; Zhao et al., 2010a; 2010b; 2009) is the database used in this work. It has one training set and two testing sets. 1200 ppi is the resolution of each image of them. The training set contains 210 images of 35 different subjects (i.e.

6 images per subject acquired in two different sessions 3 images per session). Both testing sets, labeled DBI and DBII, have 1480 images from 148 different persons (i.e. 10 images per person obtained in two different sessions 5 images per session). The images in the training and the DBI sets have 320 x 240 pixels (i.e. partial fingerprints), whereas images in the DBII has 640 x 480 pixels (i.e. full-size fingerprints). Additionally, the PolyU-HRF database holds a subset of DBI images containing the ground truth coordinates of pores (i.e. 30 images with an average of 425 pores). This ground truth is used to evaluate the true and the false detection rate of pores.

2.5.2 *Pores Detection*

We aim in this section to establish the feasibility of our method comparing it with another state of the art method. Our approach consists of thinning the input image after the binarization step. Afterward, the connected component process is stated to compute the sizes of each object of the thinned image in order to remove the regions that their sizes lower than 7 pixels. The obtained image has undergone a pruning step to remove the spur and the endpoints of each branch or ridge. Finally, the pores are extracted by the substrate of the produced image from the initial thinned image. To proving our proposed method, we use the subset of the PolyUHRF database that has the ground truth coordinates of pores to compute the true detection rate (TDR) and the false detection rate (FDR) presented by the Eqs. 2.4 and 2.5

$$TDR = \frac{TP}{Total\, number\, of\, true\, pores} \tag{2.4}$$

$$FDR = \frac{FP}{Total\, number\, of\, true\, pores} \tag{2.5}$$

where TP and FP represent the number of correctly and falsely predicted pores, respectively. The performance of our approach is provided by comparing it with some state of the art methods, as may be seen in Table 2.1.

Table 2.1 Average performance metrics in percentage and standard deviation (in parenthesis).

Methods	TDR %	FDR %
DOG (Zhao et al., 2010b)	80.8 (6.5)	22.2 (9.0)
Gabor Filter (Jain et al., 2006)	75.9 (7.5)	23 (8.2)
DAPM (Zhao et al., 2010a)	84.8 (4.5)	17.6 (6.3)
CNN (Labati et al., 2018)	84.69 (7.81)	15.31 (6.2)
(Khmila et al., 2019)	88.83 (10.1)	5.8 (5.5)
Proposed method	90.17 (9.3)	5.58 (5.8)

This table shows that our method attains the higher TDR value, and the lower FDR, which also proves the higher efficiency of our technique, compared to other well-known existing methods. Our pore detector is not required of the frequency estimation and ridge orientation. Therefore, it requires fewer computer resources. Thus, the experimental results prove that our method can detect pores more accurately and robustly. So, the proposed approach can help to enhance the verification precision of pore based fingerprint recognition systems.

2.5.3 *Recognition*

2.5.3.1 Parameters Setting

For the efficiently operating of the system, a set of parameters is used in the matching step. Those parameters are determined empirically and specially for this database.

The first the Euclidean distance ED between pores $p_{1,i}$ and $p_{2},_{j}$ must be lower than $D = 50\ pixels$.

Second the thresholds th_q of the differences of number neighbors of pores in the q zones where $q = \{1, 2, 3\}$ are equal to $th_q = \{2, 4, 6\}$. finally for the local similarity estimation, we choose the threshold of distance ξ between the neighbors of $p_{1,i}$ and neighbors of $p_{2},_{j}$ equal to 0.004. And for the orientation deviations we used $\beta = 0.002$.

2.5.3.2 Evaluation Results

To evaluate the accuracy of our proposed method we choose 100 fingerprints images acquired from 10 fingers from DBI, and we keep the following matches: For the genuine scores, each image in the first session is compared to all the other images of the same finger in the second session. For the impostor comparison trials, the first image of every finger in session 1 will be compared with all the first images of all the other fingers in the same session. The obtained scores of genuines and impostors are used to plot the distribution similarity scores as shown in Fig. 2.16.

It clearly shows the existence of two separate classes. Also the performance is evaluated using Equal Error Rate (EER), a specific point attained when FPR (False Positive Rate) and (TPR) true positive rate coincide. A graphical representation of the capabilities of the individual matchers is shown in the ROC (Receiver Operating Characteristic) plot. Ideally, the EER value is expected to be low. The lowest value is 0. Therefore, the closer the ROC plot to the upper left corner, the lower the EER, and the higher the overall accuracy of the system. The ROC curve in Fig. 2.17 illustrates the relative trade offs between true positive rate (TPR) and false positive rate (FPR) of the match scores for our dataset. Since there is a minor overlap between the distributions of genuine and impostor scores, we have a low EER value of $EER = 5\%$.

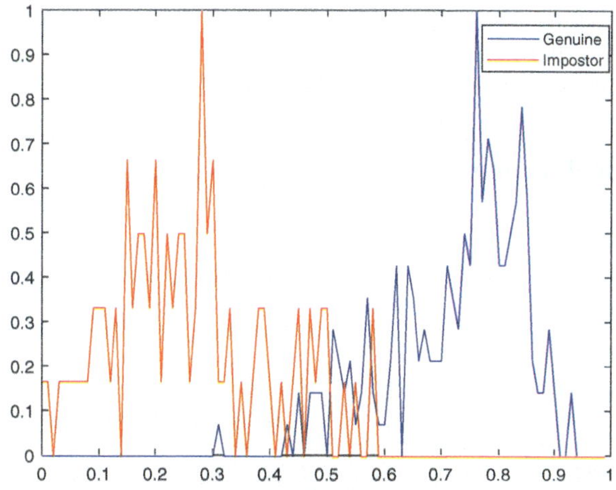

Fig. 2.16 Distribution of similarity scores

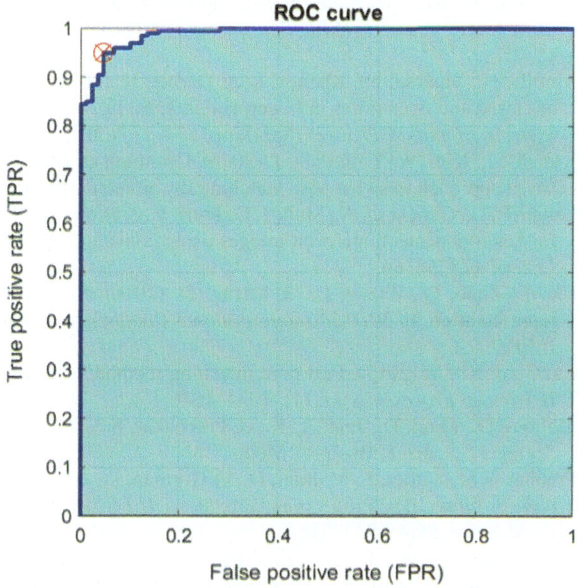

Fig. 2.17 ROC of our proposed method

2.6 Conclusion

In this chapter we have proposed two novel approaches. The first one aims at detecting pores from high-resolution fingerprint images and, the second one permits to match those pores without alignment. The detection method based on a morphological operation (skeletonization) and the labeled connect component method and the matching method is based on the local information of pore to match fingerprints without referring to the others features to align fingerprints. The experimental results show that our methods outperform the existing algorithms of pore detection and present good result of matching ($EER = 5\%$) for a small database.

References

Chen, F., Huang, X., & Zhou, J. (2013). Hierarchical minutiae matching for fingerprint and palmprint identification.*IEEE Transactions on Image Processing, 22*(12), 4964–4971.

Cui, J., Ra, M. -S., & Kim, W. -Y. (2014). Fingerprint pore matching method using polar histogram. In *The 18th IEEE International Symposium on Consumer Electronics (ISCE 2014)* (pp. 1–2). IEEE.

de Paula Lemes, R., Segundo, M. P., Bellon, O. R., & Silva, L. (2014). Dynamic pore filtering for keypoint detection applied to newborn authentication. In *2014 22nd International Conference on Pattern Recognition* (pp. 1698–1703). IEEE.

Dubois, D., Prade, H., & Sandri, S. (1993). On possibility/probability transformations. In *Fuzzy logic* (pp. 103–112). Berlin: Springer.

Jain, A. K., Chen, Y., & Demirkus, M. (2006). Pores and ridges: High-resolution fingerprint matching using level 3 features. *IEEE Transactions on Pattern Analysis and Machine Intelligence, 29*(1), 15–27.

Khmila, H., Smaoui, N., Khanfir, I., & Derbel, H. (2019). An efficient method for the extraction of closed and open pores in fingerprint images. In *2019 16th International Multi-Conference on Systems, Signals & Devices (SSD)* (pp. 154–157). IEEE.

Kim, M.-J., Kim, W.-Y., & Paik, J. (2018). Optimum geometric transformation and bipartite graph-based approach to sweat pore matching for biometric identification. *Symmetry, 10*, 175.

Labati, R. D., Genovese, A., Muñoz, E., Piuri, V., & Scotti, F. (2018). A novel pore extraction method for heterogeneous fingerprint images using convolutional neural networks. *Pattern Recognition Letters, 113*, 58–66.

Liu, F., Zhao, Q., Zhang, L., & Zhang, D. (2010). Fingerprint pore matching based on sparse representation. In *2010 20th International Conference on Pattern Recognition* (pp. 1630–1633). IEEE.

Lu, G., & Xu, Y. (2017). Fast pore matching method based on deterministic annealing algorithm. *IET Image Processing, 11*(11), 1034–1040.

Maltoni, D., Maio, D., Jain, A. K., & Prabhakar, S. (2009). *Handbook of fingerprint recognition*. Springer Science & Business Media.

Prabhakar, S., Kittler, J., Maltoni, D., O'Gorman, L., & Tan, T. (2007). Introduction to the special issue on biometrics: Progress and directions. *IEEE Transactions on Pattern Analysis and Machine Intelligence, 29*(4), 513–516.

Puri, C., Narang, K., Tiwari, A., Vatsa, M., & Singh, R. (2010). On analysis of rural and urban Indian fingerprint images. In *International Conference on Ethics and Policy of Biometrics* (pp. 55–61). Berlin: Springer.

Stosz, J. D., & Alyea, L. A. (1994). Automated system for fingerprint authentication using pores and ridge structure. In *Automatic systems for the identification and inspection of humans* (Vol. 2277, pp. 210–223). International Society for Optics and Photonics.

Xu, Y., Lu, G., Lu, Y., & Zhang, D. (2019). High resolution fingerprint recognition using pore and edge descriptors. *Pattern Recognition Letters, 125*, 773–779.

Zhao, Q., Zhang, D., Zhang, L., & Luo, N. (2010a). Adaptive fingerprint pore modeling and extraction. *Pattern Recognition, 43*(8), 2833–2844.

Zhao, Q., Zhang, D., Zhang, L., & Luo, N. (2010b). High resolution partial fingerprint alignment using pore-valley descriptors. *Pattern Recognition, 43*(3), 1050–1061.

Zhao, Q., Zhang, L., Zhang, D., & Luo, N. (2009). Direct pore matching for fingerprint recognition. In *International Conference on Biometrics* (pp. 597–606). Berlin: Springer.

Chapter 3
Fractal Analysis for Iris Multimodal Biometry

Hana Boukhriss, Améni Yengui Jammoussi, Imene Khanfir Kallel, and Nabil Derbel

Abstract Iris recognition is a new technology that ensures a high level of security which refers to the automatic recognition of individuals based on their iris characteristics. Due to its stability and high level of discrimination, the iris is the most effective biometric modality for large-scale identification and verification systems. To build up an iris recognition system, a database of iris composed for huge number of images for the training phase. In this chapter, in the first part, we present a mono-modal system based on fractal analysis. Then, a multi-modal system based on fusion at different levels. Several processes have been implemented to evaluate our proposed recognition system using PBMLT iris images and the well-known UBiris database. To conduct experiments, firstly, each image of the eye is segmented to extract only the iris from surrounding elements. Then, feature extraction process is preceded to extract relevant biometric signatures. Finally, a matching process based Hamming distance is achieved. The results obtained from segmentation part were a 77.08% segmentation rate in the visible and 87.71% in the near infrared. In the second part, EER, FAR and FRR are calculated in order to evaluate both uni-modal and multi-modal systems.

H. Boukhriss
Cybernics Team, Control and Energy Management Laboratory (CEM Lab), Electrical Department, National School of Engineering of Gabes, University of Gabes, Gabes, Tunisia

A. Y. Jammoussi (✉)
Cybernics Team, Control and Energy Management Laboratory (CEM Lab), Electrical Department, ENIS, ISBS, University of Sfax, Sfax, Tunisia

I. K. Kallel
Cybernics Team—Control and Energy Management Laboratory (CEM Lab), Electrical Department, ENIS, ISBS, University of Sfax, Sfax, Tunisia
e-mail: imen.khanfir.kallel@gmail.com

Image et Traitement de l'Information iTi Ã l'IMT Atlantique, Brest, France

N. Derbel
Control and Energy Management Laboratory (CEM Lab), National School of Engineering of Sfax, University of Sfax, Sfax, Tunisia

N. Derbel and O. Kanoun (eds.), *Advanced Methods for Human Biometrics*,
Smart Sensors, Measurement and Instrumentation 40,
https://doi.org/10.1007/978-3-030-81982-8_3

Keywords Iris recognition · PBLMTiris · Segmentation · Normalization · Encoding · Hamming distance · Uni-modal · Multi-modal · Fusion

3.1 Introduction

In general, biometrics refers to technologies that analyze and measure physical and/or behavioral human characteristics for the authentication purposes (Jain et al., 2004). A typical biometric verification system is based on two processes (Fig. 3.1):

- Enrollment: A person's biometric data is enrolled to the system's biometric database.
- Authentication: A person's newly acquired data is compared to the enrolled biometric data and a match score is generated. Based on match scores, we decide whether or not the model comes from the same person.

The human characteristics for biometric system include fingerprints, DNA, voice patterns, irises, hand measurements and facial patterns, etc.

An efficient biometric system is based on the following criteria:

- Universal, as they can be found in all individuals
- Unique features in order to differentiate one individual from another
- Permanent and invariant over time
- Small computational complexity for feature extraction
- Precise classification algorithms
- Low cost and ease of implementation.

The performance of the biometric system is usually evaluated based on some indicators. Some of them are detailed below:

- False Acceptance Rate (FAR) is the percentage of identification instances in which unauthorized persons are incorrectly accepted.
- False Rejection Rate (FRR) is the percentage of identification instances in which authorized persons are incorrectly rejected.
- Equal Error Rate (EER): as the number of false acceptances (FAR) goes down, the number of false rejections (FRR) will go up and vice versa. The point at which

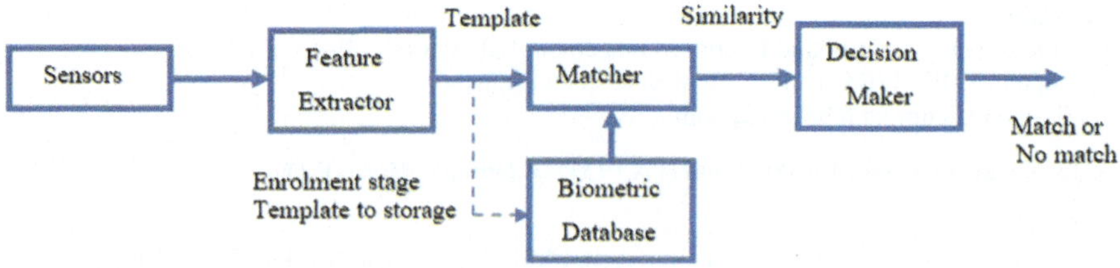

Fig. 3.1 Biometric recognition system

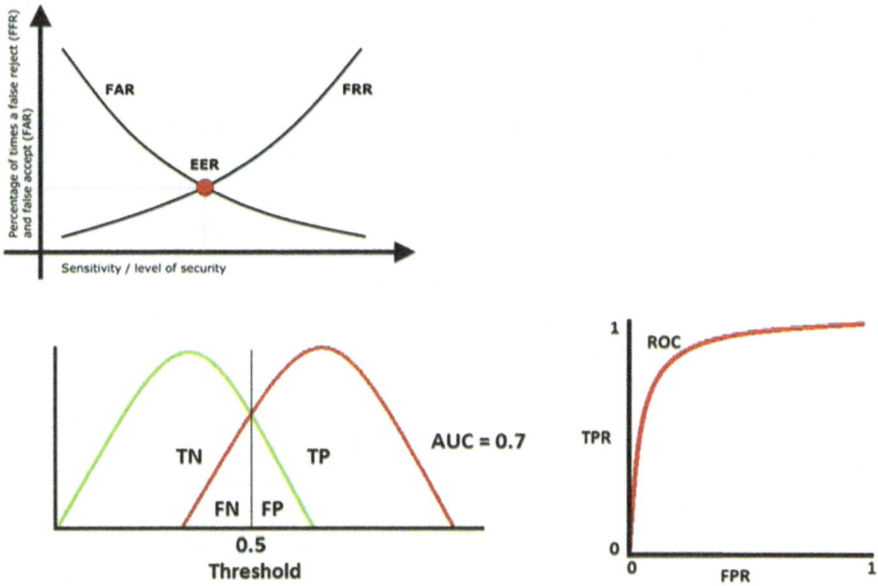

Fig. 3.2 Performances evaluation of the biometric system with different metrics

the lines intersect also has a name: the Equal Error Rate (EER). This is when the percentage of false acceptance rate equals false rejections rate.

- False Positive Rate (FPR): number of items wrongly identified as positive out of total true negatives
  $FP/(FP+TN)$
- True Positive Rate (TPR): number of items correctly identified as positive out of total number of positives
  $TP/(TP+FN)$
- ROC curve: Receiver operating characteristics curve showing performance measurement for classification problem at various thresholds settings. The ROC curve is plotted with TPR against the FPR where FPR is on the x-axis and TPR is on y-axis.
- AUC-ROC: Area Under the Curve is one of the most important evaluation metrics for checking classification performance (Fig. 3.2).

After introducing the main characteristics of an efficient recognition system, we present in the following section some works proposed in the literature. The remaining of this chapter is organized as follows. In Sect. 3.2 we review some related works to iris recognition system. In Sect. 3.3, we present our proposed method for feature extraction. Then, our uni-modal recognition system is described in detail in Sect. 3.4. An improvement of the recognition system based on fusion of information is presented in Sect. 3.5. Section 3.6 is devoted to experimental results. Finally, Sect. 3.7 presents discussions and draws some conclusions.

3.2 Related Works

The iris modality was adopted in many proposed biometric systems for its effectiveness to provide identity. The iris contains a lot of important personal identity information as well as the amount of information in the DNA (Krichen, 2007). Thus, this information is sufficient to certify the identity of an individual with great precision.

Our work falls within the implementation of an iris identification system. Iris recognition is now considered as one of the most accurate biometric recognition techniques (Llano et al., 2017). A lot of researches related to iris recognition are cited in the literature. The first approach, suggested by Daugman (2009), describes the first work methodology related to iris biometry. Daugman used an integro-differential operator to segment the circular iris and the pupillary region, as well as the arches of the upper and lower eyelids (Dubey and Madan, 2014). Noh (Seung-In et al., 2003) have proposed a new feature extraction technique, based on local and global characteristics using the wavelet transform coefficients. Another approach, presented by Xu et al. (2006), describes an improved iris recognition system based on a biometric technique for high-level applications such as border crossing systems. Also, Azizi and Pourreza (2009) proposed an alternative approach that treats the extraction of iris characteristics using the contour transform, capturing the intrinsic geometric structures of the iris image. The latter has been decomposed into sub-blocks containing all the texture information. This technique has used both the Gabor filter and the Haar wavelet.

3.3 Feature Extraction Based on Fractal Analysis

The extracted features should be significant in order to ensure the effectiveness of our identification system. In iris recognition, most of the methods found in the literature review use different types of wavelets or filtering (Gale and Salankar, 2014; Sreelekha, 2014; Yu et al., 2015). In our work, we make use of a fractal based method proposed by Khanfir in (Khanfir, 2009) which computes the local fractal dimensions on 256 contour map fragments of the iris texture using box-counting fractal dimension. The proposed technique is to quantify the evolution of texture roughness within an iris image. This quantification is done by calculating local fractal dimensions on fragments of the contours map of the texture of iris. We propose to share the texture image in 256 blocks and then calculate the fractal dimension of each block. A vector of 256 elements is then recovered. Each of them describes the irregularity of the texture within the block to which it corresponds. In order to describe the variation of roughness within the texture of the iris, the gradient of the vector of dimensions is calculated. The result is then coded in '0' and '1' according to the sign. The required 255-bit code is retained as a biometric signature. The main steps for extracting the iris signature are described below.

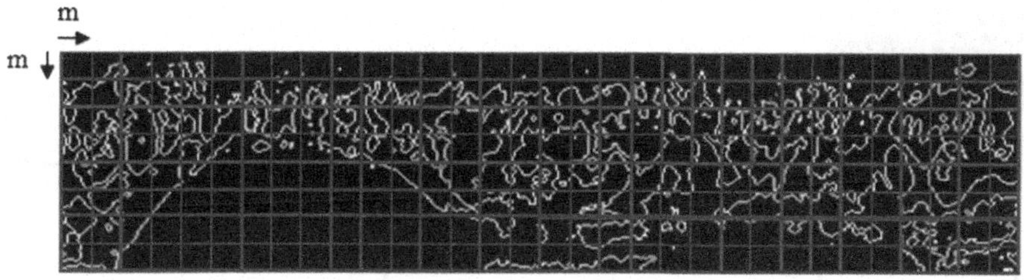

Fig. 3.3 Iris texture contours map divided into 256 blocks

Fig. 3.4 Example of an iris code (ID 023_D_10) in barcode

The contours map with the same size as the original image, $128 \times 512px$, is cut into blocks of size $(m \times m)$, with $m = 16px$ (Fig. 3.3).

Each block of dimension $(m \times m)$ is iteratively swept by cells of size $(k \times k)$:

1. We begin by $k = 2$ for the first iteration;
2. We compute the number of cells containing a fragment portion;
3. We increase k and we return to (2);
4. We stop when k reached $(m - 1)$;
5. We calculate the fractal dimension value $Df(m) = \ln(N)/\ln(k)$.

The dimensions calculated within blocks have values ranging from 0 to 2.The signature of the iris is extracted by the threshold relative to zero of the 256-dimensional gradient vector (Fig. 3.4).

3.4 Uni-Modal Recognition System

The building blocks of a recognition system are presented in the following diagram. The pre-processing step is used to localize and normalize the iris portion of the acquired image. Then, a fractal analysis is processed on each image to extract discriminative information of the iris. Finally, the iris signature is compared to the signatures in the database using the Hamming distance to calculate dissimilarity scores (Fig. 3.5).

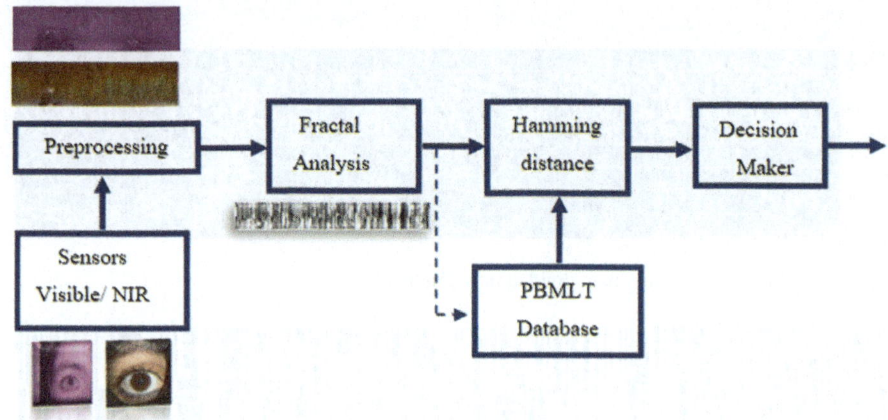

Fig. 3.5 Proposed system architecture

Table 3.1 Characteristic of PBLMTiris

Characteristic	PBMLTiris database
Sensor	Intern
Environment	Visible/Near infrared
DAPM	84.8 (4.5)
Numbers of persons	500
88.83 (10.1)	5.8 (5.5)
Session	2
Format	JPEG
Resolution (NIR)	3072 × 2304
Resolution (Visible)	2150 × 2100
Size (V)	Near 2.98MO
Size (IR)	Near 3.5MO

3.4.1 PBMLTiris Database Description

In the iris identification field, some databases of standard human eye iris images are used, the most common are the CASIA (in several versions) and UBIRIS (in other versions) databases (Roshini et al., 2017). The clarity of iris textures is the main goal in database acquisition. Iris identification has been an active topic research in our team. Therefore, an iris database for testing and evaluating our methods becomes more and more necessary. In order to satisfy this need, we intend to construct our database referred as PBMLTiris with high image quality. The characteristics of this database are depicted in Table 3.1.

PBMLT iris database (PBMLTirisV) is composed of 10000 images in the visible with V refers to visible and 10000 in the infrared (PBMLTirisNIR) with NIR

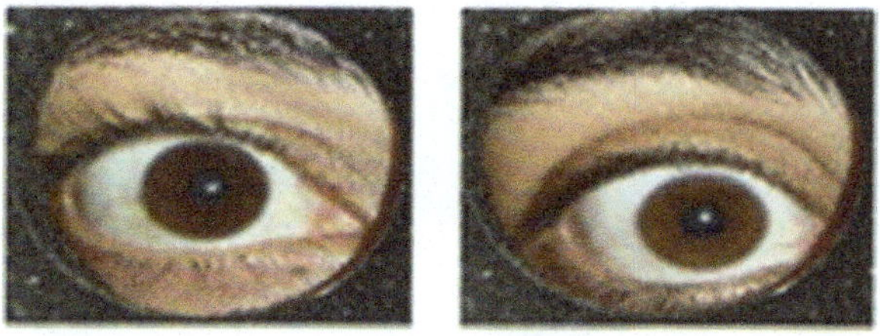

Fig. 3.6 Image examples from PBMLT irisV

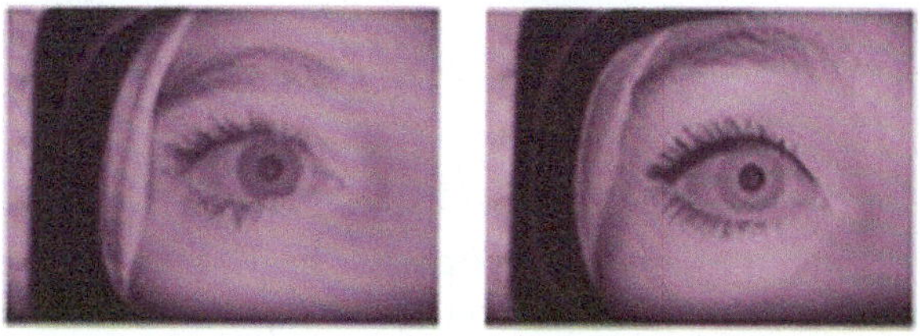

Fig. 3.7 Image examples from PBMLT irisNIR

refers to near infrared. The iris images are captured using two powerful digital cameras. PBMLTiris contains in each sub-base $(500 \times 2 \times 10) = 10000$ images of iris corresponding to 500 individuals.

These images are of very good quality with very clear iris texture details. Each person participates with 10 iris images. The same principle is applied to the images of PBMLTiris, it is taken under visible light. The images have also undergone several preprocessing before being made available to the researchers, as shown by Figs. 3.6 and 3.7.

3.4.2 Pre-processing

Before feature extraction process, a preprocessing step based on iris cropping is used. In fact, the image of the acquired eye does not only include the texture of the Iris. It is then necessary to segment and isolate this region of interest from the rest of the image. There are several methods in literature dealing with detection and cropping of human iris. The main objective is to eliminate no-essential information to recognition (outline, eyelashes, eyelids, and skin). Starting with detection of pupil center denoted as:

$$Cp = (Xp, Yp) \tag{3.1}$$

With Cp is the center of the pupil, Xp is the projection on the $x - axis$ and Yp is the projection on the $y - axis$ and we process by the following equations:

$$\begin{cases} Xp = Ri + \alpha \\ Yp = Ri + \alpha \end{cases} \tag{3.2}$$

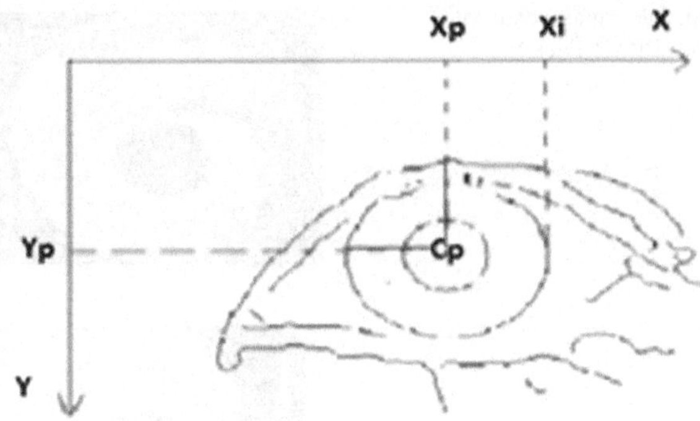

Fig. 3.8 Procedure of the pre-processing step

where α is a variable that expands the iris radius in order to cover all iris sizes in our database. After determining $Cp = (Xp, Yp)$, we can extract the radius of iris. Then, from the center of the pupil, we go as far as to find a second contour.

Figure 3.8 shows the procedure followed to locate the center of the pupil.

3.4.3 *Iris Segmentation (Daugman's Operator)*

(Daugman, 2009; Dixit and Kazi, 2015; Masšek, 2003; Thomas et al., 2016; Verma et al., 2012)

This part is very important because the performance the whole recognition system depends on how well the segmentation task performs. The main goal here is to find accurately the center coordinates and the radius of the iris and the pupil respectively using Daugman's equation. The center point of Daugman's theory of border recognition is simulated by the Daugman's Integro-Differential Operator equation as follows (Thomas et al., 2016):

$$\max(r, x_0, y_0)\, |G_\sigma(r) \times \sigma/\sigma r| \int_{(r,x_0,y_0)} I(x, y)/2\pi r\, ds \tag{3.3}$$

$I(x, y)$: the intensity of the pixel at coordinates (x, y) in the image of an iris.
(x_0, y_0): the assumed center coordinates of the iris.
r: denotes the radius of various circular regions with the center coordinates at (x_0, y_0).
σ: the standard deviation of the Gaussian distribution.
$G_\sigma(r)$: Gaussian filter of scale sigma (σ).
s: the contour of the circle given by the parameters (r, x_0, y_0) (Fig. 3.9).

Some segmentation results on the PBMLT database are presented on Figs. 3.10 and 3.11 for images acquired in the visible light and the NIR infrared. Then, in Table 3.2, the segmentation rate obtained with 240 images from PBMLT database is compared to the segmentation rate with the same number of images in the UBiris database.

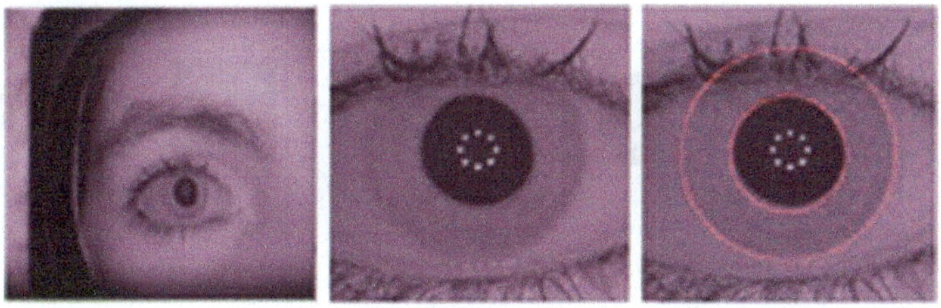

Fig. 3.9 Results of cropping and localization on PBMLTirisNIR ($ID100D01$)

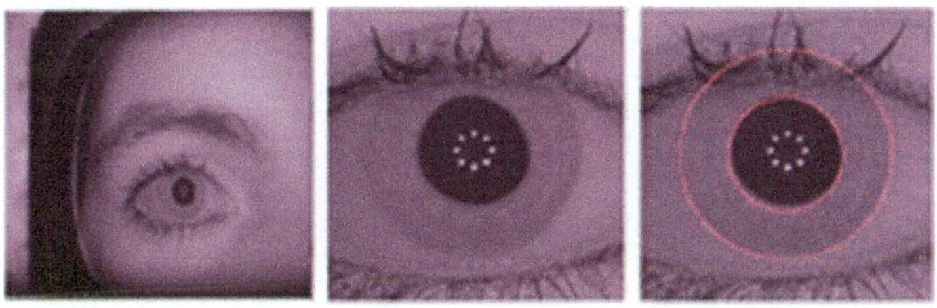

Fig. 3.10 Results of cropping and localization on PBMLTirisNIR ($ID100D01$)

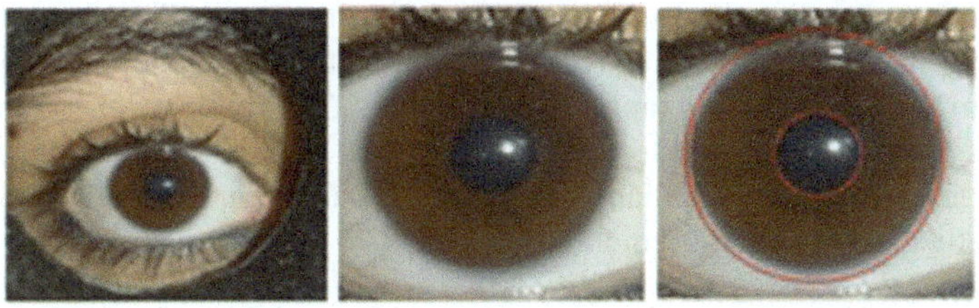

Fig. 3.11 Results of cropping and segmentation on PBMLTirisV ($ID100D01$)

Table 3.2 Segmentation rates of PBMLT and UBiris databases

	UBiris	PBLMTiris
Number of samples	240	240
Segmentation rate	95, 75%	99, 75%

3.4.4 Normalization Based on the Pseudo-Polar Method (Masšek, 2003)

The iris is an irregular disk. These irregularities are due to dilatation and contraction of the pupil. The two features that should be mentioned are: two circles that are not concentric and the non-linear changes of the texture of the iris. These features prompted Daugman to develop a method of pseudo-polar iris disk normalization

Table 3.3 Iris normalization

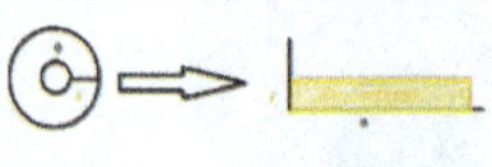		
Pseudoplar transformation	Iris contoured and iris rolled out (ID140L16-NIR)	Iris contoured and iris rolled out (ID140L16-V)

called the 'Rubber Sheet' method. This method is pseudo-polar because the two circles of the iris and the pupil are not concentric (Table 3.3).

3.4.5 *Matching*

After the feature extraction process based on the fractal analysis, we proceed to the matching step. In this stage, the iris code is compared to the codes in the database using the Hamming distance to calculate dissimilarity scores. These templates vary in size between researchers, but converting the biometric iris features to binary allows the codes to be compared very rapidly on computers. The Boolean operator exclusive OR (⊕) is typically used to carry out comparisons between the feature vectors of biometric templates. In Daugman's early work (Daugman, 2009), he presents the Hamming distance as per this equation:

$$HD = \frac{1}{N}\sum_{j=1}^{n} CodeA_j \bigoplus codeB_j \qquad (3.4)$$

The equation is a summation of the differences between the bits of the two iris codes. $CodeA_j$ and $CodeB_j$ are divided by the total number of bits N, in the iris code. For Daugman's method the total is 2048 bits, $CodeA$ is the iris code being tested, while $CodeB$ is the code that $CodeA$ is being compared to, and j is the bit number being tested, ranging between one and the total number of bits (Figs. 3.12 and 3.13).

Fig. 3.12 Example of an iris code (ID 023_D_10) in barcode

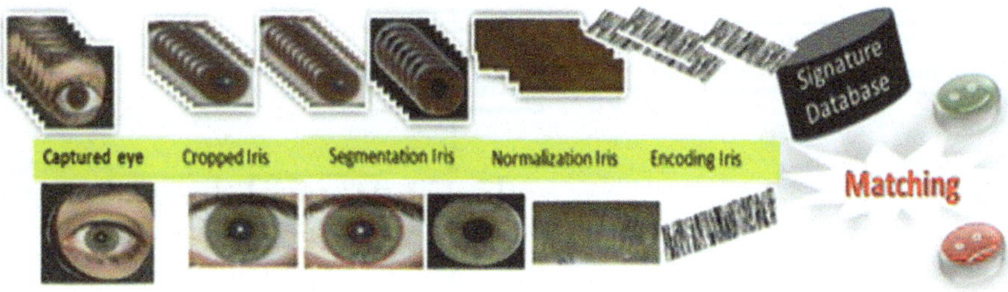

Fig. 3.13 Proposed system

3.5 Multi-modal Recognition System

3.5.1 Limitations of Uni-Modal Recognition System (Singh et al., 2019)

Uni-modal biometric system makes use of only one biometric trait to perform the recognition operation. Some limitations like poor data quality overlap between identities and illumination problem can affect the performance of such system. In our work, we will make use of a single biometric trait with different manners in order to construct a multi-biometric recognition system. In fact, the use of multiple sources often leads to an improved system reliability and enhanced recognition performance especially when the combined information is more relevant to discriminate an individual than the information obtained from a single source. Thus, combining information obtained from different sources can well contribute to the system performance.

3.5.2 Fusion Sources

3.5.2.1 Multi-Sensors (Song et al., 2019)

For iris recognition system, the image acquisition process can be achieved by using visible and near-infrared light to take a high contrast of a person's iris. With visible light, the iris texture is difficult to be discriminated especially for dark irises. Thus, the acquisition with infrared light can provide more information about the iris texture. Therefore, we can capture different information from the same modality by utilizing multiple sensors.

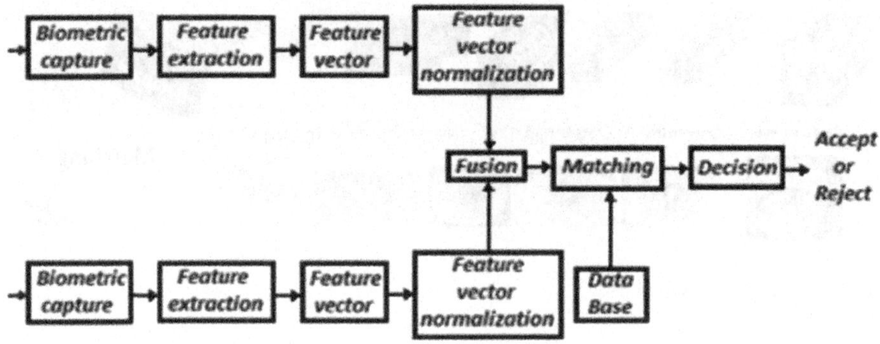

Fig. 3.14 Fusion at the feature level

3.5.2.2 Multi-Instances: Right and Left (Bharadi et al., 2018)

In the case of iris recognition system, humans have two irises. Thus, we can extract information separately from the left and right irises. Their extracted features can be used individually or in combined manner to improve the performance of the system by building multi-instance biometric system.

3.5.3 Fusion Levels

In multimodal biometric systems, there are a number of levels at which fusion can be incorporated.

3.5.3.1 Sensor Level

The data fusion is done at sensor level. Data acquired from various sensors to form one vector. Fusion of information obtained from many different sensors for a single biometric feature. The fusion is done immediately after acquisition.

3.5.3.2 Feature Level

The fusion at the feature level is achieved by different feature vectors extracted from various biometric modalities. In our case, the fusion can be done with the extraction of features of iris with different manners (Fig. 3.14).

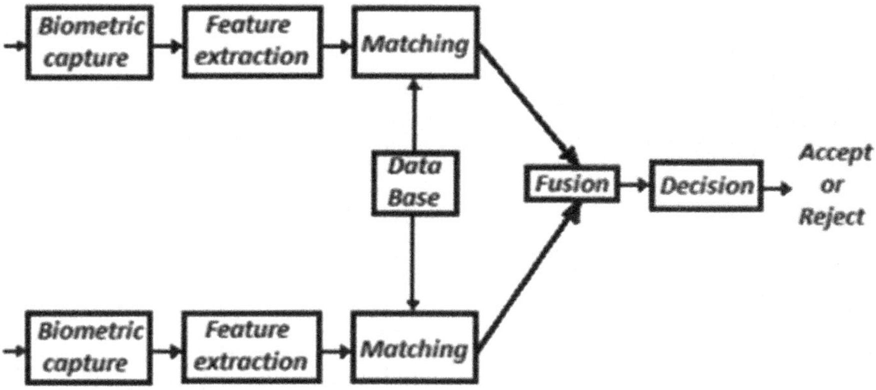

Fig. 3.15 Fusion at the score level

3.5.3.3 Score Level

In this fusion level, each unimodal biometric system implements matching algorithm to its extracted feature vectors independently. Then fusion is done between the matching results (Fig. 3.15).

3.5.3.4 Decision Level

Each uni-modal biometric system makes its decision separately. Then, these individual decisions are combined to produce the final decision: rejection or acceptance.

3.6 Experimental Results

3.6.1 Segmentation Results

The proposed segmentation method was tested on our database PBLMTiris (V and NIR) containing about 1800 images of iris in the visible environment and 1000 images of iris in the near infrared environment. To conduct a comparison study, we make use of UBIRIS database from the literature. The characteristics of this latter are presented in Table 3.4.

Table 3.5 summarizes the obtained results by the two variants of PBMLTIRIS and UBIRIS. It is clear that the segmentation rate of PBMLTirisNIR outperforms the UBIRIS segmentation rate.

Table 3.4 Characteristics of UBIRIS

Characteristic	UBiris database
Sensor	Intern
Environment	Near infra-red
Numbers of persons	240
Session	2
Format	JPEG
Resolution	2560 × 1704

Table 3.5 Comparison of the obtained results by the segmentation method

PBLMTiris (%)	PBLMTirisV (%)	PBLMTirisNIR (%)	UBIRIS (%)
Right segmentation rate	92.25	95.30	95.75
Left segmentation rate	90.25	96.56	

3.6.2 Uni-Modal System Evaluation

Our proposed recognition system was tested on the PBLMTiris database (V and NIR) in order to evaluate its performance. Our database PBLMTiris is divided into two parts (one for learning the other for testing). Firstly, we evaluated intra-class distances for all the possible combinations of genuine attempts. Secondly, we evaluated the inter-class distance for all the possible combinations of impostor attempts. We start by checking if a person is who he claims to be. This amounts to comparing the distance between two signature vectors with respect to a previously determined threshold during a learning phase. This consists of calculating the hamming distances between all the irises of the learning database and drawing up a distribution. Figures 3.16, 3.17 and 3.18) show these distributions.

These later show the learning results of the two variants of our PBLMTiris database and UBIRRIS. It is clear that the areas of uncertainty, which corresponds to the overlapping part of the two distributions for PBMLTirisNIR and UBIRIS are quite similar. We used the Receiver Operating Characteristic (ROC) curves for evaluating the performance of our proposed work. The ROC curves plot FRR (false rejection rate) versus FAR (false acceptance rate) plots. Unlike the Equal Error Rate (ERR) that considers only a pair of FAR and FRR, the ROC curve describes a variation of FRR versus FAR and thus can give an evaluation of a biometric system. Figure 3.19 shows the Receiver Operating Characteristic (ROC) curves for the two variants of our PBLMTiris database and UBIRRIS.

Table 3.6 summarizes the results obtained by used databases in the verification mode.

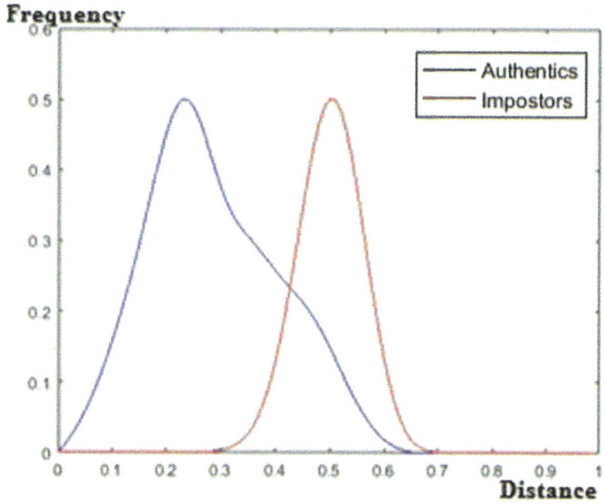

Fig. 3.16 Authentics and impostors distribution distances on PBMLTirisV

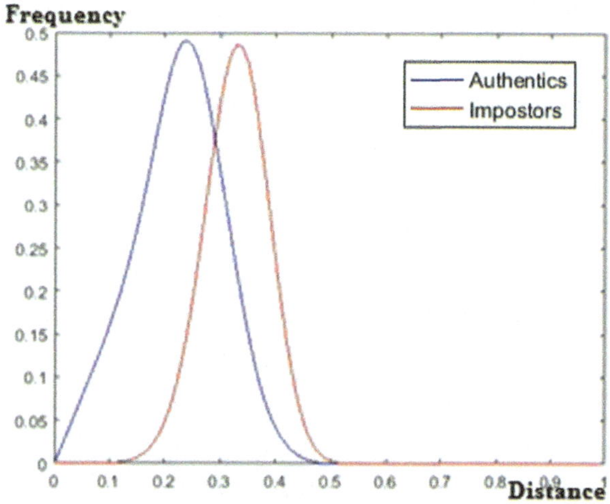

Fig. 3.17 Authentics and impostors distribution distances on PBMLTirisNIR

Table 3.6 Characteristic of Ubiris and PBLMTiris (pair of iris)

Database	EER	FRR	CRR	CAR	FAR	SURFROC
UBiris	0.29	13.92	85.81	86.08	19.19	0.93
PBLMTiris NIR	0.29	13.77	85.43	86.23	4.57	0.8985
PBLMTirisVisible	0.42	11.09	88.91	86.80	13.20	0.9415

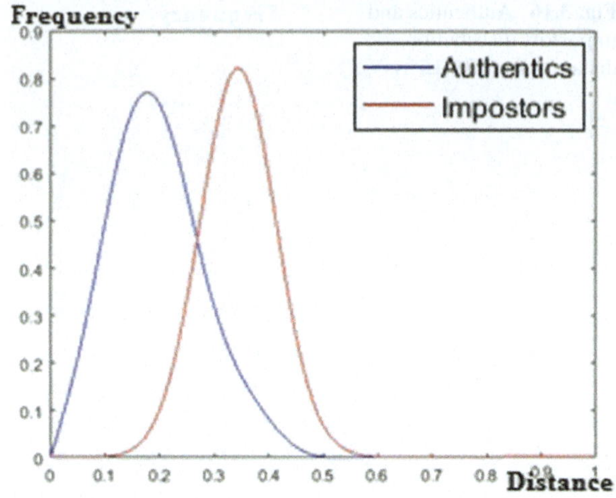

Fig. 3.18 Authentics and impostors distribution distances on UBiris

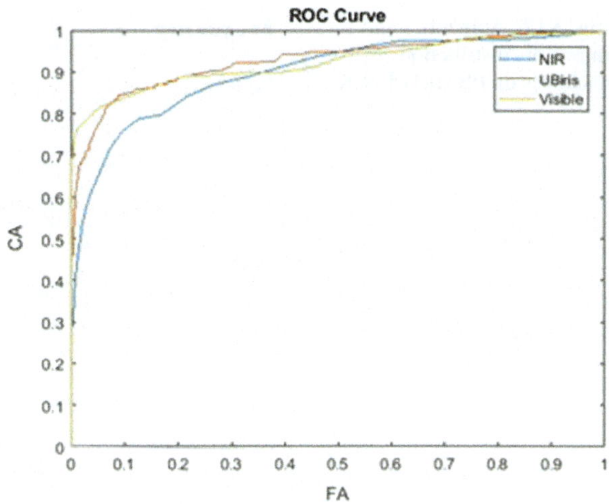

Fig. 3.19 ROC curves for different databases

3.6.3 *Feature Level Fusion Results*

In our work, features of the same nature and dimension are concatenated. The merge is performed by concatenating images of:

- The left iris and the right iris in the visible mode
- The left iris and the right iris in the near infrared mode

The obtained results are presented inf Figs. 3.20 and 3.21.

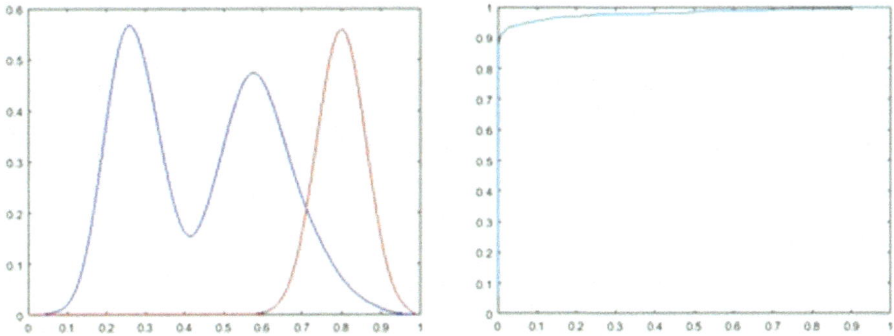

Fig. 3.20 LeftV iris and right V iris $EER = 0.7280$ $FAR = 6.5107$ $CRR = 93.4893$ $FRR = 5.4201$

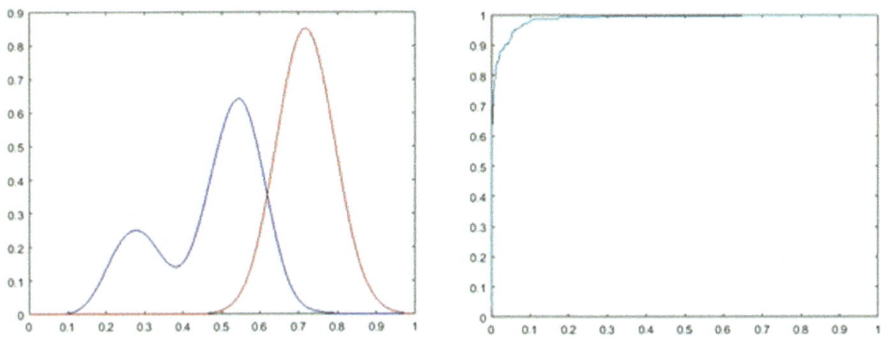

Fig. 3.21 Left NIR iris and right NIR iris $EER = 0.6240$, $CRR = 94.3825$, $FAR = 5.6175$ and $FRR = 5.4422$

3.6.4 Sensor Level Fusion Results

In our case, we have two sensors (in visible light sensor (V) and in near infrared light (NIR)). The fusion is performed by concatenating images of:

- The leftV iris and the right NIR iris
- The leftV iris and the left NIR iris
- The rightV iris and the left NIR iris
- The rightV iris and the right NIR iris.

The obtained results are depicted in Figs. 3.22, 3.23, 3.24 and 3.25.

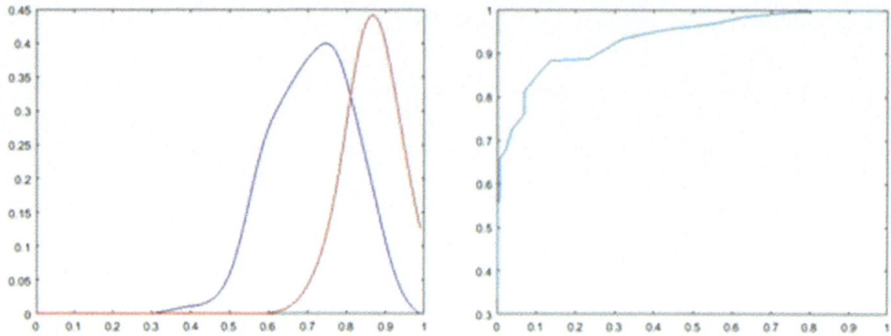

Fig. 3.22 LeftV iris and the right NIR iris, $EER = 0.8040$ $FAR = 13.4715$ $CRR = 86.5285$ $FRR = 11.7949$

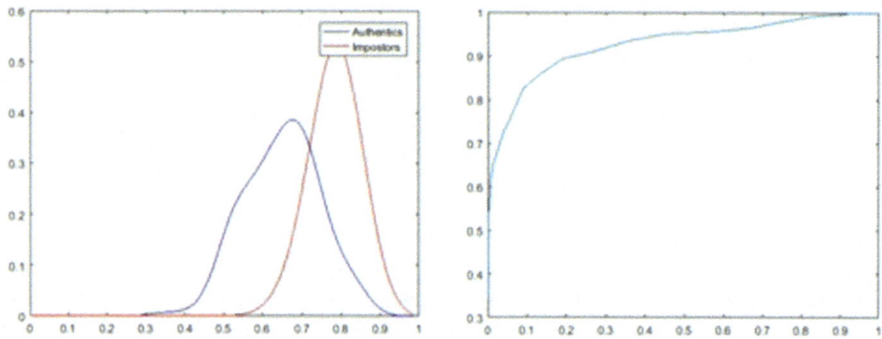

Fig. 3.23 LeftV iris and the left NIR iris $EER = 0.7170$ $FAR = 13.4636$ $CRR = 86.5364$ $FRR = 13.9918$

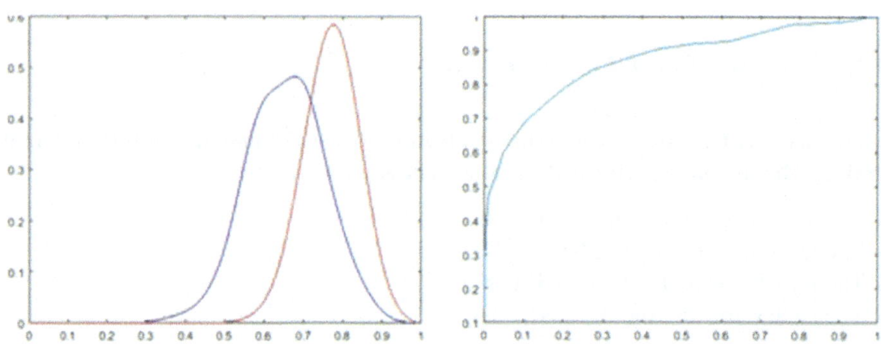

Fig. 3.24 RightV iris and the left NIR iris $EER = 0.8040$ $FAR = 13.4715$ $CRR = 86.5285$ $FRR = 11.7949$

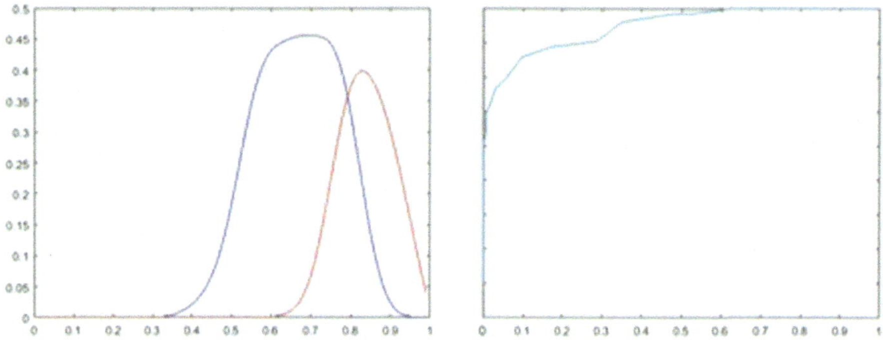

Fig. 3.25 RightV iris and the right NIR iris $EER = 0.7620$ $FAR = 9.6939$ $CRR = 90.3061$ $FRR = 14.1414$

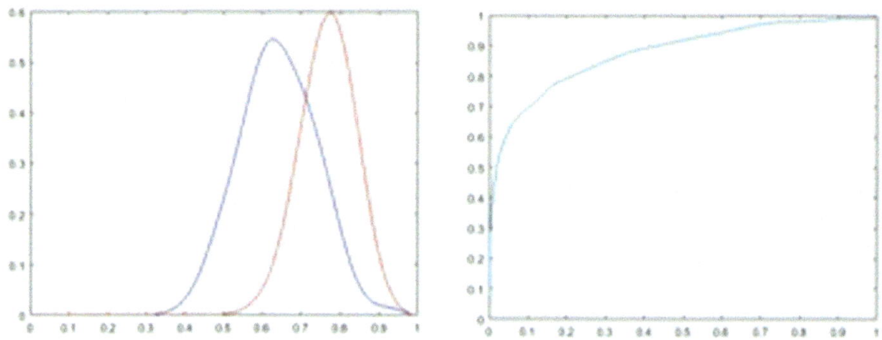

Fig. 3.26 Maximum score $EER = 0.7190$ $FAR = 22.2488$ $CRR = 77.7512$ $FRR = 19.3548$

3.6.5 Score Level Fusion Results

The combination of scores is based on mathematical methods of combination. In the combination approach, individual scores are combined to form a single score, which is then used to make the final decision. Kittler et al. (1998) developed a theoretical framework for combining identification information obtained from multiple classifiers using simple combination methods such as sum rule, product rule, max rule, min rule, and median rule (Figs. 3.26, 3.27, 3.28 and 3.29).

3.7 Discussion and Conclusion

Iris biometry is one of the technologies that provide a high level of security due to the unique characteristics of the iris texture and stability that is extended during the human life. This work falls within a national project that aims at the fight against

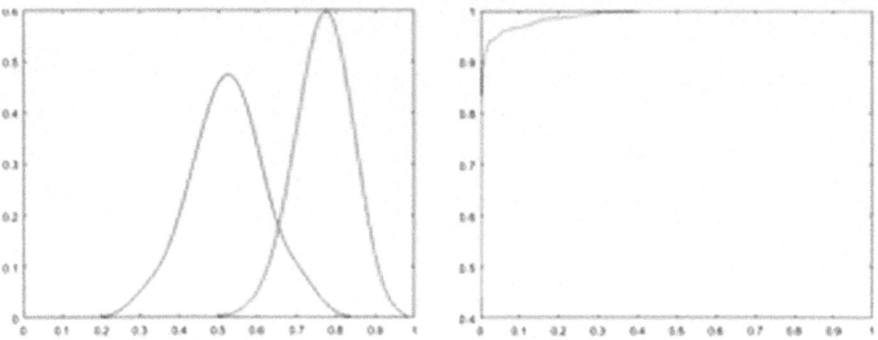

Fig. 3.27 Minimal score $EER = 0.6620$ $FAR = 5.7183$ $CRR = 94.2817$ $FRR = 4.0323$

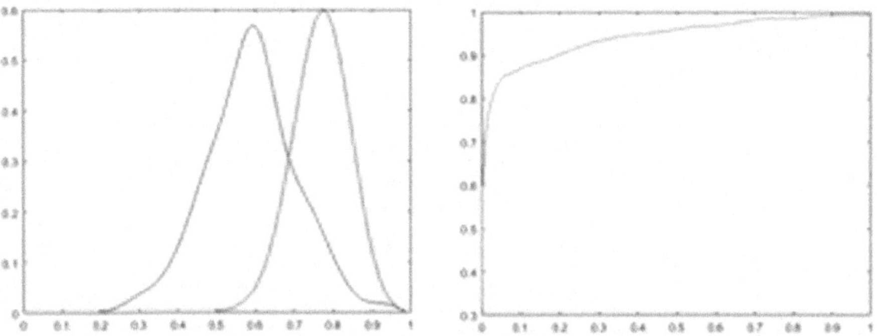

Fig. 3.28 Equal score $EER = 0.6910$ $FAR = 11.7627$ $CRR = 88.2373$ $FRR = 12.3656$

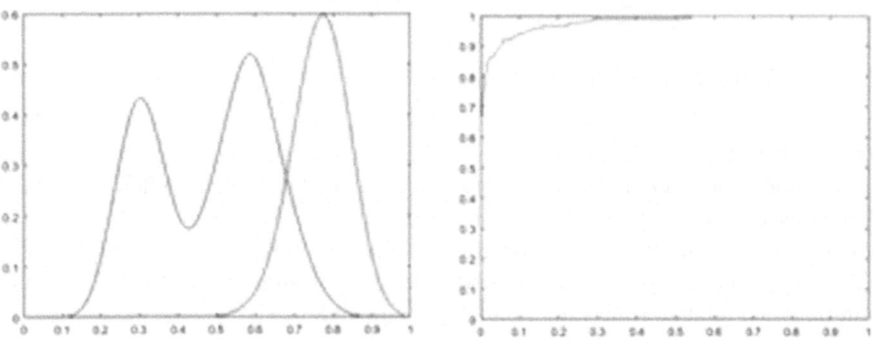

Fig. 3.29 Medium score $EER = 0.6770$, $FAR = 8.2703$, $CRR = 91.7297$, $FRR = 7.7957$

terrorism. In order to succeed and design an effective iris recognition system, a benchmark database is necessary. For this end, ideal conditions for iris images acquisition are assumed in our work. We have used Daugman method for segmentation step and fractal based method for feature extraction. Our proposed uni-modal method can give up to 83 % correct acceptation rate (CCR) of the recognition system. We noticed that the mean value of FAR and FRR with PBLMTirisNIR is less than PBLMTirisV. For the multi-modal method based on the fusion of information for the iris trait, we notice a considerable improvement in the results. In fact, the CRR exceeds 93% with the feature level merging method and the score level method.

Acknowledgment This work has been supported by the project PBMLT (Plateforme Biométrique Multimodale de Lute contre le Terrorisme) and financed by the Ministry of Higher Education, Tunisia.

References

Azizi, A., & Pourreza, H. R. (2009). Efficient iris recognition through improvement of feature extraction and subset selection. arXiv:0906.4789.

Bharadi, V., Shah, D., Thapa, N., Pandya, B. H., & Cosma, G. (2018). Multi-instance iris recognition. In *2018 Fourth International Conference on Computing Communication Control and Automation (ICCUBEA)* (pp. 1–6). IEEE.

Daugman, J. (2009). How iris recognition works. In *The essential guide to image processing* (pp. 715–739). Elsevier.

Dixit, A., & Kazi, M. K. (2015) Iris recognition by daugman's method. *International Journal of Latest Technology in Engineering, Management & Applied Science, India, IV*(VII).

Dubey, R., & Madan, A. (2014). Iris localization using daugman's intero-differential operator. *International Journal of Computer Applications, 93*(3), 06–12.

Gale, A. G., & Salankar, S. (2014). A review on advance methods of feature extraction in iris recognition system. *IOSR Journal of Electrical and Electronics Engineering (IOSR-JEEE) e-ISSN*, 1676–2278.

Jain, A. K., Ross, A., Prabhakar, S., et al. (2004). An introduction to biometric recognition. *IEEE Transactions on Circuits and Systems for Video Technology, 14*(1).

Khanfir, I. (2009). *Contribution Ã l'Identification d'Individus par l'Iris*. Ph.D. thesis, Tunisia, Electrical Engineering from the National Engineering School of Sfax.

Kittler, J., Hatef, M., Duin, R. P., & Matas, J. (1998). On combining classifiers. *IEEE Transactions on Pattern Analysis and Machine Intelligence, 20*(3), 226–239.

Krichen, E. (2007). *Reconnaissance des personnes par l'iris en mode dégradé*. Ph.D. thesis, Evry, Institut national des télécommunications.

Llano, E. G., García-Vázquez, M. S., Zamudio-Fuentes, L. M., Vargas, J. M. C., & Ramírez-Acosta, A. A. (2017). Analysis of the improvement on textural information in human iris recognition. In *VII Latin American Congress on Biomedical Engineering CLAIB 2016, Bucaramanga, Santander, Colombia, October 26–28th, 2016* (pp. 373–376). Berlin: Springer.

Mašek, L. (2003). Recognition of human iris patterns for biometric identification.

Roshini, M., Vasanth, G., Kodandaramaiah, G., & Jilani, S. (2017). Literature review on segmentation in iris recognition. *International Journal, 7*(1).

Seung-In, N., Bae, K., Park, Y., & Kim, J. (2003). A novel method to extract features for iris recognition system. In *International Conference on Audio-and Video-Based Biometric Person Authentication* (pp. 862–868). Berlin: Springer.

Singh, M., Singh, R., & Ross, A. (2019). A comprehensive overview of biometric fusion. *Information Fusion, 52*, 187–205.

Song, Y., He, Y., & Zhang, J. (2019) A survey of visible iris recognition. *AIRCC Publishing Corporation, 9*(3).

Sreelekha, K. V. R. (2014). A human iris recognition using Haar wavelet decomposition. *International Journal of Science and Engineering Research (IJOSER)*.

Thomas, T., George, A., & Devi, K. I. (2016). Effective iris recognition system. *Procedia Technology, 25*, 464–472.

Verma, P., Dubey, M., Verma, P., & Basu, S. (2012). Daughman's algorithm method for iris recognition-a biometric approach. *International Journal of Emerging Technology and Advanced Engineering, 2*(6), 177–185.

Xu, G., Zhang, Z., & Ma, Y. (2006). Improving the performance of iris recognition system using eyelids and eyelashes detection and iris image enhancement. In *2006 5th IEEE International Conference on Cognitive Informatics* (Vol. 2, pp. 871–876). IEEE.

Yu, M., Lin, Y., Tang, X., Xu, J., Schmidt, D., Wang, X., & Guo, Y. (2015). An easy iris center detection method for eye gaze tracking system.

Part II
Authentication by Biological Signals

Chapter 4
Security with ECG Biometrics

Juan Sebastian Arteaga Falconi and Abdulmotaleb El Saddik

Abstract Security with biometrics has become popular with the increase use of technology and the availability of portable gadgets. Available commercial products include face recognition, fingerprints and other biometrics to grant access to authorized users. ECG as biometrics is a relatively new technology and has advantages over traditional biometrics. This chapter describes the advantages of ECG as biometrics and provides a general overview of ECG as a biometrics technology. It describes general biometric concepts and explains terminology used in ECG and terminology used in ECG as a biometric. The chapter describe the approaches taken by research around this topic and explains each step of the process of ECG as Biometrics. It includes an overview of popular algorithms, filters, and classifiers. The chapter finishes with the evaluation process which describes arrangements, datasets and metrics used by the literature to present results.

Keywords ECG Authentication · ECG Biometrics · ECG · EKG · Electrocardiograms · Identification · Security

4.1 Biometrics Definition

Biometrics has two connotations, one for medicine and one for forensics/security. In medical terms, biometrics is the science of measuring elements related to health. These measurements can be related to a disease, a medicament or to physical characteristics of the population e.g. height and weight. In terms of forensics and security, it refers to the science of identifying an individual based on unique physical characteristics (Ross and Jai, 2015) and is widely used in modern day applications such

J. S. Arteaga Falconi · A. El Saddik (✉)
University of Ottawa, Ottawa, Canada
e-mail: elsaddik@uottawa.ca

J. S. Arteaga Falconi
e-mail: jarte060@uottawa.ca

N. Derbel and O. Kanoun (eds.), *Advanced Methods for Human Biometrics*,
Smart Sensors, Measurement and Instrumentation 40,
https://doi.org/10.1007/978-3-030-81982-8_4

as digital twins (El Saddik, 2018). This chapter will refer to biometrics within the second connotation.

The physical characteristics of an individual can be behavioural or physiological. Behavioral characteristics are unique patterns of an individual when performing a specific task such as: walking, talking, writing, typing. An individual can consciously alter-for a moment-a behavioural biometric trait, but in the long term the execution of a behavioural task will reveal the genuine trait. On the other hand, unique physiological characteristics are task independent and need a physical damage to alter the trait. Iris, Fingerprints, Face, Ear shape among others are some examples of physiological traits. People are in constant use of behavioural and physiological biometrics. They recognize each other by their unique physical aspect-face recognition-and trigger an interaction with the identified subject. Also, is common to know who is approaching before looking at the person based on the unique sound of their steps or voice if they are speaking.

The previous paragraph did not mentioned Electrocardiograms-ECG-in any category. This is because some authors consider ECG as behavioural biometrics and others as physiological. These differences arise from the fact that physical and psychological activities-such as running, drinking coffee, stress-can change the heart rate-which is present in the ECG. However, this chapter considers ECG under the physiological category because-although it can be altered-research in ECG biometrics indicates it is possible to correct these variations through a normalization procedure (Israel et al., 2005).

Nonetheless if it is behavioural or physiological, biometrics has two operation modes: identification and authentication (Wayman, 2015). It is common to see a misinterpretation of these terms. Identification is the operation mode in which a biometric system matches biometric data of a subject-call it query subject-against the stored biometric data of many other subjects-one-to-many matching scheme. As result, it will determine the identity of the query subject based on the closest match to the stored biometric data. In contrast, authentication matches the biometric data of a query subject against a biometric data of one individual called genuine. This is a one-to-one matching scheme and the objective is to determine if the query subject is a genuine or impostor user (Ross and Jai, 2015).

Biometrics have an important role in areas like forensics. Where, they have developed many biometrics identification techniques, including fingerprints. Fingerprints have been used for almost one hundred years as linking evidence to solve crimes (Maltoni, 2015). People leave fingerprint traces on the objects they touch. In forensics, investigators can recuperate these prints to match against a database of possible suspects. The vast experience with fingerprints in forensics, has made fingerprints one of the fastest and accurate biometric approaches. Nowadays, many secure applications implement biometrics as a mean to identify/authenticate people. Biometrics is displacing traditional techniques such as passwords because biometrics is not transferable and more secure-at least it demands a bigger effort to spoof it. It is common to see biometrics implemented in everyday gadgets like mobile phones or personal computers.

Advantages of biometrics in forensics might be a disadvantage for secure applications. If people leave traces of a biometric trait-like fingerprints-then it becomes available to attackers. This leaves an open door to forgery. Duplication of biometric traits is a security risk because an attacker can use it to steal personal information from a protected devices or can access to a protected location. An attacker can collect most biometric traits, face with a picture, voice with a recorder, fingerprints with latex (Roberts, 2007). There are solutions around each biometric technology to detect forgery, however the success of fake traits still high (Tan and Schuckers, 2010). The problems of biometric duplication have led to the development of new biometric approaches. One of them is ECG biometrics and next section will discuss it in more detail.

4.2 Biometrics with ECG

ECG or EKG is the abbreviation for electrocardiograms. ECG is the representation of the electrical activity of the heart and physicians use it to diagnose heart conditions. Each person posses a unique ECG shape and represents a challenge to perform diagnoses. To overcome this problem, physicians have established parameters for each ECG characteristic. If an ECG characteristic its out of the established boundaries, it will help a physician to determine a specific heart condition. This challenge in medicine, is an advantage in biometrics. The uniqueness nature of ECG makes this bio signal suitable for biometrics. Also, ECG does not leave traces of biometric traits as mentioned in the previous section and is hidden to the naked eye. This is an advantage over traditional biometrics such as fingerprints because it makes it more difficult to duplicate and therefore reduces the risk of forgery.

ECG represents an advantage to prevent forgery, but the speed and novelty of ECG biometrics is a constraint. In terms of speed, ECG is slow. Low speed is not related to the processing complexity. It is related to the collection of information to get accurate results. The resting average heart rate of person is 60–100 bpm or beats per minute-this excludes athletes that could go lower to 40 bpm. Within this range it will take 1–0.6 s to get the information for one beat. More heart beats provide more information and leads to a higher accuracy, but time will increase. This shows a direct relation between accuracy and time and next sections will discuss this aspect in more detail. But it is important to clarify that processing speed of ECG is not a major concern, the concern is the collection information for accurate results. In terms of novelty, ECG is a relatively new approach for biometrics. ECG as biometrics was first proposed by Biel et al. (2001) and the advances of today are not comparable to well established biometrics such as fingerprints. This is a constraint for immediate implementation, but it represents great research opportunities in this area.

Terminology in ECG can create confusion, specially the term "lead". Lead refers to the electrode placement position and not to the number of electrodes, although is closely related. In medicine they denote ECG lead as: I, II, III, aVF, aVR, aVL, V1, V2, V3, V4, V5, V6. Each lead offers a different perspective of the ECG signal which

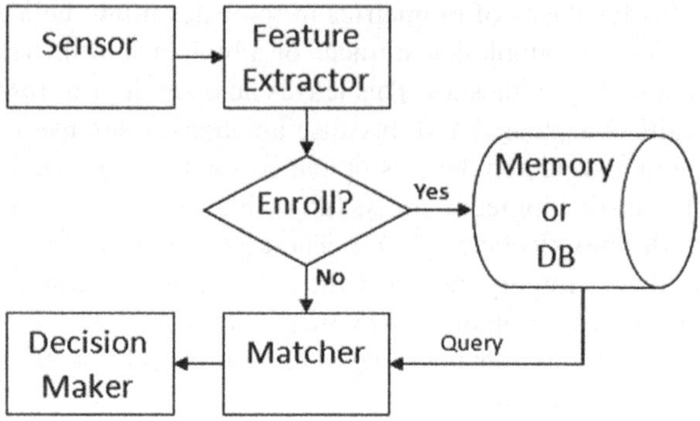

Fig. 4.1 ECG biometric modules

help physicians to determine a heart condition. Leads I, II and III use three electrodes and displays all leads at the same time. Two electrodes will display only one lead-could be any lead, not necessarily led I. A five-electrode arrangement displays led I, II, III and only one of the V1-6. A twelve-electrode arrangement will display all the leads at the same time (Goldberger, 2012). During implementation, it is important to consider two electrodes for ECG biometrics. It is more convenient for users to provide information within two electrodes. The placement of more than two electrodes will require some skills from users. With two electrodes, users can provide their ECG with both hands or a finger in each hand. Each hand or finger will touch one electrode which constitutes a lead I arrangement. This is not a fixed rule, it depends on the application requirements and its implementation will dictate the most appropriate approach.

The authentication procedures in ECG biometrics are like other biometric procedures. First it has two working phases: enrollment and verification. Enrollment phase "trains" the system with a genuine user data, generates and store his biometric template. Verification phase generates a biometric template from a querying user and compares it against the stored template of a genuine user. This phase will determine if the user is genuine or impostor.

The working phases of ECG biometrics comprise of the modules shown in Fig. 4.1: sensor, feature extractor, enrollment, storage, matcher and decision maker. The sensor is the capturing interface of the biometric information, ECG uses an electrocardiograph. It captures the signal according to the chosen lead configuration. Most of the ECG biometrics research focus on lead I because is the most convenient for an implementation, it facilitates the placement of ECG electrodes on the user. The feature extractor module processes the signal and extract the unique biometric features that will generate a biometric template. If the working phase is enrollment, then it will store the biometric template in memory. If is in the verification phase it will go to the matching module. The matching module compares the biometric template from the feature extractor against the biometric template stored in memory. It implements an algorithm which will provide a result value to the decision maker module. The decision maker module takes this value and determines if is a match or non-match. In another words, determine if the querying user is a genuine user or an impostor.

4.3 ECG Biometrics Approaches

Research in biometrics have contributed with many approaches to implement ECG as biometrics. All these approaches can be classified in 3 groups: fiducial, non-fiducial and hybrid (Odinaka et al., 2012). A fiducial approach extracts specific biometrics features such as valley to peaks distances, amplitudes, angles. In general, it detects a specific feature of the signal and performs a biometric match based on the correlation of these features. Non-fiducial approaches extract features based on the whole signal and not a specific characteristic. Wavelets or autocorrelation coefficients are common approaches in this group. Hybrid approaches use a combination of the previous two. This chapter considers hybrid approaches as a multibiometric solution. This is because hybrid approaches will use the same signal but a different algorithm to perform matching; multibiometrics classifies this solution as a multi-algorithm approach.

4.3.1 Fiducial Approaches

As mentioned earlier, Biel et al. (2001) where the first proposing ECG as biometrics-to the best of the knowledge of the authors of this chapter. They proposed the extraction of ECG biometric features based on time, amplitude, and slope. See Fig. 4.2. They experimented with several leads and they conclude that lead arrangement of three electrodes is enough to perform authentication.

Singh and Singh (2012) authenticate user by calculating the Euclidean distance on 20 ECG features extracted from the fiducial points. They use 3 min ECG to obtain 82% True Acceptance Rate-TAR-and 7% False Acceptance Rate-FAR.

In another approach, Israel et al. (2005) studies the effects in ECG authentication by changes in the heart rate due to physical or psychological conditions. They found heart rate changes in the ECG fiducial features are linear and they corrected with a normalization procedure. They use ECG samples lengths of 20 s.

Arteaga-Falconi et al. (2016) studies ECG authentication time. They perform authentication aiming for mobile devices. To achieve it, they reduced the authentication time to 4 s by an independent threshold adjustment on the biometric features-these features are based on fiducial points-and the use of an R peak detection algorithm with a lower average time error (Arteaga-Falconi et al., 2015). They maintain a TAR of 84.93% while decreased the FAR to 1.29%.

Many other studies achieve better results, but it is important to tell if they use a multibiometric approach. A unibiometric approach-one biometric-cannot compare

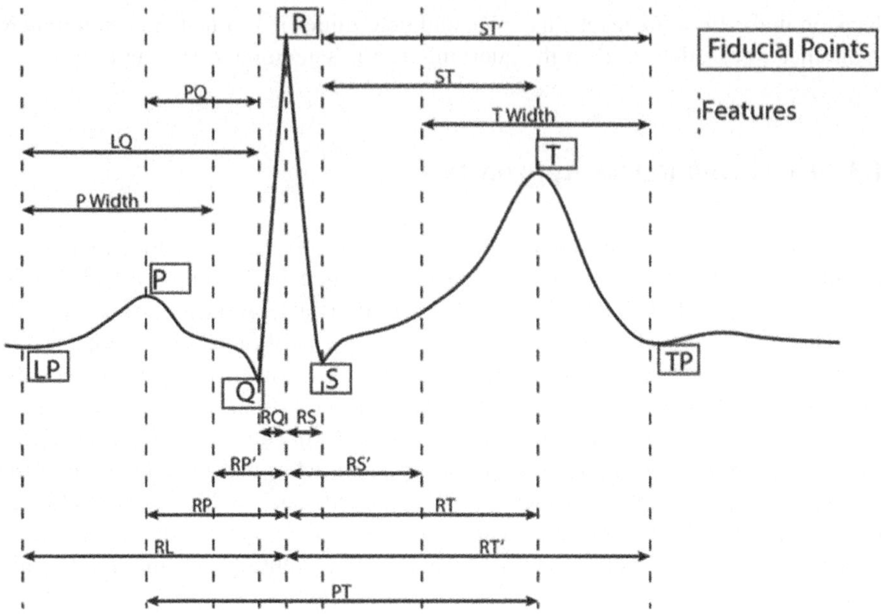

Fig. 4.2 ECG Features from fiducial points

results with a multibiometric approach. Multibiometric approaches provide more information and the outcome has a better accuracy.

Multibiometrics in ECG is a wide topic and a different work should analyze this topic and provide more specific details for a better comprehension of the topic.

4.3.2 Non-fiducial Approaches

Chan et al. (2008) does not extract features based on fiducial points. They segment the ECG signal by heart beats and align them to each other. They use wavelet coefficients as their features to perform matching. They perform their experiment with lead I ECG signals collected from the fingertip of the users.

Another non-fiducial approach (Wübbeler et al., 2007) places the heart beats in vectors and measures the distance between them to establish a match. They use 10 s of ECG signal in lead I, II and III placements. They with data of different dates ranges from months to years. Based on their experiments they conclude stability-validity-of ECG biometric features up to 2727 d-no alterations of ECG biometric features. This time is not the maximum in which ECG still stable, it is the maximum time in which they perform their experiments. The validity of ECG still not clear and it needs more research.

ECG biometrics with autocorrelation is another non-fiducial approach and is proposed by Plataniotis et al. (2006). They use an autocorrelation to extract coefficients from the ECG signal. They reduce the number of coefficients with a Direct Cosine Transform-DCT. They used 10 s ECG signal and perform matching with Euclidean distances and Gaussian log likelihood metrics.

4.4 ECG Signal Filters

A high-quality ECG signal is essential to have high accuracy in ECG biometrics. The quality of the signal depends on the quality of the electrocardiograph instrument. Due to health regulations electrocardiographs used in medicine provides the best ECG signals. However, those are for medical purposes and they might not be suitable-due economical reason and mass production-to use them in biometrics. Independent of the equipment quality, signal filtration will improve ECG signals. Obviously, non-medical equipment requires a better filtration of the ECG signal.

There are three main source of noise in ECG signals: muscle, power source and baseline wander (AlMahamdy and Riley, 2014). The source of muscle noise is the electrical stimulation of muscles to create body movements. Power source noise comes from the power line, which 60 Hz in North America 50 Hz in Europe. Part of the power source noise includes other sources such as electronic devices which induces noise in other band frequencies. Baseline wander noise is cause by the respiration movements and contact with the ECG electrodes.

The basic filters used in biometrics are high pass, low pass, band pass and smooth filters. High pass filters remove the muscle noise and the cut frequency is usual 0.5–1 Hz. Low pass filter removes the high frequency components generated by another electronic device. Band pass filters remove the power source noise and the cut frequency varies from each region. After the filtration noise might still be present in the signal. Smooth filters help to remove any residue of noise. It provides a cleaner ECG signal by removing any interference left in the ECG.

Many filter solutions remove noise from ECG signals. The appropriate selection of each filter depends on many factors such as: equipment, environment, and application. Some filters might work well in some scenarios and might not work in others.

Common filters for ECG are wavelets, adaptive filters and the Savitzky-Golay filter. Any of these filters-except the last one-can remove high, low or a specific band of frequencies. The Savitzky-Golay filter is a smoothing filter and is widely use in ECG.

ECG uses wavelets filters because has the advantage of removing frequency components at a specified time. It works on time and frequency components. Filtration of ECG signals use the Discrete Wavelet Transform-DWT. This is because modern ECG signals are digital, and this approach is appropriate for this scenario. Although, ECG filtration can use Continuous Wavelet Transform, it might not be as efficient as DWT in terms of complexity.

Adaptive Filters are also common for ECG filtration as well. However, for ECG biometrics might not be the best choice. The adaptation of the signal to a pattern might cause to lose some vital information for biometrics. ECG biometrics cannot completely discard the use of these filters, but it must implement it with caution. It might lead to a reduction of accuracy or incorrect evaluation results. ECG biometrics can use Least Mean Square-LMS-or Recursive Least Squares-RLS-approaches. LMS are simpler and presents an advantage to approximate unknown signals. RLS requires less interaction to converge, however each interaction is more complex.

The Savitzky-Golay filter smooths the ECG signal by removing any remaining noise. This filter uses polynomial approximations in a specified window length. The fitting polynomial should have coefficients in an order less than the window size. The advantage of Savitzky-Golay is these filters have a constant coefficient through the moving windows which makes it simple to apply.

In general, ECG signals can apply many filters to remove the noise. However, ECG biometrics always considers filters that has less affection on the biometric information. This is in amplitude and time components. As mentioned earlier; the use of filters will depend on many conditions and noise will always affect the biometric information on the signal. The application should look at the filters that affects the less on their purpose. For biometrics is the uniqueness of the ECG signal.

4.5 ECG Biometric Classifiers

The ECG biometric classifier holds two modules from the general view presented in Fig. 4.1. The matcher and the decisions maker are part of the biometric classifier. The matcher generates the scores and the decision maker decides if the scores are in the boundaries of a genuine or an imposter.

In ECG biometrics the information is proportional to the signal time length. A longer ECG signal provides more information and as a result the classifier will have a higher accuracy. The time length in the classifier will depend on the application. If the application is for high security purposes a longer ECG signal is the appropriate approach. If the application requires constant access-e.g. mobile phones-then a shorter time will be the best approach, but this should not set the accuracy bellow an accepted security standard.

In the event an application cannot alter the signal length and requires a higher accuracy; the selection of an appropriate classifier improves the accuracy-as mentioned earlier, filtration stage is also important. Literature provides many classifiers approach, each one with advantages for a specific application and disadvantages for other applications. This section will describe these approaches in a general manner.

ECG biometric classifiers have two categories: manual and machine learning classifiers. Both approaches require a learning or enrollment stage. As mentioned earlier, the enrollment stage registers a genuine user by generating a biometric template and stores it in memory for further comparison. The difference is that machine learning uses training data to automatically generates a boundary between genuine and impostors. While the manual approach requires to manually specify the boundaries. Manual approaches use training data to register an ECG biometric template. Then, the application generates specific rules for classification based on the distance to the registered biometric template. These rules do not change unless the algorithm changes.

Manual classifications of ECG biometrics use fiducial points from ECG signals and generate features. Manual classification approaches applies a thresholds on the distance between the enrolled template and the querying user template. Arteaga-Falconi et al. (2016) studied individual features and established a specific threshold value to each feature. This approach maintains a similar accuracy but improved the acquisition time.

The most used Machine learning classifiers in ECG biometrics are: Support Vector Machines-SVM-, K nearest neighbours-KNN-, Linear Discriminant Analysis Classifiers-LDA-, Neural Networks. The last one can only implement ECG identification. This is because authentication solves a one-class classification problem and those require data for all the classes to perform classification. As mentioned earlier, authentication is a one-to-many approach. As a result, an authentication approach has access just to the genuine user data and not the impostors.

Therefore, it can only train with one class of data, anything else is consider an impostor. Support Vector Machines or SVM use multidimensional space. Each ECG biometric feature represents one dimension in the SVM classification space. SVM automatically finds a hyperplane by using support vectors. These vectors are points closer to the hyperplane. SVM optimally finds vectors that maximize the classifier margin. SVM use different kernels for classification. For authentication, a Radian Basis function-also known as RBF or Gaussian kernel-allow to perform one class classification (Arteaga-Falconi, 2018).

To classify a specific point as part of a class, K Nearest Neighbors or KNN uses a parameter called k to find the k closest values to the point in question. KNN classifier founds k points which belong to different classes. The class with most k points determines the class of the point in question. As SVM, KNN is multidimensional and each feature belongs to a dimension and creates a multidimensional space. KNN can be efficient with small amounts of data, but complexity increases with more data. An example of implementing ECG biometrics with KNN is the work presented by Agrafioti and Hatzinakos (2009). KNN can implement ECG identification or authentication because KNN works to solve a one-class classification problem.

ECG biometrics authenticates with Linear Discriminant Analysis-LDA (Israel et al., 2005). LDA reduces the number of features and uses only with the most representative features. To achieve this, LDA measures the separation between classes known as between-class variance. Between-class variances are the distance between the mean of each class. Then it calculates the within-class variance which is the distance between the mean of a class and each feature within the class. LDA reduces the dimensions with features that maximizes Between-class variance and minimize the within-class variance.

Classification of ECG biometrics also implements many varieties of neural networks. Some of the most common are the perceptron and the special class known as convolutional neural network-CNN. The perceptron classifies data similarly to other machine learning approaches, except uses an online training approach. An online training approach reduces the error each time data arrives to the perceptron. In contrast to other machine learning approaches, where training creates a fixed model. Perceptron becomes more complex as the number of features increase. CNN solves this problem by adding a filter that removes redundant features. This filter is a convolutional filter, therefore the name of Convolutional Neural Networks.

This chapter does not describe in detail the function of neural networks, instead gives a very abstract view of these machine learning approaches. Fortunately, there is vast literature available-not just in Biometrics-which explains these approaches in more detail.

4.6 Evaluation of ECG Biometrics

The evaluation procedure provides a quality measure of ECG biometrics. It determines the accuracy of the system for comparison purposes or to determine if the system meets with a specific accuracy requirement of an application.

The evaluation procedure must specify the source of ECG signals for the evaluation. The source of ECG signals can be databases or data acquired with the same system. When the evaluation compares results with another ECG biometric system, most likely uses databases. Literature on ECG biometrics uses public and private databases for their evaluation. A well known public ECG database is Physionet (Goldberger et al., 2000). Physionet is a collection of datasets for different purposes and ECG biometrics can use them as well. Before using a dataset from physionet the biometric system should consider the special conditions of each dataset. Some datasets will contain ECG signals that contains a medical condition such as arrythmias, cardiac arrests and more. Physionet has the ECG-ID dataset (Nemirko and Lugovaya, 2005) for ECG biometrics. This dataset has ECG signal lengths of 20 s and evaluation should take this in consideration in case the ECG biometric system needs longer ECG signal lengths for the enrollment process. Another public available database is the CYBHi (da Silva et al., 2014), it is a collection of ECG signals acquired with non-medical grade equipment from the palms and fingers of users. Has two collection sets, one collects ECG signals within a short time frame between each acquisition and the other collect ECG signals within a longer period between acquisition. Another ECG database is the UofTDB (Pouryayevali et al., 2014). It is a private database and it need to contact the authors to get access to it. It collects data from the fingertips in different postures with sessions over a period of 6 months.

In a different evaluation scenario, when the purpose is to provide an accuracy specification value, it is better to use data acquired with the same system. ECG data source for evaluation purposes affects the results. Traditionally, ECG from databases collects data with high end medical equipment. This represent a much higher quality of the signal and will give better results. When the equipment is for non-medical purposes, the ECG signal quality will be lower. This will affect the results with lower accuracy. To have a better evaluation of ECG biometric algorithms is better to use the signal source for a fair comparison. Otherwise, each algorithm might work well with one type of data and will not perform as well with others.

Evaluation measurements of ECG biometrics have a close relation to measurements of classification systems. It uses False Acceptance Rate-FAR-, True Rejection Rate-TRR-, True Acceptance Rate-TAR-and False Rejection Rate-FRR. False Acceptance Rate or FAR refers to the quantity of false positives in the system. This is a measurement of how many impostors the ECG biometric system wrongly admits as genuine users. Equation (4.1) expresses this relation.

$$\mathrm{FAR} = \frac{\text{Accepted Impostors}}{\text{Impostor Attempts}} \tag{4.1}$$

TRR or True Rejection Rate measures the number of impostors the system correctly rejects. Equation (4.2) expresses this concept.

$$\mathrm{TRR} = \frac{\text{Rejected Impostors}}{\text{Impostor Attempts}} \tag{4.2}$$

TRR is closely related to FAR because both uses impostors for the evaluation. Equation (4.3) shows the relation.

$$\mathrm{FAR} = 1 - \mathrm{TRR} \tag{4.3}$$

True Acceptance Rate or TAR measures the number of genuine users correctly accepted by the ECG biometric system. Equation (4.4) shows this relation.

$$\mathrm{TAR} = \frac{\text{Accepted Genuine}}{\text{Genuine Attempts}} \tag{4.4}$$

False Rejection Rate or FRR measures the number of genuine users incorrectly rejected by the ECG biometric system. Calculation of FRR uses the equation (4.5).

$$\mathrm{FRR} = \frac{\text{RejectedGenuine}}{\text{GenuineAttempts}} \tag{4.5}$$

FRR relates to TAR because both use genuine users to evaluate the rate. Equation (4.6) express the relation between FRR and TAR.

$$\mathrm{TAR} = 1 - \mathrm{FRR} \tag{4.6}$$

Although FAR and TRR are related through Eq. (4.3) and TAR and FRR are related through Eq. (4.6), FAR and TAR are indirectly related. This relation is indirect because each rate is independent, equations cannot calculate FAR based on TAR and the same goes for the opposite direction. However, the relation exists because an

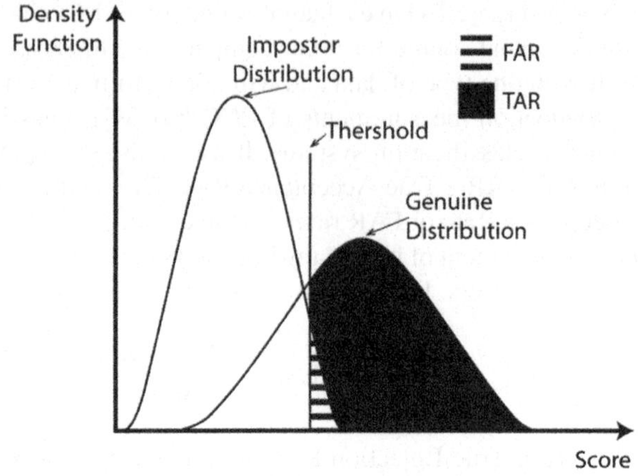

Fig. 4.3 Threshold relation between TAR and FAR rates

ideal system will have a TAR of 1 with a FAR of 0. Ideal systems only exist in theory and an actual implementation does not achieve those rates. An actual implementation with accurate rates will have a TAR value close to 1 and a FAR value close to 0. If the system wants to improve TAR it will affect FAR. If the system wants to improve FAR it will affect TAR. A threshold value is the parameter that affects the relation between FAR and TAR. Modification of thresholds increases TAR and decreases FAR or increases FAR and decreases TAR. Figure 4.3 express the indirect relation between these rates.

Thresholds also affects the relation between TRR and FRR, as mentioned earlier those are related to FAR and TAR respectively. Therefore, modification of thresholds also increases TRR and decreases FRR or increases FRR and decreases TRR.

Evaluation also uses combination of rates. For instance, it can evaluate FAR and FRR or TAR and TRR. In general, the most common approach to evaluate are: TAR against FAR and FAR against FRR. When the evaluation uses TAR and FAR, it generates a graph and calls it Receiver Operating Characteristics Curve or ROC Curve. This is valid approach-many authors use it-to measure the system response in terms of Accuracy -correct classification of genuine users-against error-the misclassification of impostors as genuine. Figure 4.4 shows a general view of the ROC curve. The ideal working point is when TAR is 1 and FAR is 0.

When evaluation uses FAR and FRR it generates a Detection Error Trade-off or DET graph. This graph represents the classification errors of the ECG biometric system. It is a plot of the misclassification of impostors as genuine-FAR-against the misclassification of genuine as impostors-FRR. This is also a widely adopted approach for evaluation of ECB biometrics. This chapter recommends the use of this approach because it provides a point for evaluation against other ECG biometric classifiers. The name of this point is Equal Error Rate-EER-and is the point in which both errors-FAR and FRR-are the same. Some evaluation approaches provide values of TAR and FAR without threshold values. Which is valid because thresholds will be different for each approach. However, it represents a challenge to compare with

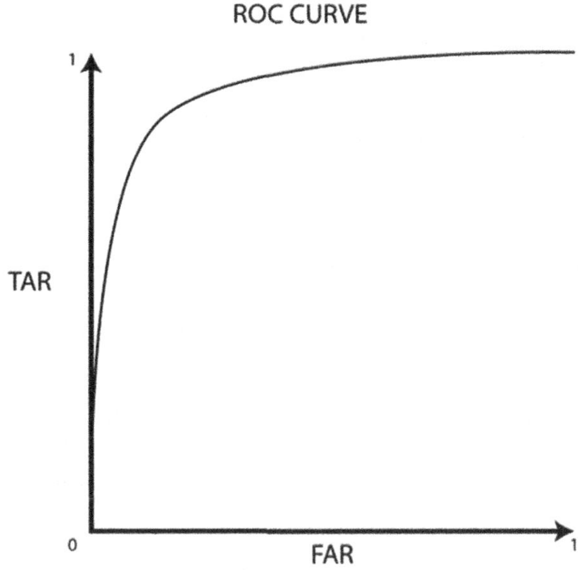

Fig. 4.4 Receiver operating characteristics curve—ROC

other algorithms, because each algorithm must match either TAR, FAR, FRR or TRR for a fair assessment. When the evaluation provides EER, it facilitates the comparison because other algorithms does not have match different rate values for a fair comparison. They need to provide the value in which errors are equal. Because EER is a measure of errors; the most accurate system provides the lowest EER value. An ideal system will provide an EER of 0. Because of the previously mentioned relations between FAR, TAR, FRR and TRR; an EER of 0 represents a TAR of 1 with a FAR of 0 or a TRR of 1 with a FRR of 0. Figure 4.5 shows a DET graph with its corresponding EER.

An ECB biometric system sometimes uses multiple thresholds as in Arteaga-Falconi et al. (2016). When the system involves multiple thresholds-multivariate-the evaluation with the DET curve involves a different approach proposed in Arteaga-Falconi et al. (2018). This approach detects the outliers of the many points generated by the multiple thresholds and generates the DET curve and estimates the EER based on the intersection with a line with a slope value of 1.

This section recommends evaluation approaches and ECG information sources based on general purposes. The final application of ECG biometric will define the most appropriated evaluation approach and the most appropriate source of ECG information for the evaluation. Please take the information on this section as a recommendation but never as must-do instructions.

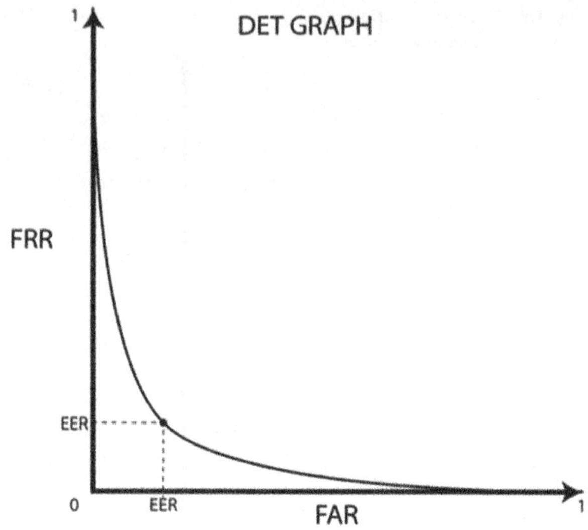

Fig. 4.5 Detection error trade-off (DET) graph and equal error rate (EER)

4.7 Conclusion

ECG as biometrics has advantages and disadvantages over traditional biometrics. ECG as biometric has a better protection against spoofing attacks but requires a longer acquisition time of the biometric trait. This leaves the door open to a broad field for future research in this topic.

This chapter presented a general overview of ECG as biometrics and is written in the same order as the implementation of ECG biometrics. The purpose is to facilitate a starting point to further research in this field, which will help to propagate the use of this technology. Some statements presented in this chapter are clearly identified recommendations, because ECG biometrics could be used in many different applications and just the final application will determine the final path of the recommendations which this chapter presents.

References

Agrafioti, F., & Hatzinakos, D. (2009). ECG biometric analysis in cardiac irregularity conditions. *Signal, Image Video Process, 3*(4), 329–343.

AlMahamdy, M., & Riley, H. B. (2014). Performance study of different denoising methods for ECG signals. *Procedia Computer Science, 37*, 325–332.

Arteaga-Falconi, J. S., Al Osman, H., & El Saddik, A. (2015). R-peak detection algorithm based on differentiation. *IEEE 9th International Symposium on Intelligent Signal Processing (WISP)* (pp. 1–4).

Arteaga-Falconi, J. S., Al Osman, H., & El Saddik, A. (2016). ECG authentication for mobile devices. *IEEE Transactions on Instrumentation and Measurement, 65*(3), 591–600.

Arteaga-Falconi, J. S., Al Osman, H., El Saddik, A. (2018). ECG and fingerprint bimodal authentication. *Sustainable Cities and Society, 40*, 274–283.

Arteaga-Falconi, J., Tobón, D. P., & Saddik, A. E. (2018). EER calculation and DET approximation in a multi-threshold biometric system. *IEEE International Symposium on Medical Measurements and Applications (MeMeA)* (pp. 1–6).

Biel, L., Pettersson, O., Philipson, L., & Wide, P. (2001). ECG analysis: A new approach in human identification. *IEEE Transactions on Instrumentation and Measurement, 50*(3), 808–812.

Chan, A. D. C., Hamdy, M. M., Badre, A., & Badee, V. (2008). Wavelet Distance Measure for Person Identification Using Electrocardiograms. *IEEE Transactions on Instrumentation and Measurement, 57*(2), 248–253.

da Silva, H. P., Lourenço, A., Fred, A., Raposo, N., & Aires-de-Sousa, M. (2014). Check your biosignals here: A new dataset for off-the-person ECG biometrics. *Computer Methods and Programs in Biomedicine, 113*(2), 503–514.

El Saddik, A. (2018). Digital twins: The convergence of multimedia technologies. *IEEE Multimedia, 25*(2), 87–92.

Goldberger, A. L., Amaral, L. A., Glass, L., Hausdorff, J. M., Ivanov, P. C., Mark, R. G., Mietus, J. E., Moody, G. B., Peng, C. K., & Stanley, H. E. (2000). PhysioBank, PhysioToolkit, and PhysioNet?: Components of a new research resource for complex physiologic signals. *Circulation, 101*(23), e215–e220.

Goldberger, A. L. (2012). *Clinical electrocardiography: A simplified approach*, 8th ed. Mosby Elsevier

Israel, S. A., Irvine, J. M., Cheng, A., Wiederhold, M. D., & Wiederhold, B. K. (2005). ECG to identify individuals. *Pattern Recognition, 38*(1), 133–142. http://www.sciencedirect.com/science/article/pii/S0031320304002419.

Maltoni, D. (2015). Fingerprint recognition, overview. In Li, S. Z., & Jain, A. K.(Eds.), *Encyclopedia of biometrics* (pp. 664–668). Boston, MA: Springer US.

Nemirko, A. P., & Lugovaya, T. S. (2005). Biometric human identification based on electrocardiogram. In *XII-th Russian Conference on Mathematical Methods Pattern Recognition* (pp. 387–390).

Odinaka, I., Lai, P. H., Kaplan, A. D., O'Sullivan, J. A., Sirevaag, E. J., & Rohrbaugh, J. W. (2012). ECG biometric recognition: A comparative analysis. *IEEE Transactions on Information Forensics and Security, 7*(6), 1812–1824.

Plataniotis, K. N., Hatzinakos, D., & Lee, J. K. M. (2006). ECG biometric recognition without fiducial detection. In *Biometrics Symposium: Special Session on Research at the Biometric Consortium Conference* (pp. 1–6). https://doi.org/10.1109/BCC.2006.4341628

Pouryayevali, S., Wahabi, S., Hari, S., & Hatzinakos, D. (2014). On establishing evaluation standards for ECG biometrics. *IEEE International Conference on Acoustics, Speech and Signal Processing (ICASSP)* (pp. 3774–3778).

Roberts, C. (2007). Biometric attack vectors and defences. *Computers & Security*, *26*(1), 14–25.

Ross, A., & Jain, A. K. (2015). Biometrics, overview. In Li, S. Z., & Jain, A. K. (Eds.) *Encyclopedia of biometrics* (pp. 289–294). Boston, MA: Springer US

Singh, Y. N., & Singh, S. K. (2012). Evaluation of electrocardiogram for biometric authentication. *Journal of Information Security, 3*(1), 39–48.

Tan, B., & Schuckers, S. (2010). Spoofing protection for fingerprint scanner by fusing ridge signal and valley noise. *Pattern Recognition, 43*(8), 2845–2857

Wayman, J. L. (2015). Biometric verification/identification/authentication/recognition: The terminology. In Li, S. Z., Jain, A. K., (Eds.), *Encyclopedia of Biometrics* (pp. 263–268). Boston, MA: Springer US

Wübbeler, G., Stavridis, M., Kreiseler, D., Bousseljot, R. -D., & Elster, C. (2007). Verification of humans using the electrocardiogram. *Pattern Recognition Letter, 28*(10), 1172–1175. http://www.sciencedirect.com/science/article/pii/S0167865507000463.

Chapter 5
ECG Biometric System for Human Recognition Based on the Possibility Theory

Nesrine Keskes, Amal Charfi, Sameh Fakhfakh, Imene Khanfir Kallel, and Nabil Derbel

Abstract Biometrics are attracting widespread interest due to the increasing need for systems security everywhere. Researchers contributed to this field by applying various biometric modalities such fingerprint, face and iris recognition. In this chapter, a new recognition biometric system based on ECG signal is proposed. One of the main issues that could face the use of ECG as a biometric trait is its variability over time. Therefore, in this study, the possibility theory is introduced to take into account the information alteration caused by undesired factors and mainly by the variation of the signal. It is important to mention that, until now, a biometric system based on possibility modeling approach has only been applied for few modalities like iris and fingerprint. This chapter presents the first work that investigates its efficiency to classify ECG signals. The proposed method is applied on signals taken from the PTB database provided by the PhysioBank archives. A pre-processing step is presented, then features are extracted based on the AC/DCT method.After features transition to the possibility space, a fusion methodology, at the classification level, is performed to improve the system efficiency. For the evaluation of the proposed system, experiments show significant results, the area under curve deducted from the ROC curve has the value of 0.96, an equal error rate of 6.67% which are superior to some related works,the accuracy rate achieve a 93.22% as well a 93.12% for sensitivity rate and 93.32% for specificity rate.

N. Keskes (✉) · A. Charfi · S. Fakhfakh · I. K. Kallel · N. Derbel
Control and Energy Management Laboratory, National Engineering School of Sfax, University of Sfax, Sfax, Tunisia
e-mail: nesrine.keskes@gmail.com

A. Charfi
e-mail: charfi.amal@gmail.com

S. Fakhfakh
e-mail: fakhfakhsameh@yahoo.fr

I. K. Kallel
e-mail: imen.khanfir.kallel@gmail.com

N. Derbel
e-mail: n.derbel@enis.rnu.tn

N. Derbel and O. Kanoun (eds.), *Advanced Methods for Human Biometrics*,
Smart Sensors, Measurement and Instrumentation 40,
https://doi.org/10.1007/978-3-030-81982-8_5

Keywords Biometric recognition system · ECG signal · Feature extraction · Classification · Possibility theory · Possibility distribution · Information imperfections · Possibilistic modeling · Small sample size · Similarity measure

5.1 Introduction

Recently, the technology world has witnessed a growing spread in such a way that security become a primary request everywhere. Many systems and devices like smartphones, smart houses, banks, and airports...that rely on authentication or identification process need high level of security. In such situations, the science of biometry has been introduced as an efficient tool for people recognition instead of classical means like keys, passwords and cards that could be easily stolen, copied or lost. A biometric system can be classified according to its ability to responds to different requirements such as robustness, availability and acceptability (Wayman, Jain, Maltoni, & Maio, 2005). Among the most widely used biometric modalities, in this study we are dealing with the (ECGs) one. It is recognized that it is not expensive like the DNA and not falsifiable such as the fingerprint. The ECG signal is a graphic presentation of the heart electrical activity (Trobec, Tomašić, Rashkovska, Depolli, & Avbelj, 2018), which is generated by a polarization and depolarization of the ventricles and atria. The useful information taken away from this signal are localized in its peaks amplitude, waves and intervals. It is a periodical signal, each period or heartbeat is constituted of different parameters known as the P wave, the QRS complex and T wave (Fig. 5.1).

Feature extraction and classification are the key levels of an authentication/identification process: The efficiency at those two levels determines the performances of the biometric system. In the literature, many researchers have addressed to this topic so that various techniques were adopted to apply the ECG for human recognition. Two categories exist to extract ECG features.The first one employs the fiducial points (temporal domain) such as peaks, intervals and amplitudes (Keskes, Ghribi,

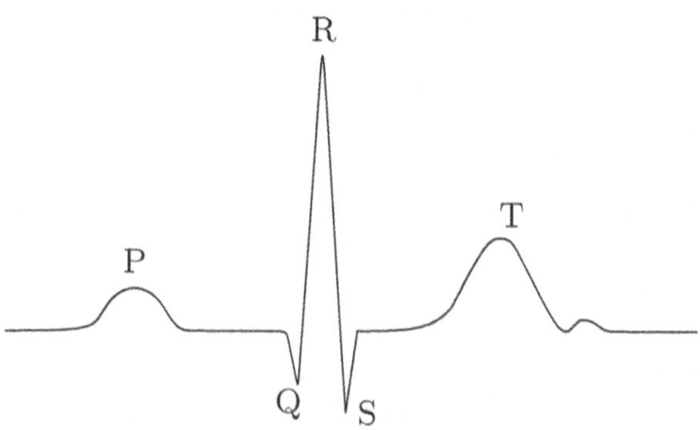

Fig. 5.1 Normal ECG heartbeat

Table 5.1 Feature extraction techniques

References	Proposed method
Biel, Pettersson, Philipson, and Wide (2001)	Fiducial based method + features extracted are the QRS complex, T wave and P wave
Kyoso and Uchiyama (2001)	Fiducial based method + 4 features: P wave duration, PQ interval, QRS and QT durations.
Plataniotis, Hatzinakos, and Lee (2006)	Non fiducial based method. + autocorrelation of the ECG windows + Discrete Cosine Transform to reduce the dimension
Wübbeler, Stavridis, Kreiseler, Bousseljot, and Elster (2007)	Non fiducial method. + examining the distance between two ECG heartbeats
Irvine, Israel, Scruggs, and Worek (2008)	Non fiducial based method. + application of the PCA for feature extraction. + Euclidean distance measurement for final decision
Chiu, Chuang, and Hsu (2008)	Discrete Wavelet Transform (DWT) for feature extraction + Euclidean distance applied to measure similarity

Barioul, & Derbel, 2018). The second is based on non-fiducial points (frequency domain) (Jung & Lee, 2017). Table 5.1 represents the main methods (Pouryayevali, 2015), applied for feature extraction.

In order to achieve a high level of performance, several studies have employed robust classifiers such as K-nearest-neighbor (KNN), Neural network (NN), support vector machine (SVM) and other classification approaches. In current related researches, such in (Shdefat, Joo, Choi, and Hee-Cheol, 2018) moving average filters to denoise the signal has been used. Then it has been segmented into windows of 30 s. Correction techniques have been applied for the detected ectopic intervals. Then correlation based on a QRS template matching algorithm has been used with linear SVM for classification task. In Vuksanovic and Alhamdi (2014), the researcher has tested his method on PTB, MIT-BIH and Milano databases, which consist of extracting analytic and modelling features (time, width, amplitudes and AR parameters) then he has applied the KNN classifier for subject recognition. The author in Jung and Lee (2017) has utilized the wavelet transform for pre-processing step, for feature extraction and selection he has applied the autocorrelation, the discrete cosine transform and the method of window removing to eliminate unnecessary windows. For the classification step, he has used the nearest-neighbor (NN), the linear discriminant analysis (LDA) and the support vector machine (SVM). The work in Singh and Singh (2012) is based on a fiducial feature extraction method and for the final decision a matching algorithm has been used between the database template and the query sample, the proposed system has been applied on the ECG modality at first and then on a combined ECG, face and fingerprint modalities. However, and as a fact of knowledge, the electrocardiogram is a bio-signal measured on alive people. Taking into consideration, effectively, the undesirable facts that can affect the system, is of great interest. Moreover the ECG signal is characterized by its changeability and variability. A considerable amount of research have dealt with those factors but few ones have focused on the uncertainty of the information extracted from this signal. Thus, the challenge in this work, is to concept an ECG biometric system that takes into account the uncertainty of received data, which can be resulted from the way of taking the signal (position, emotion, body movements…), the quality of the ECG device (electrodes or fingertip) and other additional noises. Therefore, the possibility theory is introduced as the well appropriate framework for making an accurate decision with ECG data. The possiblistic based modeling approach has proved, again, its efficiency for biometric recognition context which motivates its future application, to other modalities such the EEG and the EMG. The rest of this chapter is organized in the following way. Section 5.2 explains the possibility theory. Section 5.3 describes the proposed methodology. It allocates for the experimental results and discussion. Finally, concluding remarks were cited in Sect. 5.4.

5.2 Possibility Theory

Based on the fuzzy set theory, the possibility theory was proposed since the 70s by Zadeh (1978), then was developed with other authors like Prade and Dubois (RAMER, 1989). Unlike the probability, which focus on the randomness to deal with uncertainty, the possibility theory can deal with additional types of imperfection such as the uncertainty, imprecision or ambiguity.It offers an efficient representation for imperfect data through possibility distributions. This theory is characterized by two complementary measures; the possibility and necessity N. Several methodologies have been developed in previous works to build the possibilistic measure, some of them are based on transformation from probability distribution to possibility distribution (Bounhas, Mellouli, Prade, & Serrurier, 2010) and others are based on a direct evaluation of the data.

5.2.1 Possibility Distribution

In the possibility theory, the possibility distribution of the collected data presents an important concept for data analyses and classification. It is an essential tool to represent uncertain or imprecise information. The possibility distribution noted π, is an application from the universe Ω to [0, 1],where $\Omega = \{\omega_1, \omega_1, \ldots, \omega_k\}$. Conventionally, a state is said impossible if $\pi(\omega_i) = 0$, and it is totally possible if $\pi(\omega_i) = 1$.

For a distribution, extreme forms of knowledge existing could be defined as:

- Complete knowledge: $\exists\, \omega_0, \pi(\omega_0) = 1$ and $\forall\, \omega_n \neq \omega_0 \in \Omega, \pi(\omega_n) = 0$
- Complete ignorance: $\forall\, \omega_n \in \Omega, \pi(\omega_n) = 1$

A possibility distribution is characterized by its height, its support and its core as it is explained in Fig. 5.2.

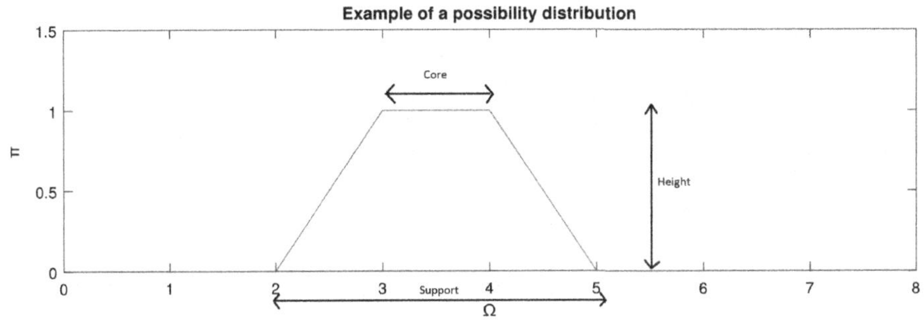

Fig. 5.2 Characteristics of a possibility distribution

5.2.2 Transformation from Probability Distribution to Possibility Distribution

It is useful to transform a probability distribution into possibility distribution in case of dealing with a weak source of information or if applying possibility calculation is more efficient than probability calculation (Dubois & Prade, 2006). Based on some consistency principles, many transformations techniques exist in the literature that we can cite those proposed by Dubois and Prade, Klir, Masson and others. To apply this theory, let consider that the probability measure presented as $p = (p_1, p_2, \ldots, p_n, p_{n+1})$, the possibility measure presented as $\pi = (\pi_1, \pi_2, \ldots, \pi_n, \pi_{n+1})$ which are organized in decreasing order and $p_{n+1} = 0$, $\pi_{n+1} = 0$.

5.2.2.1 Klir Transformation

Klir has examined the uncertainty preservation principle using two scales given as following:
Linear scale:

$$\pi_i = \frac{p_i}{p_1} \tag{5.1}$$

Logarithmic scale:

$$\pi_i = \left(\frac{p_i}{p_1}\right)^{\alpha} \tag{5.2}$$

5.2.2.2 Dubois and Prade Transformation

They have proposed two probability-possibility transformations, defined as following:

Symmetric transformation:

$$\pi_i = \sum_{j=1}^{n} \min(p_i, p_j) \quad (5.3)$$

Asymmetric transformation:

$$\pi_i = \sum_{j/p_j \le p_i} p_j \quad (5.4)$$

5.2.2.3 VT Transformation

This transformation is developed by Mouchaweh, Bouguelid, Billaudel, and RiERA (2006) expressed as following:

$$\pi_i = \left(\frac{p_i}{p_1}\right)^{k\, p_i} \quad (5.5)$$

where k must respect the following condition to guarantee the principle of consistency:

$$0 \le k \le \frac{\log p_n}{(1 - p_n)\, \log \dfrac{p_n}{p_1}} \quad (5.6)$$

5.3 Methodology

In this section, we explain different steps of the proposed ECG biometric system. Figure 5.3 shows the whole block diagram of the used algorithm, starting with pre-processing phase based on a band-pass filter, then the signal is segmented into non overlapping windows of 4s each one, and AC/DCT method is applied to extract non-fiducial features. For better results, a discriminant power technique has been employed for feature selection. Finally, in the classification phase, a possibility distribution is performed for each selected feature then a similarity measure is effected and fused for the authentication. More details are explained in following subsections.

5.3.1 ECG Signal Pre-processing

The electrocardiogram is captured through electrodes set on the skin (Trobec et al., 2018). It transforms the activity of the heart into electrical signal. During this task, different existing factors can affect on the quality of the signal. From these

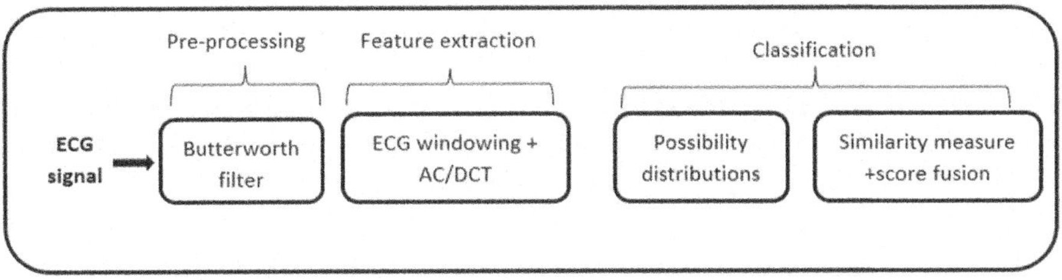

Fig. 5.3 Methodology

undesirable sources of noise, the power line interface, which is around the 50 Hz and the baseline wander caused by the respiration, which is between 0.15 Hz and 0.3 Hz frequencies. Also, the signal can be degraded with the presence of motion artifact and due to an Electromyography (EMG) noise or even a loss of electrode contact (Satija, Ramkumar, & Manikandan, 2018). Thus to denoise the acquired signal, a pre-processing step is essentially needed. For that a band-pass Butterworth filter of order 4 for cut-off frequencies at 0.5 Hz and 40 Hz is applied based on empirical results (Agrafioti, Gao, & Atzinakos, 2011; Vuksanovic & Alhamdi, 2014; Wang, Agrafioti, Hatzinakos, & Plataniotis, 2007).

5.3.2 *Feature Extraction*

Feature extraction stage is the task of abstracting the different characteristic points from the electrocardiogram, which are the key to distinguish between individuals, since every person has its unique parameters. The applied method is based on non-fiducial extraction features. After filtering the signal, it is segmented it into non-overlapped windows, each one has a 4 s length and the origin is any arbitrary point, it must contain at least one heartbeat period (Pummer, 2016), and this is which make it easier than extracting temporal features in terms of the computational complexity and time duration. Then, an auto-correlation is applied on each window since we are dealing with a periodical signal, so that we put the accent on the important shapes existing in the window. In other words, the main peaks such as the P, Q, R, S and T are emphasized when the signal is auto-correlated, whereas small waves or noise near the zero are neglected. A normalization step is then applied in order to eliminate the variation of the amplitudes of each wave even existing for the same person. It is presented in the following equation:

$$Rxx(m) = \frac{1}{Rxx(0)} \sum_{i=0}^{N-|m|-1} x(i)x(i+m) \tag{5.7}$$

where $x(i)$ is a windowed signal, $x(i+m)$ is the shifted window for a lag time equal to $m = 0, 1, \ldots, M-1$.

Results after the auto-correlation are considered highly dimensioned, thus to avoid wasting time and to discard the non-discriminant features we used the Discrete Cosinus Transform (DCT) (Jung & Lee, 2017). The features then are extracted as power coefficients, and samples near the zero value are neglected since they are non useful for the classification task. So if M points of the DCT points were calculated, only C points ($C \ll M$) are taken to constitute the feature vector.

5.3.3 Possibility Theory Based ECG Classification

In this part, we explain the possibilistic classification approach which is illustrated in Fig. 5.4.

As a first step in this approach, the extracted features are transformed into probability distribution relying on the histogram method. Possibility distribution is then inferred with the Symmetric transformation of Dubois–Prade (Dubois, Prade, & Sandri, 1993). Then we apply the Symmetric transformation. Once we get the possibility distribution for all the features, we select the most distinctive ones and later we measure the similarity between the obtained possibility distributions based on Info-Closeness technique. By calculating the average score of similarity resulted from each selected feature, we fusion all the measures of similarity. Finally, according to a choice of an optimal decision threshold, classification task is accomplished.

5.3.3.1 Generating Probability Distribution

The histogram presents a descriptive graphical tool to express the data distribution. It indicates the frequently occurrence of the data in each class (Mouchaweh & Billaudel, 2002).

Thus, the histogram is calculated by dividing an interval $(y_2 - y_1)$ of a feature into h equal length sub-intervals named bins. The bin width is defined by:

$$B = \frac{y_2 - y_2}{h}$$

with y_1 and y_2 are the minimal and maximal values of the features. The number n_k of occurrences of the data samples presents the height of a bin k calculated in its interval. Then the probability p_k affected to the bin k is defined as following:

$$p_k = \frac{n_k}{n}$$

where n is the total number of data samples.

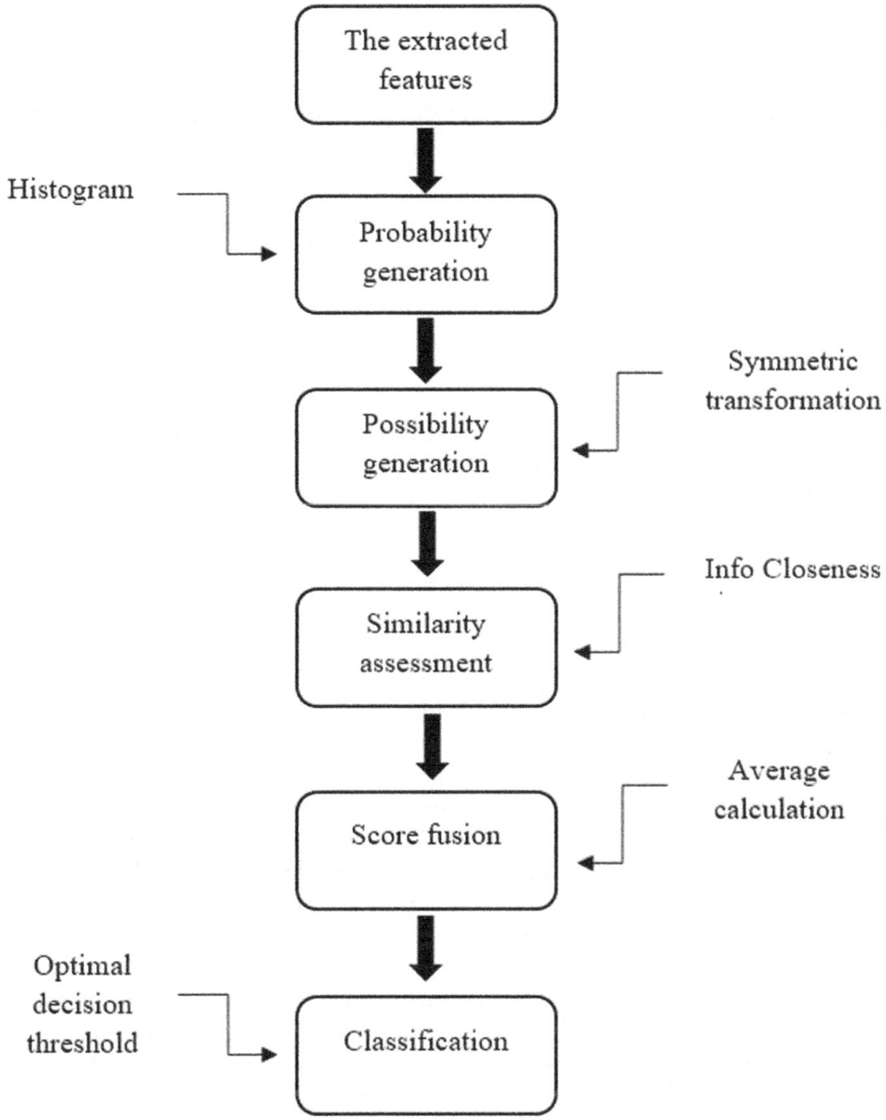

Fig. 5.4 Classification based possibility process

5.3.3.2 Generating Possibility Distribution

Several techniques were proposed in the literature to ensure the transition from the probabilistic representation to possibilistic one. In this evaluation, the symmetric probability-possibility transformation proposed by Dubois and Prade (Dubois, Prade, & Sandri, 1993), is employed to infer possibility distributions. It is used to be one from the most efficient methods, in term of generated possibility distribution, as well as time computation.

5.3.3.3 Feature Selection

In order to ameliorate the classification results, a power discriminant factor is measured for every possibility distribution of each feature vector that reflect its capability to distinguish between different classes. For each feature possibility distribution, the power discriminant is computed according to their intra-class and inter-class variances. It is expressed as following:

$$P_{dis} = \frac{V_{inter}}{V_{inter} + V_{intra}} \tag{5.8}$$

5.3.3.4 Template Matching and Decision

Similarity assessment represents a fundamental stage in different fields especially for data analysis. In possibility theory, similarity measure refers to the formula enabling to determine the proximity between the compared possibility distributions. The choice of a similarity measure is of great impact on the decision making. However, the selection of the most appropriate measure, in the possibility framework, remains an open issue and it is quite related to the context of application. The similarity measure, in this work, is the tool allowing to determine the resemblance between the compared templates. It is useful to evaluate the similarity between templates belonging to the same class to build the genuine distribution and the similarity between templates belonging to different classes to build the imposters distributions (Fig. 5.8). The adequate measure should detect high degree for genuine distributions and low degrees for imposters ones.

If we considerate that π_1 and π_2 are two possibility distributions, then the similarity measure between them, S (π_1, π_2) is an application, where its values are between 0 and 1 and it respects the following properties:

1. Positivity: $\forall\, \pi_1, \pi_2,\ S(\pi_1, \pi_2) \geqslant 0$.
2. Symmetric: $\forall\, \pi_1, \pi_2,\ S(\pi_1, \pi_2) = S(\pi_2, \pi_1)$.
3. Identity: $\forall\, \pi,\ S(\pi, \pi) = 1$.

Divers methods were proposed to estimate the similarity between possibility distributions. While some researches confirm the reliability of the use of classic distance metrics, others investigate the use of uncertainty measure in order to evaluate the difference between distributions. In fact, uncertainty measures were introduced in the possibility theory literature, in order to qualify the information amount inherent to each possibility distribution. Information closeness, that we rely on our proposed model, is one of the leader similarity assessment works based on uncertainty comparison. It is defined as following (Higashi & Klir, 1983):

$$G(\pi_1, \pi_2) = g(\pi_1, \pi_1 \vee \pi_2) + g(\pi_{12}, \pi_1 \vee \pi_2) \tag{5.9}$$

where $g(\pi_i, \pi_j) = U(\pi_j) - U(\pi_i)$, $\vee$ represents the maximum operator.

The U-uncertainty measure is applied on an ordered possibility distribution such that: $\pi_1 > \pi_2 > \pi_n$

$$U(\pi) = \sum_{i=2}^{n} (\pi_i - \pi_{i+1}) \log_2 i + (1 - \pi_1) \log_2 n \tag{5.10}$$

To have output values on the interval [0, 1], the normalized Information Closeness measure transformed to:

$$\text{InfoCloseness} = 1 - \frac{g(\pi_i, \pi_j)}{G_{max}} \tag{5.11}$$

where:

$$G_{max} = 2\log_2(n) - \log_2(n-1) = \log_2 \frac{n^2}{n-1}$$

5.3.3.5 Score Fusion

Relatively, data fusion presents a recent procedure in the scientific literature. It is the operation that integrates multi datum in order to extract a new information which is more representative of the whole data. Several level of data fusion were appeared in previous works. Those levels are divided into:

- Sensor level: its interest is to collect the various data captured from different sources or sensors and then fuse them. In our case, this level cannot be applied since the ECG data are extracted from the same source.
- Feature level: it is the combination of the extracted characteristics after processing task.
- Score level: it is the most used level, which leads to more precise classification. In the presented work, it is the fusion of similarity measures.
- Decision level: it is the simplest level, which is based on the fusion of each resulted decision (Yes/No).

After the application of each defined data fusion level, the fusion on the level of score has proved the most efficient results. The process of score fusion for each subject is described as following: Since we are dealing with 4-feature vector, at the beginning, a possibility distribution is performed for each feature, then genuine and imposter are generated relying on the similarity measure, so that we get 4-scores. In the proposed algorithm, the fusion is effected by a mathematical combination (5.12). The scores generated from each feature are fused to form a unique score for each subject which is use later for the decision stage.

$$S = \frac{1}{n} \sum_{i=1}^{n} s_i \tag{5.12}$$

where:
n: the number of selected feature
s_i: the similarity measure for a feature i

According to the calculation of genuine and imposter of this biometric system, a threshold is delivered by obtaining the minimum equal error rate. Thus, for the identification task, the resulted similarity is compared to the optimal threshold.

5.3.4 Experimental Results and Discussion

We have tested our biometric system on signals taken from the PTB database (Goldberger et al., 2000). It is formed by 549 records from 290 person, men and women aged 17–87. Each individual has 1–5 records. The sampling frequency is at 1000 Hz. For the experiment, we have applied our approach on 20 subjects with 3 records taken in different time for each one, 800 samples have been considered. The database is divided into a training set formed by 600 samples and the testing set formed by 200 samples.

In a first step, ECG signal is denoised using a band-pass Butterworth filter of order 4 for cut-off frequencies at 0.5 Hz and 40 Hz based on empirical results. Figure 5.5 presents an example of a filtered signal.

In a second step, the signal is segmented into non-overlapping windows with 4s length for each one. After that, features are extracted based on the AC/DCT method. Each window is auto-correlated by applying the Eq. (5.7), in order to emphasis the main parameter of the signal. Figure 5.6 explains different phases for the feature extraction method.

The DCT application reduce the feature dimension into 8 significant features. In order to get better classification results, a step for feature selection is included in this work. In the training phase, 3 models of possibility distribution is performed for each feature of each person, Fig. 5.7 shows an example where the models are almost confound since they belong to the same class. And according to the variation of all the features, the number of bin was fixed to 8, in order to get the probability distribution and later the possibility distribution.

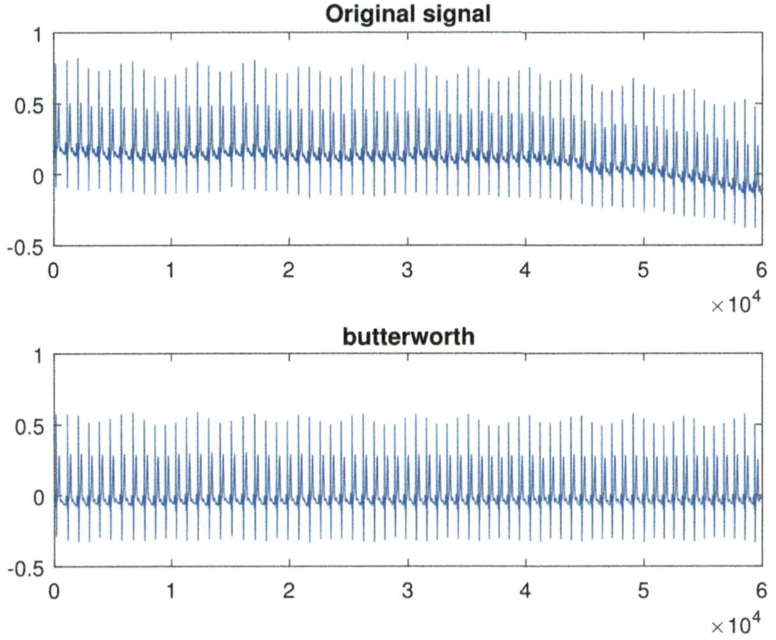

Fig. 5.5 Denoised ECG signal

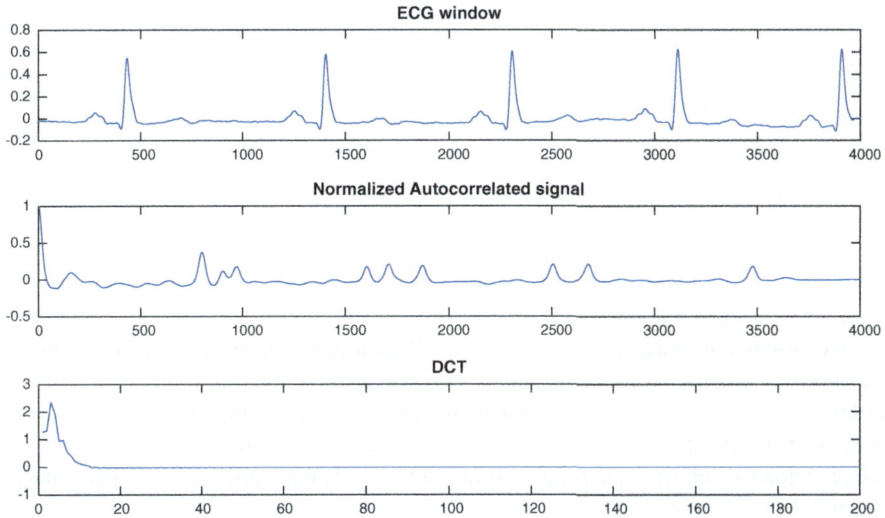

Fig. 5.6 Feature extraction steps

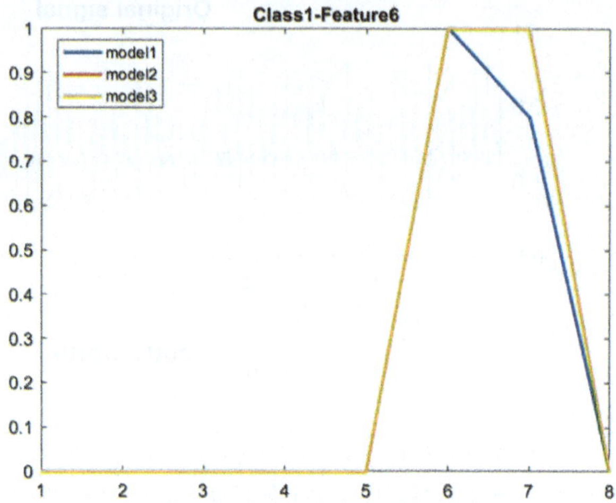

Fig. 5.7 Class 1 possibility distribution for the feature 6

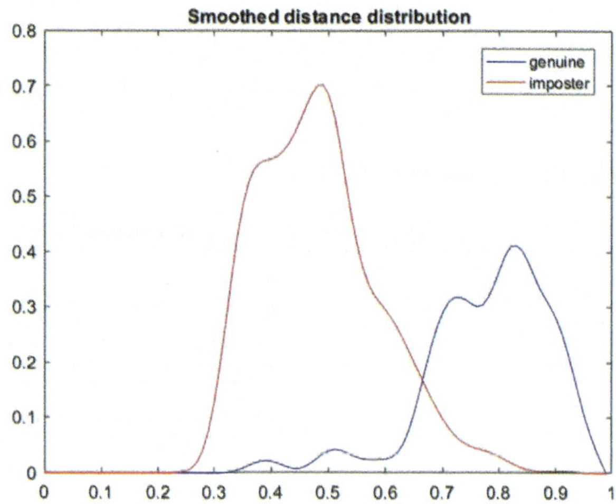

Fig. 5.8 Genuine and Imposter curve

To evaluate our biometric system for identification, we have calculated the genuine (intra-class) and imposter (inter-class) scores for the different possibility distributions and before applying the similarity measure, a power discriminant factor is computed to choose the most selective features. From 8 features, feature 2, 3, 6 and 8 were selected due to their high power discriminant value. The genuine and imposter curves are plotted in Fig. 5.8.

The receiver operating characteristics (ROC) curve is widely used as performance predictor in classification problems. It is introduced to evaluate the ability of the proposed model to classify genuine and imposter cases. It has two parameters; the false positive rate and the true positive rate. It indicates that the system has a good performance when the curve is closed to 1. As it is shown in Fig. 5.9, from the ROC

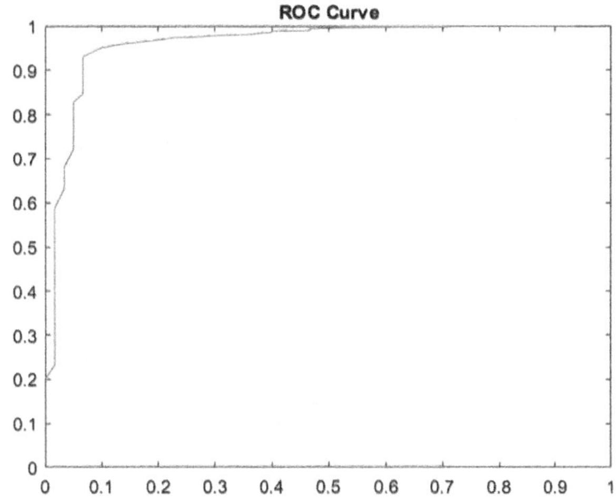

Fig. 5.9 ROC curve

curve we can calculate the area under curve (AUC = 0.96) which is considered as an interesting value.

The false acceptance rate (FAR) and the false rejected rate (FRR) are two errors types that can inform about the classification system efficiency. In biometric system, FAR indicates the rate of incorrectly acceptance of unauthorized persons and FRR indicates the probability of rejection of legitimate candidates.

$$FRR = \frac{\text{Genuine scores falling below threshold}}{\text{All genuine score}}$$

$$FAR = \frac{\text{Imposter scores exceeding threshold}}{\text{All imposter score}}$$

The false acceptance rate (FAR) and the false rejected rate (FRR) in Fig. 5.10 show good results since it is near to zero.

The equal error rate (EER) is the intersection point between the FAR and FRR, the less value it has the more efficient classification we get. In this experiment, as it is clear in Fig. 5.11 the EER is equal to 6,67% which is improved compared with previous works: EER = 14% in Agrafioti et al. (2011), and EER = 16.4% in Zemzemi et al. (2018). Moreover,the accuracy, the sensitivity and the specificity of the proposed biometric system are calculated for the classification performance, according to the undermentioned equations:

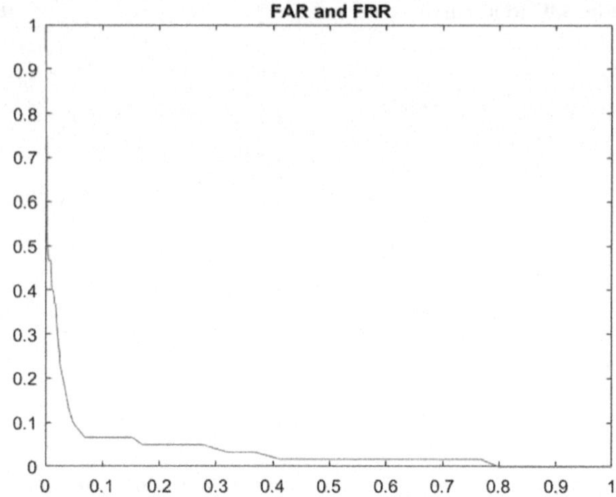

Fig. 5.10 FAR and FRR curve

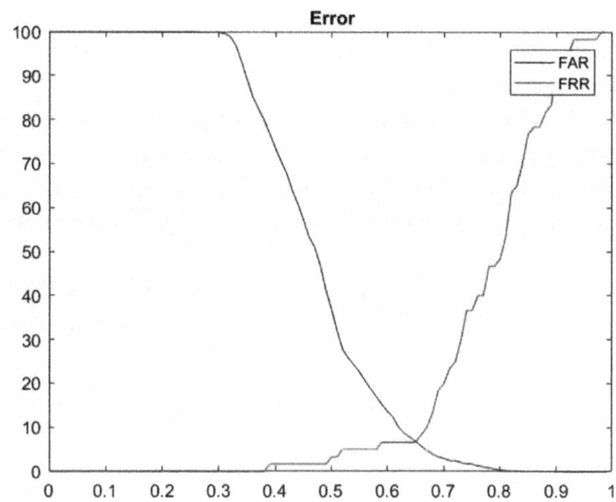

Fig. 5.11 Error curve

$$\text{Accuracy} = \frac{TP + TN}{TP + TN + FP + FN} = 93.22\% \tag{5.13}$$

$$\text{Sensitivity} = \frac{TP}{TP + FN} = 93.12\% \tag{5.14}$$

$$\text{Specificity} = \frac{TN}{TN + FP} = 93.32\% \tag{5.15}$$

where:
TP: Number of true positive samples
TN: Number of true negative samples
FP: Number of false positive samples
FN: Number of false negative samples

Table 5.2 Comparison of results

References	No. subject	Accuracy (in %)	EER (in %)
Singh and Singh (2012)	73	82	10.8
Zemzemi et al. (2018)	10	82	16.4
Proposed method	20	93.22	6.67

In Table 5.2 results are compared to previous related work according to the number of subject, the accuracy and the equal error rate.

The limitation in the proposed model,compared to Singh's work (Singh & Singh, 2012) is that the PTB database provides few ECG records for each subject, that's why we get few segmented ECG windows. This affect the efficiency of possiblistic modeling and then it may adversely influenced the classification performance. Thus, it should be noted that although the performance of the proposed biometric system is important, those results could be improved if we augment the number of samples for each individual, so that the possibility distribution becomes more accurate and efficient.

5.4 Conclusion

This chapter has investigated an ECG recognition system. The process for the identification purpose starts with filtering the acquired ECG signal using the Butterworth band-pass filter. Then a non-fiducial feature extraction method is applied on each 4 s ECG window based on the AC/DCT technique. Moreover, the application of the possibility theory is implemented through generating a probability distribution then a possibility distribution based on the symmetric transformation. Finally, in order to perform person identification, the classification task relied on the information closeness as a measure of similarity. Significant results have been elaborated from this study, the error rate is reduced to 6.67% compared to previous work, air under curve achieved the value of 0.96, an accuracy rate of 93.22%,a sensitivity rate of 93.12% and a specificity rate of 93.32%.

Despite the interesting obtained results, the current study was limited by the number of ECG samples provided by the PTB database: Possibility modeling accuracy is very related to the samples representativeness.

Since the actual results are promising, we intent to build our specific ECG database in order to validate our ECG based recognition model on a larger number of samples.

References

Agrafioti, F., Gao, J., & Atzinakos, D. (2011). Heart biometrics: Theory, methods and applications. In *Biometrics*. IntechOpen

Biel, L., Pettersson, O., Philipson, L., & Wide, P. (2001). Ecg analysis: A new approach in human identification. *IEEE Transactions on Instrumentation and Measurement, 50*(3), 808–812.

Bounhas, M., Mellouli, K., Prade, H., & Serrurier, M. (2010). From bayesian classifiers to possibilistic classifiers for numerical data. In *International Conference on Scalable Uncertainty Management* (pp. 112–125). Springer.

Chiu, C.-C., Chuang, C.-M., & Hsu, C.-Y. (2008). A novel personal identity verification approach using a discrete wavelet transform of the ecg signal. In *2008 International Conference on Multimedia and Ubiquitous Engineering (MUE 2008)* (pp. 201–206). IEEE.

Dubois, D., & Prade, H. (2006). Représentations formelles de l'incertain et de l'imprécis. *Concepts et méthodes pour l'aide à la décision, 1*, 11–165.

Dubois, D., Prade, H., & Sandri, S. (1993). On possibility/probability transformations. In *Fuzzy logic* (pp. 103–112). Springer.

Goldberger, A., Amaral, L., Glass, L., Hausdorff, J., Ivanov, P., Mark, R., Mietus, J., Moody, G., Peng, C., & Stanley, H. (2000). Physiobank, physiotoolkit, and physionet: Components of a new research resource for complex physiologic signals. *Circulation* [online], *101*(23), e215–e220.

Higashi, M., & Klir, G. J. (1983). On the notion of distance representing information closeness: Possibility and probability distributions. *International Journal of General Systems, 9*(2), 103–115.

Irvine, J. M., Israel, S. A., Scruggs, W. T., & Worek, W. J. (2008). Eigenpulse: Robust human identification from cardiovascular function. *Pattern Recognition, 41*(11), 3427–3435.

Jung, W.-H., & Lee, S.-G. (2017). Ecg identification based on non-fiducial feature extraction using window removal method. *Applied Sciences, 7*(11), 1205.

Keskes, N., Ghribi, S. F., Barioul, R., & Derbel, N. (2018). Parameter extraction of ecg using labview. In *2018 15th International Multi-Conference on Systems, Signals & Devices (SSD)* (pp. 49–54). IEEE.

Kyoso, M., & Uchiyama, A. (2001). Development of an ecg identification system. In *2001 Conference Proceedings of the 23rd Annual International Conference of the IEEE Engineering in Medicine and Biology Society* (Vol. 4, pp. 3721–3723). IEEE.

Mouchaweh, M. S., & Billaudel, P. (2002). Influence of the choice of histogram parameters at fuzzy pattern matching performance. *WSEAS Transactions on Systems, 1*(2), 260–266.

Mouchaweh, M. S., Bouguelid, M. S., Billaudel, P., & RiERA, B. (2006). Variable probability-possibility transformation. *International Journal of Approximate Reasoning*.

Plataniotis, K. N., Hatzinakos, D., & Lee, J. K. (2006). Ecg biometric recognition without fiducial detection. In *2006 Biometrics Symposium: Special Session on Research at the Biometric Consortium Conference* (pp 1–6). IEEE.

Pouryayevali, S. (2015). ECG biometrics: New algorithm and multimodal biometric system. Ph.D. thesis.

Pummer, C. (2016). Continuous biometric authentication using electrocardiographic (ECG) data. Ph.D. thesis, MSc thesis, 2016, Hagenberg.

RAMER, A. (1989). Possibility theory: An approach to computerized processing of uncertainty, by didier dubois and henri prade, plenum press, New York 1988. xvi+ 263 pages. *International Journal of General Systems, 15*(2), 168–170.

Satija, U., Ramkumar, B., & Manikandan, M. S. (2018). A review of signal processing techniques for electrocardiogram signal quality assessment. *IEEE Reviews in Biomedical Engineering, 11*, 36–52.

Shdefat, A. Y., Joo, M.-I., Choi, S.-H., & Hee-Cheol, K. (2018). Utilizing ecg waveform features as new biometric authentication method. *International Journal of Electrical and Computer Engineering, 8*(2), 658.

Singh, Y. N., & Singh, S. K. (2012). Evaluation of electrocardiogram for biometric authentication. *Journal Information Security, 3*(1), 39–48.

Trobec, R., Tomašić, I., Rashkovska, A., Depolli, M., & Avbelj, V. (2018). *Body sensors and electrocardiography*. Berlin: Springer.

Vuksanovic, B., & Alhamdi, M. (2014). Analysis of human electrocardiogram for biometric recognition using analytic and ar modeling extracted parameters. *International Journal of Information and Electronics Engineering*, *4*(6), 428.

Wang, Y., Agrafioti, F., Hatzinakos, D., & Plataniotis, K. N. (2007). Analysis of human electrocardiogram for biometric recognition. *EURASIP Journal on Advances in Signal Processing*, *2008*(1), 148658.

Wayman, J., Jain, A., Maltoni, D., & Maio, D. (2005). An introduction to biometric authentication systems. In *Biometric systems* (pp. 1–20). Springer.

Wübbeler, G., Stavridis, M., Kreiseler, D., Bousseljot, R.-D., & Elster, C. (2007). Verification of humans using the electrocardiogram. *Pattern Recognition Letters, 28*(10), 1172–1175.

Zadeh, L. A. (1978). Fuzzy sets as a basis for a theory of possibility. *Fuzzy Sets and Systems, 1*(1), 3–28.

Zemzemi, M., Keskes, N., Barioul, R., Fakhfakh, S., Kallel, I. K., Derbel, N., & Kanoun, O. (2018). Toward a low cost, high performance ecg based biometrics: A preliminary work. In *2018 15th Interantional Multi-Conference on Systems, Signals & Devices (SSD)* (pp. 55–59). IEEE.

Chapter 6
Surface EMG Based Biometric Person Authentication by a Grasshopper Optimized SVM Algorithm

R. Barioul, S. Fakhfakh Ghribi, S. Suresh, H. Ben Jmaa Derbel, and O. Kanoun

Abstract Biometric techniques can provide extremely accurate and secured access to information compared to traditional person authentication methods through the use of users unique characteristic data. One of the newest and a promising physiological entity used in this field is the Surface electromyography (sEMG). However, machine learning based optimization are needed to have more accurate and precise results. In this context, the aim of this chapter is to introduce the Grasshopper optimized support vector machine to the field of person authentication based on sEMG. In this chapter, a first investigation of the introduced optimized algorithm accuracy will be conducted by the identification of 10 persons from the sEMG signals collected on their forearms via two Myoware sensors. 27 features in time and frequency domain were extracted from the collected database and graphically evaluated fo their person distinguish ability before being considered as an input to a feature selection stage. Three conventional feature selection methods which are Neighbourhood component analysis (NCA), Principal component analysis (PCA) and Relief-f where compared by their performances in combination with support vector machine (SVM) to perform the person identification. Furthermore, to insure a good performance of the person authentication, an optimization of the support vector machine by the grasshopper optimization algorithm has been implemented. A person recognition accuracy of 80% is reached by the GOA-SVM with 20 search agents and 100 iterations of GOA

R. Barioul (✉) · S. Suresh · O. Kanoun
Chair of Measurement and Sensor Technology, Technische Universität Chemnitz, Chemnitz, Germany
e-mail: rim.barioul@etit.tu-chemnitz.de

O. Kanoun
e-mail: kanoun@ieee.org

S. F. Ghribi · H. B. J. Derbel
CEM Research Laboratory, National School of Engineer of Sfax, University of Sfax, Sfax, Tunisia
e-mail: fakhfakhsameh@yahoo.fr

H. B. J. Derbel
e-mail: h.derbel@fss.rnu.tn

N. Derbel and O. Kanoun (eds.), *Advanced Methods for Human Biometrics*, Smart Sensors, Measurement and Instrumentation 40,
https://doi.org/10.1007/978-3-030-81982-8_6

even without any additional feature selection process while the best performance was shown by GOA-SVM with NCA as 88% of person recognition accuracy.

Keywords Support vector machine · Surface electromyography · Grasshopper optimization algorithm · Biometric person authentication

6.1 Introduction

Traditional methods for person authentication include knowledge-based PIN or password, proof incorporated information, and ownership-based ID card methods. These conventional techniques involve inconveniences of remembering the passwords, risk of password counterfeiting, risk of cards loss and theft. To overcome these problems, the need for biometrics is increasing regularly. In biometric techniques, basically persons unique characteristics are captured and recorded in a database. So whenever an identity verification is required, a new record is captured and compared with previously stored records in database and only if the patterns match, the person's identity is confirmed. Furthermore, advances in feature extraction and selection techniques, classification and optimization algorithms, make the design of an accurate and precise person identification model based on person biometrics such as DNA, ECG, EEG or EMG a potential possible field of research (Kim & Pan, 2017). Furthermore, biometric based algorithms are today worthy not only in confidential financial transactions in Banks and business sectors but also for personal data privacy and military applications (Kalyani, 2017).

The surface electromyography (sEMG) signal is the biomedical signal that measures the difference in electric potential collected at the surface of the muscles and generated during its contraction or relaxation. The sEMG signal is a convoluted signal with a range depending on the anatomical and physiological properties of the measured muscle (Kim & Pan, 2017; Lee, Ryu, & Youn, 2017). Moreover, this signal was proved to contain informations about the identity of subjects (Krishnamohan & Holi, 2011). This study aims to conduct a biometric person authentication from sEMG signals. For this, first a collection of signals and a feature extraction and evaluation for person distinguishing abilities needed to be conducted. Also, as the focus of this study is the software parts of the sEMG based person authentication, signals were collected from the forearm muscles with a reduced number of commercial Myoware sensors. In addition, to insure the conformity of the data collection conditions, all subjects were asked to perform the same hand gesture, which was a hard fist during the signal collection while the two sensors were recording simultaneously. After that features from time and feature domains are extracted, graphically represented and evaluated via scatter plots for person distinguishing ability. These features are then scaled through a min-max normalization. Furthermore a person recognition is done by classifying the signals using the basic Support Vector Machine algorithm with and without the feature selection step to choose the best selection method in this study case. Also, in order to insure a robust and reliable classification process with avoiding local optima convergences of the modal parameters, the nature based meta-heuristic algorithm called Grasshopper Optimization Algorithm(GOA) was used to create an optimized version of SVM. The GAO-SVM version which was presented and proved to outperform other optimized SVM versions with benchmark data on Ahanch et al. (2018) was here implemented and introduced to the field of sEMG based person authentication.

This chapter includes a presentation of related works to EMG based person authentication in Sect. 6.2, theoretical background of the GOA-SVM algorithm in Sect. 6.3. The extracted features and there evaluation are included in Sect. 6.4 with the person identification experimental results and discussion before the conclusion.

6.2 Biometry Based on sEMG Signals

Surface EMG as a biometric technique for person authentication is one of recent fields of research which getting more attention recently as the surface EMG had been proved to be efficient in other fields of sensor applications in term of computational reliability, device ability and subject safety. In Krishnamohan and Holi (2011) researchers collected the data on 3 days from 2 EMG sensors: One on forearm and the other on the wrist. Personal features were extracted by a non-uniform filter bank and the collected data were reduced through vector quantization modelling and classified using Gaussian Mixture Modelling (GMM). In this study 49 person were identified with a recognition rate average performance of 83.94%. In another study, Kim et al., used 12 time domain features to extract personal information from 20 persons wearing 4 four EMG sensors during gait analysis activity such as walking and sitting. The ANN was used in this investigation and subjects could be identified using the EMG signal applied to gait based human identification. When all 12 features were used, the average identification rate was 99.7% while it was 96% when only three muscles and five features were used (Kim, Ryu, Lee, & Kim, 2016). Another investigation of biometric person authentication based on gait analysis by surface EMG signals was performed by the Center for Bionics, Korea Institute of Science and Technology. In this study researchers proved that the EMG based identification methods have a good security features due to the fact that people have unique characteristics that could be seen in their walking behaviour including walking stride, speed, duration of burst and magnitude of the EMG signal from their muscles. This study involved the collection of EMG signals from the low-limb muscles of subjects and five-dimensional training and test data extracted from the values of root mean square (RMS), mean absolute value (MAV), dominant frequency (DF), integrated EMG signals (iEMG), and onset time of muscle firing (OS). A linear discriminant analysis (LDA) classifier algorithm was implemented for person recognition with am accuracy of 93% (Lee et al., 2017).

In Kim and Pan (2017), authors analysed EMG-based biometrics and implemented a personal identification and motion recognition system. The developed system extracted features using non-uniform filter bank and Waveform Length (WL), and reduced the dimension using Principal Component Analysis (PCA) and Linear Discriminant Analysis (LDA). The collected EMG signals were measured at two channels on the forearm of four persons, and the motions performed repeatedly were: Fist, rotating the wrist and raising the hand. The system performance was compared with the UCIs sEMG for Basic Hand movements Dataset in term of person identification ability from hand gestures. In the UCIs database the EMG signals were measured at the same muscles of the hands of five person aged between 20 and 22, in order to recognize six motions of the hands. The person classification was performed on the extracted features using Euclidean Distance (ED), Support Vector Machine (SVM), and K Nearest Neighbors (KNN). The collected database showed the highest personal identification rate of 95% on average for the fist. The average personal identification rate using Hook or Snap operation of UCI data was the highest at 93.89% (Kim & Pan, 2017). Moreover, in a study by Jingwei et al., on Feature Selection Problem in EMG Signals Classification with Binary Grey Wolf Optimizer, EMG database (DB4) from Non-Invasive Adaptive Prosthetic (NinaPro) was used. It consists of 10 healthy subjects with 17 Hand movement types recorded with 12 EMG electrodes. Feature extraction was implemented using STFT(Short Term Fourier Transforms) and 10 features such as Renyi entropy, Spectral Entropy, Shannon Entropy, Singular Value Decomposition-Based Entropy, Concentration Measure, Mean and Median Frequency, Two-Dimensional Mean, Variance, and Coefficient of Variation. A K-nearest Neighbor algorithm was implemented with $k = 1$. On applying a competitive Binary Grey Wolf Optimization with the KNN an accuracy of 92.69% was obtained for person identification (Too, Abdullah, Saad, Nursabillilah, & Weihown, 2018).

For previous investigations on sEMG based person authentication it could be remarked that the best results included experiments with either gate analysis or hand gestures with a big number of sensors in both cases. However, due to their complexity those methods can not be considered as user friendly and their suitability for embedded systems is questionable. To solve those limitation this chapter propose a system with only 2 EMG sensors and one performed statistic gesture for person authentication.

6.3 Hybrid Grasshopper Optimization Algorithm and Support Vector Machine (GOA-SVM)

The support vector machine (SVM) method has many advantages compared to other methods such as excellent generalization performance, as it is able to generate high-quality decision boundaries based on a small subset of training data points. In addition, the SVM method has a high ability to model complex and non-linear relations (Tanveer, 2015; Xi, Tang, Miran, & Zhizeng, 2013). But it also present some limitations such as its sensitivity to parameters initialization. Parameters including the cost (C) factor and the kernel parameters such as the gamma (γ) parameter in the radial basis function (RBF). Non adequate selection of such parameters can reduce the generalization performance of the SVM method (Klaus, 2005). In addition to this limitation and similar to other machine learning algorithms, SVM performance depends on the selected data set features, which is critically important in improving the generalization performance, increasing the computational efficiency and reducing running time, and produce very accurate classification model (Nguyen & de la Torre, 2010; Staelin, 2003). In this chapter, the grasshopper optimization algorithm of the Support vector machine (GOA-SVM), which was introduced on 2018 in Ahanch et al. (2018) and suggested the optimization of SVM parameter's tuning by the grasshopper optimization algorithm(GOA) is implemented for sEMG based person authentication.

6.3.1 Grasshopper Optimization Algorithm (GOA)

Grasshoppers are insects which eat plants and their behavior in groups or swarms builds the basis for the following algorithm. The life cycle of a grasshopper consists of three stages, namely the egg, the nymph and the adult grasshopper. As a nymph, the grasshopper has a slow pace and motion and as adults, they take swift long jumps. Searching for food is the primary objective of the swarm of grasshoppers (Mafarja et al., 2017; Saremi, Mirjalili, & Lewis, 2017). Exploration and Exploitation are key concepts to be considered in nature based algorithms to improve the convergence speed and to avoid local optima in the search space. Exploration also known as diversification leads the search agents to move abruptly to large distances. In contrast, exploitation or intensification results in short movements locally in the search space. Grasshoppers, along perform exploration and exploitation as they search for food or target (Ibrahim, Mazher, Ucan, & Bayat, 2019; Mafarja et al., 2017; Saremi et al., 2017).

The swarming behavior of grasshoppers can be described by,

$$X_i = r_1 S_i + r_2 G_i + r_3 A_i \tag{6.1}$$

where X_i defines the position of the ith grasshopper, S_i is the social interaction defined in Eq. 6.2), G_i is the gravity force on the ith grasshopper (see Eq. 6.4), A_i shows the wind advection (see Eq. 6.5), and r_1,r_2, and r_3 are random numbers in the interval [0, 1].

$$S_i = \sum_{j=1, j\neq i}^{N} s(d_{ij})\widehat{d_{ij}} \tag{6.2}$$

where d_{ij} is the Euclidean distance between the ith and the jth grasshopper that is calculated as $d_{ij} = |x_j - x_i|$. The $\widehat{d_{ij}}$ is a unit vector from the ith grasshopper to the jth grasshopper. 's' is defined as the strength of social forces given by,

$$s(r) = f \exp\left(-\frac{r}{l}\right) - \exp(-r) \tag{6.3}$$

where f indicates the intensity of attraction, and l is the attractive length scale. With different values of l and f it was found that the repulsion occurs between any two grasshoppers if the distance between them is in the interval [0, 2.079]. If a grasshopper is 2.079 units away from other grasshoppers, it enters the comfort zone.

$$G_i = -g\,\widehat{e_g} \tag{6.4}$$

where g is the gravitational constant and $\widehat{e_g}$ shows a unity vector towards the centre of the earth.

$$A_i = u\,\widehat{e_w} \tag{6.5}$$

where u is a constant drift and $\widehat{e_w}$ is a unit vector in the direction of the wind. Nymph grasshoppers have no wings, so their movements are highly correlated with wind direction.

$$X_i^d = c\left[\sum_{j=1 j\neq 1}^{G} c\frac{u_{bd} - l_{bd}}{2} s\left(\left|X_j^d(t) - X_i^d(t)\right|\right)\frac{X_j(t) - X_i(t)}{d_{ij}}\right] + \widehat{T_d} \tag{6.6}$$

$$s(d) = f\ \exp\left(-\frac{d}{l}\right) - \exp(-d) \tag{6.7}$$

where u_{bd} is the upper bound of the Dth dimension and l_{bd} is the lower bound of the Dth dimension and c is a decreasing coefficient to shrink the comfort zone, attraction region and repulsion area. The outer c_1 is used to reduce the search coverage toward the target grasshopper as the iteration count increases, while the inner c_2 has been used to reduce the effect of the attraction and repulsion forces between grasshoppers proportionally to the number of iterations (Mafarja et al., 2017; Saremi et al., 2017).

For controlling the degree of exploratory and exploitative behaviours in GOA, the parameter c should be decreased proportionally to the number of executed iterations. This mechanism increases the degree of exploitation as the iteration number increases. It reduces the comfort zone proportionally to the number of iterations as well. Parameter c is calculated as in equation below (Hamad, Houssein, Hassanien, & Fahmy, 2018; Ibrahim et al., 2019; Mafarja et al., 2017; Saremi et al., 2017; Tharwat, Houssein, Ahmed, Hassanien, & Gabel, 2018).

$$c(t) = c_{\max} - \frac{t}{t_{\max}}(c_{\max} - c_{\min}) \tag{6.8}$$

Furthermore, in each iteration, the parameter c is calculated and the distances between grasshoppers are normalized between 1 and 4. Updating the position of grasshoppers is performed iteratively until the stopping criterion is satisfied. Finally, the best grasshopper (target) is returned, which represents the best approximation for the global optimum (Mafarja et al., 2017; Saremi et al., 2017). Figure 6.1 shows the diagramatic representation of the working principle of the Grasshopper Optimization Algorithm (Ahanch et al., 2017).

6.3.2 GOA-SVM

As the optimization of SVM parameter's tuning is needed to improve its performance for sEMG based person authentication (Kim & Pan, 2017) and the implementation of the meta heuristic grasshopper optimization algorithm for this goal had shown a better efficiency than other optimization methods while tested with the benchmark datasets in Ahanch et al. (2018), the hybrid Grasshopper Optimization Algorithm and Support Vector Machine (GOA-SVM) was chosen for this study to be implement on collected sEMG signals for biometric person identification. In order to apply any meta heuristic algorithm for optimizing a problem, there are two formulation issues that should be addressed. The first is how the individual (i.e. solution) is represented, while the second is the selection of a fitness function used for evaluating the quality of the generated solutions. After deciding on these two design issues, the processes of running the GOA-SVM model can be described as follows (Ahanch et al., 2018):

1. Initialization: The GOA-SVM model starts with an initial random population of candidate solutions where each solution consists of the C and γ parameters and a number of flags corresponding to the features in the data set.
2. Update and reproduction: Reproduction operators of the GOA are applied in this step on all individuals to form a new population of candidate solutions.
3. Fitness evaluation: The quality of every generated individual is evaluated using the fitness function, which is described as

$$Fitness I_{ij} = \frac{1}{K}\sum_{k=1}^{K}\left(\frac{1}{N}\sum_{j=1}^{N}\delta[c(x_j), y_k]\right) \tag{6.9}$$

4. Termination: The search process for the best solution ends when the maximum number of iterations is reached. Then, the GOA-SVM model reports the best SVM parameters along with the subset of features that was found to have the highest classification accuracy.
5. Testing: For verification, the final step is to test the developed SVM model with the selected parameters and features based on an unseen and new part of the original data set (Ahanch et al., 2018)

Fig. 6.1 Flowchart of Grasshopper Optimization Algorithm (Ahanch, Asasi, & Amiri, 2017)

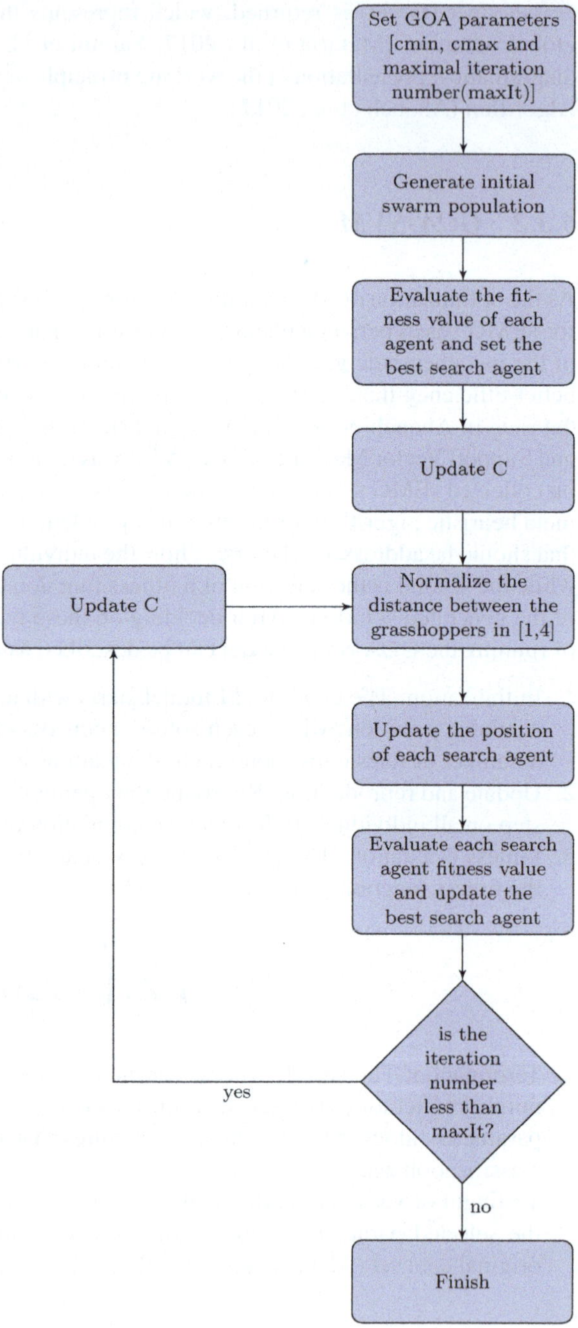

6.4 Experimental Results

In this study 2 commercial Myoware sensors were used simultaneously to collect the sEMG signal from ten healthy subject forearms. The collected database is including 200 signals of 6s duration each. For extracting the features from the sEMG signal, several in-built and user defined Matlab functions were used. The collection of the widely used features in sEMG based applications including hand gesture recognition and person identification in the literature with a total of 27 time domain and frequency domain features were extracted from each signal (Altin & Er, 2016; Arief, Sulistijono, & Ardiansyah, 2015; Karabulut, Ortes, Arslan, & Adli, 2017; Kim & Pan, 2017; Phinyomark et al., 2013; Raez, Hussain, & Mohd-Yasin, 2006), as described in Table 6.1.

To verify each feature ability to distinguish between persons from the collected database a preliminary graphical evaluation with scatter plots of sensor one versus sensor two was performed on all extracted features. Figures 6.2, 6.3 and 6.4 are showing some of the generated scatter plots as examples of the resulted irrelevant, relevant and redundant features for person authentication.

It was remarked that most of the features that posses a good person separability power such as AR2 and Mean Frequency in Fig. 6.3 were evaluated by researchers (Phinyomark, Hirunviriya, Limsakul, & Phukpattaranont, 2010; Barioul, Fakhfakh, Derbel, & Kanoun, 2019) as bad hand gestures recognition features while other features that was indicated as a good gestures separator such as the AAC in Fig. 6.2 are presenting a poor person distinguishing ability. Also, Graphically some evident redundancy i.e. mean absolute values and root mean square in Fig. 6.4 could be visualized.

The big similarity level between those features reduces their relevance for classification. Therefore, selecting the features providing a low inter-class correlation and high intra-class correlation is needed. To perform an adequate feature selection to this case of study, three well-known feature selection methods which are the Neighbourhood Component Analysis (NCA), Principal Component Analysis (PCA) and Relief-f are compared by their performances in combination with Support Vector Machine (SVM) to perform the person authentication in Table 6.2 with the best 10 selected features in each method. From Table 6.2 it could be seen that, for the collected data from the two Myoware sensors during the hard fist performance, the 10 persons where identified in the best case from the 10 features selected by the NCA with an average classification test accuracy of 75%. It is also remarked that for PCA with only 2 principle components, which refer to the 10 best features, the accuracy drooped even in comparative with an SVM without feature selection step.

This could be solved by adding more principle components, but it involves more features also, which increases the computation complexity of the algorithm and reduces it ability for being embedded.It is also remarked that in comparison with previous investigations of sEMG based person authentication this study presents less number of sensors, less number of gestures or a bigger number of persons with presenting a comparable person identification accuracy. However, considering

Table 6.1 Extrated features

Features	Formula and descriptions
Mean Absolute Value	$MAV(\mu) = \frac{1}{N}\sum_{n=1}^{N} \lvert x_n \rvert$ where N is the length of the EMG signal and xn represents the EMG signal. Mean is the most common and easy implemented feature of the time domain. It only finds the mean of EMG amplitude values over sample length of the signal and measure contraction level of the EMG signals
Root Mean Square	$RMS = \sqrt{\frac{1}{N}\sum_{n=1}^{N}(x_n)^2}$ RMS of EMG signal is the average value of the contraction level and is calculated. where N is the length of the EMG signal and xn represents the EMG signal
Variance	$VAR = \frac{1}{N-1}\sum_{n=1}^{N}(x_n - \mu)^2$ Variance is also most common statistical method for time domain feature extraction. It measures the density power of EMG signals
Standard Deviation	$STD_DEV(\sigma) = \sqrt{\frac{1}{N-1}\sum_{n=1}^{N}(x_n - \mu)^2}$ Variance is also most common statistical method for time domain feature extraction. It measures the density power of EMG signals
Kurtosis	$KURT = \frac{1}{\sigma^4} \times \frac{1}{N}\sum_{n=1}^{N}(x_n - \mu)^4$ Kurtosis is measure of peakness of probability distribution or measure of fourth order cumulative of the EMG signal. It is based on the distribution of EMG samples
Skewness	$SKEW = \frac{1}{\sigma^3} \times \frac{1}{N}\sum_{n=1}^{N}(x_n - \mu)^3$ Skewness is measure of asymmetry of a signal or measure of third order cumulative of the EMG signal It is based on the distribution of EMG samples
Willison Amplitude	$WAMP = \sum_{i=1}^{N} f(\lvert x_{i+1} - x_i \rvert - T)$ This is defined by the number of iteration or count times that the difference between two consecutive amplitudes of the EMG signal exceeds or overshoot a particular threshold specified by the programmer during processing. It is the cumulative variation that can indicate the degree variations of EMG signals. It can be formulated as, where T value is considered as 0.5
Number of Turns	$NT = \sum_{i=1}^{N-1} u(-x_i x_{i+1})$ NT is essentially defined as the number of changes in the sign of the slope, in other better terms, the number of EMG signal peaks

(continued)

Table 6.1 (continued)

Features	Formula and descriptions
Zero Crossing	$ZC = \sum_{i=1}^{N-2} u[(x_{i+1} - x_i)(x_{i+1} - x_{i+2})]$ Zero crossing is a representation of frequency information of the signal at time domain. Zero crossing is count of the times that amplitude values of the EMG signal cross zero amplitude level. In order to eliminate voltage fluctuations or background noises, threshold can be used in the calculation
Waveform Length	$WL = \sum_{n=1}^{N-1} \|x_{n+1} - x_n\|$ Waveform length is a measure of complexity of the EMG signal. It can be called as the cumulative length of the EMG waveform under consideration over the time axis segment i.e with respect to time
Mean Frequency	$MEAN_F REQ = \frac{\sum_{j=1}^{M} f_j P_j}{\sum_{j-1}^{M} P_j}$ Mean frequency is defined as the average of all frequency components which are obtained as sum of product(SOP) of the EMG power spectrum and the frequency divided by the sum total of the spectrum intensity
Median Frequency	$\sum_{j=1}^{MDF} P_j = \sum_{j=MDF}^{M} P_j = \frac{1}{2} \sum_{j=1}^{M} P_j$ Median frequency is a frequency at which the spectrum is divided into two regions with equal amplitude. It is a characteristic feature of the EMG signal
Signal to Noise Ratio	Signal to Noise Ratio is a ratio of the signal power and noise power. The signal power and noise power are estimated separately for EMG Signals
Mean Absolute Deviation	$MAD = \frac{1}{N} \sum_{n=1}^{N} \|x_n - ORT\|$ The average of the absolute deviations of data distribution of the EMG signal from their mean
Median Absolute Deviation	The median absolute deviation(MAD) is a robust measure of how spread out a set of data distribution of the EMG signal is
Myopulse Percentage Rate	$MYOP = \frac{1}{N} \sum_{i=1}^{N} [f(x_i)]$ MYOP is the average of the number of times that the EMG signal exceeds a predefined threshold of 0.01
Simple Square Integral	$SSI = \sum_{n=1}^{N} (x_n)^2$ SSI is measure of the contraction of the EMG signal. // While Mean Absolute Value calculates the averaged value, Simple squared integral calculates the complete value of contraction
Average Amplitude Change	$ACC = \frac{1}{N} \sum_{n=1}^{N-1} \|x_{n+1} - x_n\|$ AAC is the summation of difference of consecutive data points in the distribution of EMG signal

(continued)

Table 6.1 (continued)

Features	Formula and descriptions
Temporal Moment 3	$TM3 = \frac{1}{N}\sum_{n=1}^{N}(x_n)^3$ The third temporal moment is the summation of the data distribution raised to the third order
Auto-regression Coefficient	$x[n] = -\sum_{k=1}^{p}(a_k x[n-k] + e[n])$ In the Auto-Regressive model which is a widely useful feature extraction component, the samples of the signal are calculated by the linear combination of all the previous samples at a particular instance. This process essentially computes linear regression coefficients. It has been proven that the EMG spectrum varies with respect to muscular movements like contraction and relaxation that directly results in the variation in AR coefficients. AR coefficients are popular feature extraction method for biological signals. Only a few AR coefficients are usually relevant and contain useful information. Here p is degree of the AR model, $x[n]$ is data signal which composed of n data points, a_k is real valued AR coefficients and $e[n]$ is white noise term which is independent from previous samples. Choosing AR model order is an important issue. Because of the reason that lower order of AR model cannot represent signal exactly, higher order of AR model overfit the signal and represent too much noise
DASD Value	$DASDV = \sqrt{\frac{1}{N-1}\sum_{n=1}^{N-1}(x_{n+1} - x_n)^2}$ Difference in Absolute Standard Deviation in the wavelength is a derived feature from ACC
V-Order	$V_ORD = \sqrt[v]{\frac{1}{N}\sum_{i=1}^{N}(X_i)^v}$ The mathematical expression for V feature of an EMG signal which defines the order of the equation of the specified EMG signal
Temporal Moment 5	$TM5 = \frac{1}{N}\sum_{n=1}^{N}(x_n)^5$ The fifth temporal moment is the summation of the data distribution raised to the fifth order

that the SVM outperformed other classifiers for person authentication from surface EMG signals and GOA-SVM is outperforming other optimized version of SVM for benchmark data, it is expected that the GOA-SVM will outperform other methods in the field. This hypotheses was tested in the last experiment presented in this chapter and the results are presented on Table 6.3.

Considering Table 6.3, it can be remarked that NCA feature selection followed by the GOA-SVM has a similar accuracy rate with previous studies but with less sensors and more identified persons by less gestures. Therefore is could be concluded that the GOA-SVM over-perform other methods for surface EMG based person authentication.

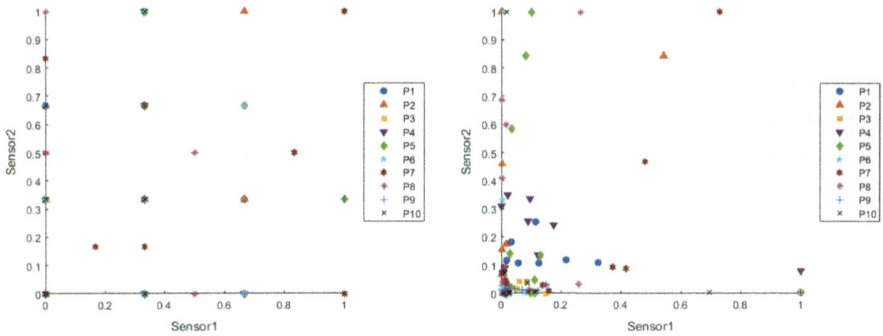

(a) Average Amplitude Change (b) Variance

Fig. 6.2 Scatter plots of selected non relevant features for person authentication

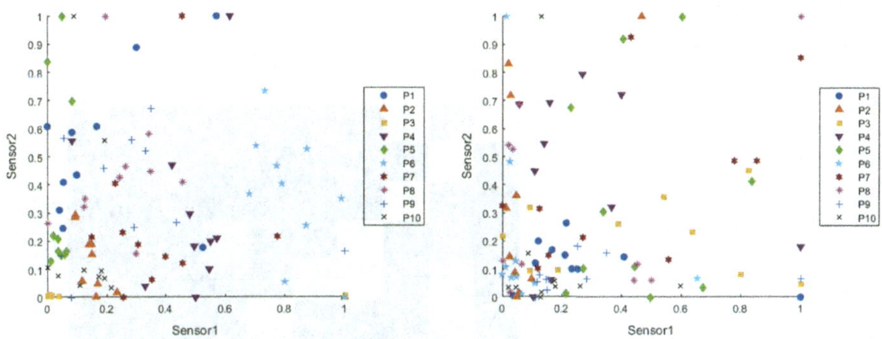

(a) Auto-regression Coefficient 2 (b) Mean Frequency

Fig. 6.3 Scatter plots of selected relevant features for person authentication

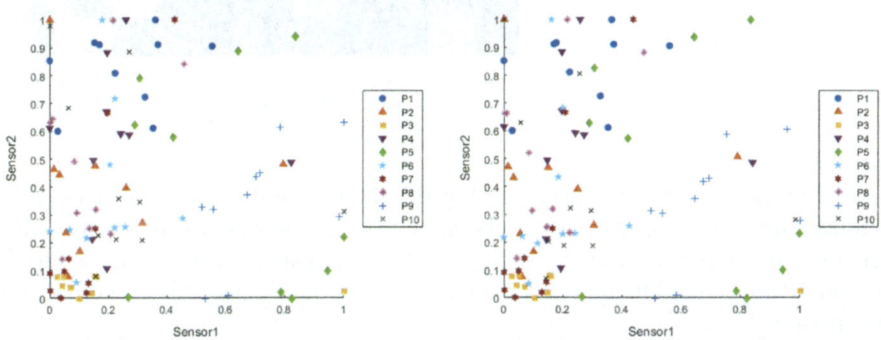

(a) Mean Absolute Value (b) Root Mean Square

Fig. 6.4 Scatter plots of selected relevant but redundant features for person authentication

Table 6.2 Influences of various selection methods on SVM accuracy

Method	Training accuracy (%)	Testing Accuracy (%)
SVM + 27 features	73.75	72.50
SVM + NCA	76.37	75
SVM + Relief f	72.50	62.50
SVM + PCA	78.75	70

Table 6.3 Influences of GOA optimization on SVM and various selection methods accuracy

Method	Training accuracy (%)	Testing accuracy (%)
GOA-SVM + 27 features	99.44	80
GOA-SVM + NCA	99.68	88
GOA-SVM + Relief f	99.93	72.50
GOA-SVM + PCA	99.37	85

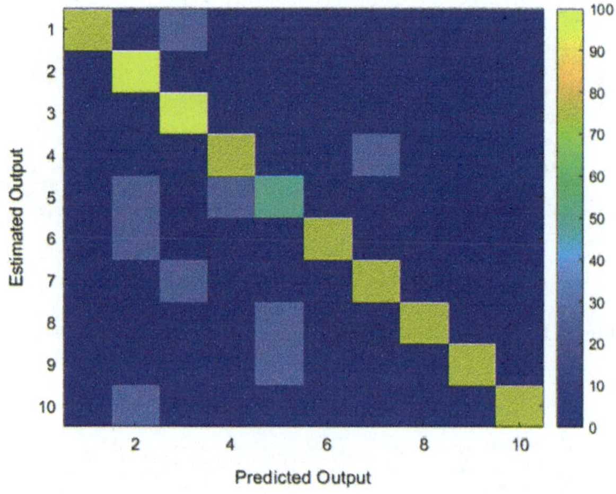

Fig. 6.5 GOA-SVM with NCA confusion matrix for model testing

Moreover, the confusion matrix in Fig. 6.5 shows that the trained model is presenting a sensitivity of 100% for 2 persons and 80% for most of the others. It is also remarked that most of the errors in classification are related the data collected from person 5 which could be probably caused by errors during the data recording with this person.

6.5 Conclusion

The aim of this chapter was to perform a sEMG based person authentication with limited number of sensors, limited number of gestures and optimized version of SVM with evaluating different widely used features in various sEMG applications for their person distinguishing abilities and testing conventional feature selections methods performances for this application. For that, in this study the hard fist gesture were performed by 10 healthy subjects for 10 times each while wearing two commercial surface EMG sensors in the forearm linked through synchronized ADC's of a teensy board version 3.6 to a personal computer. The collected signals were exploited to perform a person authentication with introducing the GOA-SVM to this field. In total 27 time and frequency domain features were extracted from the collected database and evaluated via scatter plots of sensor 1 versus sensor 2. It was remarked that the relevant features for person identification were reported to be non relevant for hand gesture recognition applications which could give a hint on further features to test for this application. Moreover it was concluded that the NCA is more adequate to this application as a feature selection method providing the best accuracy with less features which include less computation complexity. The GOA-SVM combined with the NCA resulted and average accuracy of 88% of the trained model after 20 runs which is a promising result showing the power of this algorithm in the field.

Acknowledgements This research is funded by the Deutscher Akademischer Austauschdienst (DAAD) and the German Research Foundation (DFG).

References

Ahanch, M., Asasi, M. S., & Amiri, M. S. (2017). A grasshopper optimization algorithm to solve optimal distribution system reconfiguration and distributed generation placement problem. In *IEEE 4th International Conference on Knowledge-Based Engineering and Innovation (KBEI)*, December 22.

Aljarah, I., Al-Zoubi, A. M., Faris, H., Hassonah, M. A., Mirjalili, S., & Saadeh, H. (2018). Simultaneous feature selection and support vector machine optimization using the grasshopper optimization algorithm. *Cognitive Computation, 10*, 478–495.

Altın, C., & Er, O. (2016). Comparison of different time and frequency domain feature extraction methods on Elbow Gesture's EMG. Bozok University, Yozgat, European Journal of Interdisciplinary Studies Turkey, Vol. 2, No. 3, May-August.

Arief, Z., Sulistijono, I. A., & Ardiansyah, R. A. (2015). Comparison of five time series emg features extractions using Myo Armband. International Electronics Symposium (IES).

Barioul, R., Fakhfakh, S., Derbel, H., & Kanoun, O. (2019). Evaluation of EMG signal features for hand gesture distinction. In *16th International Multi-Conference on Systems, Signals & Devices (SSD)*, July.

Hamad, A., Houssein, E. H., Hassanien, A. E., & Fahmy, A. A. (2018). Hybrid grasshopper optimization algorithm and support vector machines for automatic seizure detection in EEG signals. In *International Conference on Advanced Machine Learning Technologies and Applications*. Part of the advances in intelligent systems and computing book series Springer (AISC, Vol. 723).

Huang, G., Zhang, D., Zheng, X., & Zhu, X. (2010). An EMG-based handwriting recognition through dynamic time warping. In *2010 Annual International Conference of the IEEE Engineering in Medicine and Biology*, 31 Aug.–4 Sept.

Ibrahim, H.T., Mazher, W.J., Ucan, O. N., Bayat, O. (2019). A grasshopper optimizer approach for feature selection and optimizing SVM parameters utilizing real biomedical data sets. Neural Computing and Applications, Vol. 31, pp. 5965–5974.

Kalyani, C. H. (2017). Various biometric authentication techniques: A review. *Journal of Biometrics and Biostatics, 8*(5), October.

Karabulut, D., Ortes, F., Arslan, Y. Z., & Adli, M. A. (2017). Comparative evaluation of EMG signal features for myoelectric controlled human arm prosthetics. Nalecz Institute of Biocybernetics and Biomedical Engineering of the Polish Academy of Sciences, Elseiver.

Kim, S.-H., Ryu, J.-H., Lee, B.-H., & Kim, D.-H. (2016). Human identification using emg signal based artificial neural network. *Journal of the Institute of Electronics and Information Engineers, 53*(4), 142–148, April.

Kim, J. S., & Pan, S. B. (2017). A study on EMG-based biometrics. *Journal of Internet Services and Information Security (JISIS), 7*(2), 19–31, May.

Klaus, S. (2005). Optimal parameter selection in support vector machines. Journal of Industrial and Management Optimization, vol 1, no 4, p. 465.

Krishnamohan, P., & Holi, M. S. (2011). Gmm modeling of person information from emg signals. In *IEEE Recent Advances in Intelligent Computational Systems (RAICS'11)*, Trivandrum, India (pp. 712–717), September.

Lee, M., Ryu, J., & Youn, I. (2017). Biometric personal identification based on gait analysis using surface emg signals. Center for Bionics, Korea Institute of Science and Technology, South Korea. In *2nd IEEE International Conference on Computational Intelligence and Applications*.

Mafarja, M., Aljarah, I., Faris, H., Hammouri, A.I., Al-Zoubi, A.M., Mirjalili, S. (2017). Binary Grasshopper Optimisation Algorithm Approaches for Feature Selection Problems. Expert Systems with Applications, vol. 117, no. 1, pp. 267–286.

Negi, S., Kumar, Y., & Mishra, V. M. (2016). Feature extraction and classification for EMG signals using linear discriminant analysis. In *2nd International Conference on Advances in Computing* (pp. 1–6).

Nguyen, M.H., and de la Torre, F. (2010). Optimal feature selection for support vector machines. Pattern Recognition, 43(3), 584–591.

Phinyomark, A., Hirunviriya, S., Limsakul, C., & Phukpattaranont, P. (2010). Evaluation of EMG feature extraction for hand movement recognition based on euclidean distance and standard deviation. In *IEEE International Conference on Electrical Engineering/Electronics, Computer, Telecommunications and Information Technology*, May.

Phinyomark, A., Quaine, F., Charbonnier, S., Serviere, C., Tarpin-Bernard, F., & Laurillau, Y. (2013). EMG feature evaluation for improving myoelectric pattern recognition robustness. *GIPSA and LIG Laboratory, 40*(12), 4832–4840.

Raez, M. B. I., Hussain, M. S., & Mohd-Yasin, F. (2006). Techniques of EMG signal analysis: detection, processing, classification and applications. *Biological Procedures Online, 8*(163), 11–35.

Saremi, S., Mirjalili, S., & Lewis, A. (2017). Grasshopper optimisation algorithm: Theory and application (Vol. 105, pp. 30–47).

Staelin, C. (2003). Parameter selection for support vector machines. Hewlett-Packard Company, Technical Report HPL-2002-354R1.

Tanveer, M. (2015). Robust and sparse linear programming twin support vector machines. Cognitive Computation, Vol. 7, no. 1, pp. 137–149.

Tharwat, A., Houssein, E. H., Ahmed, M. M., Hassanien, A.E., Gabel, T. (2018). MOGOA algorithm for constrained and unconstrained multi-objective optimization problems. Applied Intelligence, Vol. 48, pp. 2268–2283.

Too, J., Abdullah, A. R., Saad, N. M., Nursabillilah, M. A., & Weihown, T. (2018). A new competitive binary GreyWolf optimizer to solve the feature selection problem in EMG signals classification. *MDPI Journal*.

Xu, Y., Guo, R., Wang, L. (2013). A twin multi-class classification support vector machine. Cognitive Computation, Vol. 5, no. 4, pp. 580–588.

Part III
Algorithm Based Methods of Multimodal Authentication

Chapter 7
Tracklet and Signature Representation Using Part Appearance Mixture Approach in the Context of Multi-shot Person Re-Identification

Salwa Baabou, Furqan M. Khan, François Bremond, Awatef Ben Fradj, Mohamed Amine Ben Farah, and Abdennaceur Kachouri

Abstract Recognizing persons in a video surveillance scene in the real world is attractive and is now showing an increasing interest. The task of person Re-Identification (Re-ID) consists in assigning the same identifier to all instances of a particular individual captured in a series of images or videos, even after the occurrence of significant gaps over time or space. This chapter addresses the problem of high variance in a person's appearance by represent it as a set of multi-modal feature distributions modeled by Gaussian Mixture Model (GMM). We tackle also the Re-ID challenges that consist in occlusions, illumination changes and person's orientation/pose. To this end, we extract the tracklets, i.e trajectories of persons. From these tracklets, we compute the signature and represent it based on the approach of Part Appearance Mixture (PAM) in the context of Multi-Shot person Re-Identification. Experiments and results on two public datasets and on our own dataset show good performance. An evaluation of the quality of this signature representation is also described.

S. Baabou (✉)
National Engineering School of Gabes, University of Gabes, Gabes, Tunisia
e-mail: baabousalwa@gmail.com

F. M. Khan · F. Bremond
INRIA Sophia Antipolis, Biot, France
e-mail: furqan.khan@inria.fr

F. Bremond
e-mail: francois.bremond@inria.fr

A. B. Fradj · M. A. B. Farah · A. Kachouri
Laboratory of Electronics and Information Technology (LETI), National Engineering School of Sfax, University of Sfax, Sfax, Tunisia
e-mail: benfradj_awatef@yahoo.fr

M. A. B. Farah
e-mail: med.farah@yahoo.fr

A. Kachouri
e-mail: abdennaceur.kachouri@enis.rnu.tn

N. Derbel and O. Kanoun (eds.), *Advanced Methods for Human Biometrics*,
Smart Sensors, Measurement and Instrumentation 40,
https://doi.org/10.1007/978-3-030-81982-8_7

Keywords Person Re-Identification (Re-ID) · Tracklet · Signature representation · Part Appearance Mixture Approach (PAM)

7.1 Introduction

Person Re-Identification (Re-ID) aims to match individuals appearing across non-overlapping camera networks at distinct times and locations. A person's appearance across different cameras is very variable, which makes the recognition of individuals more and more challenging.

A typical Re-ID system may have an image (single shot) or a video (multi-shot) as input for feature extraction and signature generation. Thus, the first step in Re-ID is to learn a person's visual signature or model and then compare the two models to get either a match or a non-match. Extracting a reliable signature depends on the availability of good observations. Besides, faulty trajectory estimation and incorrect detections introduce errors in signature generation and extraction that affect the Re-ID quality. The most obvious and simplest signature of a person is characterized by low-level features like color, texture and shape. However, these features are hardly unique, not descriptive enough and prone to variations. Color/texture varies due to cross view illumination variations, pose variations, view angle or scale changes in multi-camera settings.

A subject may be fully or partially occluded by other subjects or carrying items that lead to errors in matching between tracklets. Furthermore, some works in person Re-ID used body-parts methods (such as SDALF, MPMC...) (Pala, Satta, Fumera, & Roli, 2016) to solve the issue of signature alignment but this problem is still difficult and not efficient as these methods require real detections and many annotations. All these issues may affect the performance of person Re-ID which is still not robust enough to guarantee high accuracy in practice.

To sum up, the contribution of this chapter is: How to pre-process the tracklets to make them good for computing the signature and then represent it for multi-shot Re-ID based on PAM approach (Khan & Bremond, 2017). This may cater the high variance in a person's appearance and discriminate between persons with similar appearances. A Mahalanobis based distance is defined to compute similarity between two signatures.

This chapter is organized as follows: the main challenges of person Re-ID as well as the Re-ID related work are given is Sects. 7.2 and 7.3, respectively. The Re-ID process which contains person detection and tracking is described in Sect. 7.4.

Section 7.5 introduces the Part Appearance Mixture Approach PAM by presenting the signature representation and computing the similarity between these latter using metric learning algorithms. Finally, we evaluate the quality of our signature representation based on the realized experiments and results before concluding.

7.2 Main Challenges of Person Re-ID

There are many problems while automating Re-ID in real world surveillance systems. One of the greatest is the variation in pose and appearance of the person in various cameras and in time. This variability happens due to the different camera orientation and changes in subject pose, camera resolution or visual appearance of the person itself. In addition to that, inter-camera variations in lighting conditions, the changes in the scene illumination, different camera parameters, occlusion of the body parts. Occlusions could be either caused by other people or objects of the scene, or self-occlusions caused by own body parts. To tackle these issues, we need to have a robust feature descriptor, which is scale, pose and illumination invariant. Since generally in Re-ID, data is acquired in uncontrolled environments, i.e., without the user collaboration and varying backgrounds, the extraction of coherent discriminative feature vector demands great effort. Other complementary aspects of the Re-ID also require the collection of contextual cues such as spatio-temporal topology and the situation context as well. Many works in this field are only a decade old, and all of the challenges above makes Re-ID still an open problem in computer vision.

7.3 Related Works

Earlier studies have focused on the appearance matching problem in the context of person recognition. Person Re-Identification approaches concentrate either on signature modeling or on matching function leaning, while producing a distinctive and invariant representation for appearance matching. Signature modeling approaches focus on creating invariant and discriminative representations for individual's appearance, which are robust to viewpoint, illumination, and camera changes, as well as, occlusions and localization errors. On the other hand, matching function learning approaches [Chen, Yuan, Hua, Zheng, & Wang, 2015; Dikmen, Akbas, Huang, & Ahuja, 2010; Hirzer, Roth, Köstinger, & Bischof, 2012; Koestinger, Hirzer, Wohlhart, Roth, & Bischof, 2012; Li, Wu, Mukunoki, Kuang, & Minoh, 2015; Liao, Hu, Zhu, & Li, 2015; Liu, Ma, Zhang, & Huang, 2015; Martinel, Das, Micheloni, & Roy-Chowdhury, 2016; Prosser, Zheng, Gong, Xiang, & Mary, 2010; Shen et al., 2015; Su et al., 2015; Wang, Gong, Zhu, & Wang, 2014; You, Wu, Li, & Zheng, 2016; Zhao, Ouyang, & Wang, 2014] focus on improving Re-ID performance regardless of the underlying signature representation used to model appearance of individuals.

Single-shot approaches such as [Ahmed, Jones, & Marks, 2015; Gray & Tao, 2008; Layne, Hospedales, Gong, & Mary, 2012; Liu, Gong, Loy, & Lin, 2012a; Martinel et al., 2016; Matsukawa, Okabe, Suzuki, & Sato, 2016; Schwartz & Davis, 2009; Varior, Haloi, & Wang, 2016; Wang, Doretto, Sebastian, Rittscher, & Tu, 2007; Zhao, Ouyang, & Wang, 2013; Zhao et al., 2014] construct signature using only one image. These approaches use a mixture of color, texture, edge or "deep" features to identify persons given only one image. Hence they do not explicitly use additional information available in multi-shot case. Other approaches such as Bazzani, Cristani, Perina, Farenzena, and Murino (2010), Farenzena, Bazzani, Perina, Murino, and Cristani (2010), Karanam, Li, and Radke (2015), Liao et al. (2015), Liu et al. (2012b), Ma, Su, and Jurie (2012), Yan et al. (2016), Zeng, Wu, Tian, Zhang, and Hu (2015, Zeng et al. (2016) are used for multi-shot Re-ID; however, they extract features from each image independently and then aggregate information. For instance, after extracting descriptors of input images independently, uses Karcher mean to accumulate information (Zheng et al., 2016) uses max-pooling and (Yan et al., 2016) uses LSTM to aggregate information from multiple images of a person into a single feature vector. Hence, they fail to capture multi-modality of an individual's appearance. Other methods represent a multi-shot signature as a set of image descriptors acquired from different images of a person and use a set based similarity metric such as minimum point-wise distance (MPD), or average point-wise distance (APD). For computational and storage efficiency, instead of using all images, a smaller subset is selected. Bak, Kumar, and Brémond (2014), Liao et al. (2015), Zeng et al. (2015) use a fixed number of randomly selected images; however, uninformed random sampling is prone to losing valuable information unless a large number of samples are drawn. Alternatively, we propose that appearance modalities in multi-shot case should be discovered: (i) independently for each descriptor space (feature or body-part), and (ii) using variance of respective features as cues instead of "external" ones like orientation or pose, because most low-level features are not robust to arbitrary transformations, such as pose or orientation changes; thus variance based cues subsume pose and orientation cues. We achieve these goals by independently learning appearance model of each body region and feature descriptor given the set of images of a person. Furthermore, by modeling appearance as parametric distribution we retain more information about appearance of a person with high compression that allows us to better discriminate between persons with similar appearance.

7.4 Person Re-ID Process

The advances in computer vision, as well as machine learning techniques in the recent years, have ameliorated this expedition towards smart surveillance at a fast pace and as a result, a plethora of algorithms for the automatic analysis of the video sequences have been proposed. They include, for instance, person detection, person tracking, activity monitoring, and person Re-Identification. Some survey papers such as Paul, Haque, and Chakraborty (2013), Vezzani, Baltieri, and Cucchiara (2013), Bedagkar-Gala and Shah (2014), Wang et al. (2014), Saghafi, Hussain, Zaman, and Saad (2015) have presented them in detail.

Figure 7.1 illustrates the Re-ID process, i.e the diagram of person Re-ID system, containing the different steps that we will follow and explain it later. It starts with automatic person detection. In recent years, most of the existing person re-identification works have ignored this step and assume perfect pedestrian detection. However, perfect detection is impossible in real scenarios and misalignment can seriously affect the person Re-ID performance. Therefore, this factor should be carefully studied in future studies. In order to build a strong visual signature of people appearances, persons have to be accurately detected and tracked, so the step of person tracking should be also taken into consideration. However, person detection and multiple person tracking are difficult problems with their own hurdles. Significant amount of work has gone into the problem of person detection over the years as well as Multiple Object Tracking (MOT) (Nguyen, Chau, & Bremond, 2015; Nguyen, Bremond, & Trojanova, 2016) within a single camera's Field-Of-View (FOV) as well as multiple cameras which has also been widely researched but it remains an open problem. For feature extraction and descriptor generation, the most commonly used features are color, shape, position, texture, and soft-biometry. The adopted feature is determined by different factors. On one side, the signature should be unique and discriminative enough which can lead to the selection of biometry or soft-biometry features. On the other side, camera resolution, computational load and other implementation issues can prevent or limit their usage and more generic features are required. It is worth noting, that the Re-ID system; in which our work is focusing; as appearing in the relevant literature, turns out to be divided as we said into two distinctive steps (see Fig. 7.1): (i) extracting distinctive visual features to represent the human appearance and (ii) establishing correspondence by learning or discovering an optimal metric that can maximize the distance between samples from different classes whilst minimizing the distance between those belonging to the same class.

All these steps will be described in details in the following section.

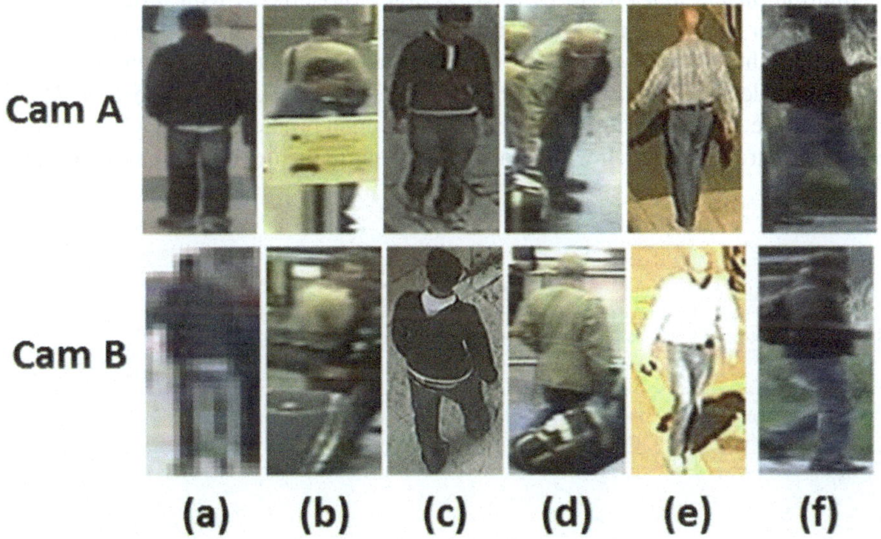

Fig. 7.1 Person Re-ID diagram

7.4.1 Detection

Person detection is the process of detecting and localizing each subject in the images, represented via bounding boxes which is by itself an intensive research field.

Subject detection can be considered also as classification process; First intuitive idea is to deal with detection as classification of all possible bounding boxes in the image, and classify them as different subjects. To take this way, we need a sliding window with certain step to span the whole image. In addition, we need the windows to be with different sizes, and scales. At the end, we have the bounding box of the subject, and a score (confidence) of the classification. Theoretically, if we have a very fast classifier, this can work. But in reality, the sliding window is slow, we need too many windows to guarantee that all possible regions are tested. To solve the problem of sliding window, instead of looking at all possible positions, we can have a smarter system that can find some "interesting" regions, and tell the classifier where to look. Example of region proposals are selective search, and Edge-boxes.

There are many subject detectors that we can cite: R-CNN, fast/er R-CNN (Ren, He, Girshick, & Sun, 2015) which are two-stage detectors; first, they propose regions, then they apply classification and bounding box regression. The modern detectors deal with the whole detection process as bounding box regression, so they are much faster which are YOLO (Redmon, Divvala, Girshick, & Farhadi, 2016) and SSD (Liu et al., 2016).

In this contribution, we will use the SSD detector (Liu et al., 2016). Figure 7.2 shows a visualization of a sample from CHU Nice dataset of the detection results.

In fact, the SSD detector differs from other single shot detectors due to the usage of multiple layers that provide a finer accuracy on subjects with different scales. (Each deeper layer will see bigger subjects). It starts with a VGG pre-trained model. Then after the image is passed on the VGG network, some convolutional (conv) layers are added producing feature maps of sizes 19×19, 10×10, 5×5, 3×3, 1×1. These, together with the 38×38 feature map produced by VGG's conv 4-3, are the feature maps which will be used to predict bounding boxes. Using the same

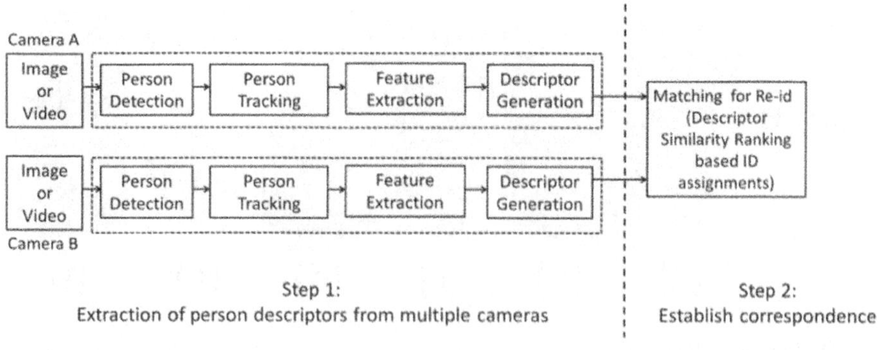

Fig. 7.2 Visualization of a sample from CHU Nice dataset of the Detection Results

concept as anchor boxes in Faster R-CNN, and the idea of dividing the image to grid in YOLO, they apply different boxes with different sizes and scales to different feature maps. For each default box on each cell, the network output are the following:

- A probability vector of length c, where c are the number of classes plus the background class that indicates no subject.
- A vector with 4 elements (x, y, width, height) representing the offset required move the default box position to the real subject.

7.4.2 *Multi-object Tracking*

Insprired by Multi-object tracking approach in Anh, Khan, Negin, and Bremond (2017), the tracking process is based on the method of a robust online multi-object tracking which combines a local and global tracker. In the local tracking step, we use the frame-to-frame association to generate the tracklets (*object trajectories*; which are represented by a set of multi-modal feature distributions modeled by the GMMs. In the global tracking step, the tracklet bipartite association method is used based on learning Mahalanobis metric between GMM components using KISSME (Koestinger et al., 2012) metric learning algorithm. The local tracker's objective is to find correct object trajectories in the past, while, the global tracker tries to find object associations between aggregated tracklets. In the first step, the tracklets are

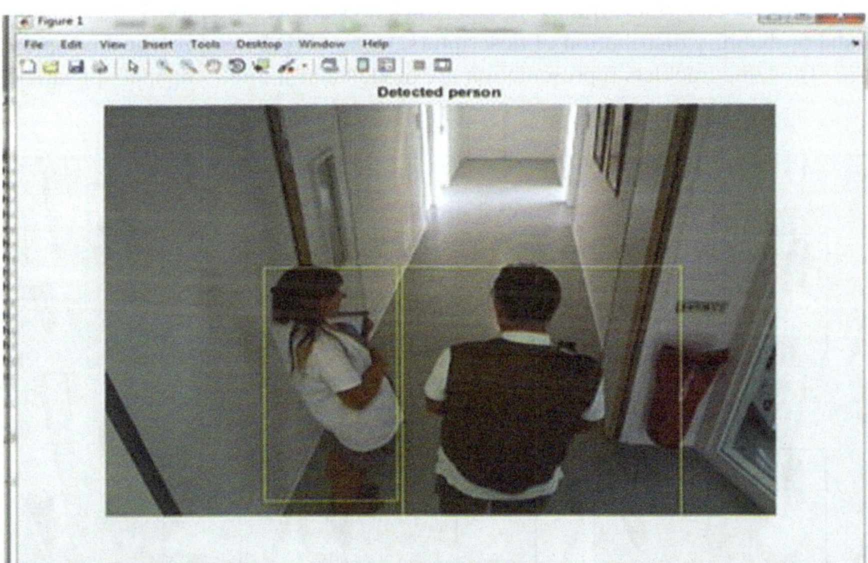

Fig. 7.3 Visualization of a sample of frames from CHU Nice dataset of the Tracking Results

constructed by putting together frame-to-frame tracker's output. For a reliable tracklet, tracklet filtering is applied by splitting spatially disconnected or occluded tracks and filtering out noisy tracklets. In the second step, in every video segment Δt, the global tracker carries out data association and performs on line tracklet matching. Association and matching process happens based on Mahalanobis metric among representations of tracklets stacked in two previous video segments ($2\Delta t$).

Figure 7.3 presents a visualization of a sample of consecutive frames from CHU Nice dataset of the tracking results.

7.5 Part Appearance Mixture (PAM) Approach

7.5.1 Signature Representation

We define a tracklet Tr_i between two frames m and n as a sequence of tracked subject's bounding-boxes as follows:

$$Tr_i = (N_i^m, N_i^{m+1}, \ldots, N_i^{n-1}, N_i^n) \tag{7.1}$$

where N represents the subject bounding-box and i is the subject ID.

Figure 7.4 shows a representation of some samples of tracklets of a person from CHU Nice dataset.

Inspired by Part Appearance Mixture PAM approach in Khan and Bremond (2017), and to cater the variance in a person's appearance, we model it as a multi-modal probability distribution of descriptors, using GMM to represent this appearance. Thus, the tracklets representation are modeled as a multi-channel appearance

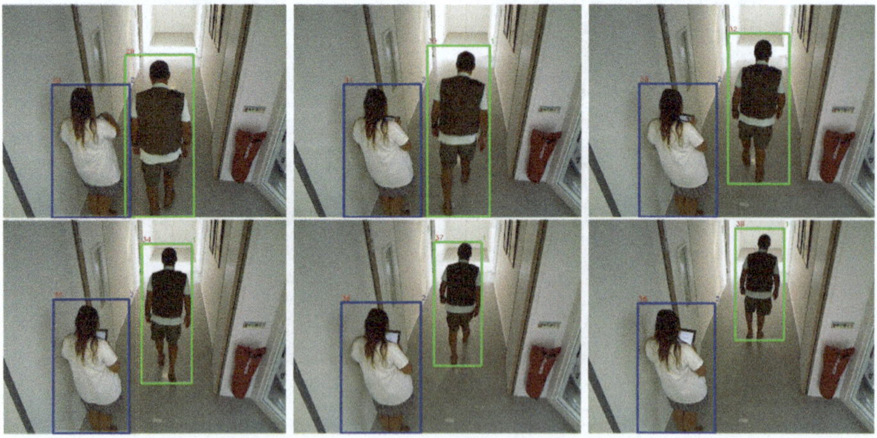

Fig. 7.4 Sample of tracklets representation from CHU Nice Dataset

mixture (appearance model). The representation divides body into three parts: full, upper and lower. Each channel in the mixture model corresponds to a particular body part. Given a set of nodes (detection bounding-boxes) belonging to a tracklet Tr_i, its PAM signature representation Q is defined as a set of appearance models $M_i^p : Q = \{M_q^p \mid p \in \{full, upper, lower\}\}$; one for each part p of person q. Each appearance model in the set is a multivariate GMM distribution of low-level features of part p. Appearance models help to overcome occlusion, pose variation and illumination problems.

To describe an object, we use appearance features that are locally computed on spatial grid of object detection bounding-boxes; the features are computed efficiently to be shared between the parts (upper and lower body regions are defined as 60% of bounding-box of the person) including: HOG (Dalal & Triggs, 2005), LOMO (Liao, Hu, & Li, 2014), HSCD (Zeng et al., 2015) and Color histogram features. While the framework exploits HOG feature as a shape based feature to overcome difficulties of pose variation, it benefits from other features to cope with illumination and appearance changes happening in long occlusions.

The similarity between two signatures is partially based on computing Mahalanobis distance between means of GMM components. People appearing in a video have different appearance and produce GMMs with variable number of components. Therefore, the number of components are not a priori determined and need to be retrieved. In order to infer the number of GMM components for each appearance model automatically, Akaike Information Criterion (AIC) model selection is used. Knowing fixed number of the components, the parameters of a GMM could be learned conveniently using Expectation-Maximization method.

7.5.2 Similarity Metric for Signature Representation

As mentioned before, a signature is a set of part appearance mixtures. Similarity between two signatures G_i and G_j is defined as the sum of similarities between the corresponding appearance models. Given the distance between two appearance mixtures d(M_1,M_2), we can convert this distance into similarity using Gaussian similarity kernel:

$$Sim(G_i, G_j) = \sum_{p \in \mathrm{P}} \exp\left(-\frac{\overline{d(M_p^q, M_p^g)} - \gamma_{p,g}}{\frac{1}{3}(\beta_{p,g} - \gamma_{p,g})}\right) \tag{7.2}$$

where $\mathrm{P} = \{full, upper, lower\}$, $\overline{d(M_p^p, M_p^g)}$ is max normalized distance between a query person q and a gallery person g of part p. $\beta_{p,g}$ and $\gamma_{p,g}$ are the maximum and minimum normalized distances, respectively, between person g in gallery and any other person q in query set. The factor $\frac{1}{3}$ in formula makes Gaussian similarity kernel goes to zero for q that has maximum normalized distance from g. We define the distance between two GMMs as he distance between their components weighted by their prior probabilities:

$$d(M_1, M_2) = \sum_{i=1:K, j=1:K} \pi_{1i}\pi_{2j} d(G_{1i}, G_{2j}) \tag{7.3}$$

where G_{nk} is the component k of GMM $M_{n\in\{1,2\}}$ with corresponding prior π_{nk} and $d(G_i, G_j) = JDiv(G_i, G_j)$ which is defined in the following section.

7.5.3 Distance Computation Between Signatures

To define similarity between two signatures, f-divergence based distances; in particular Jeffry's Divergence (JDiv) is used, and since we restrict covariance matrices to be diagonal, it can be computed as follows:

$$JDiv(G_i, G_j) = \frac{1}{2}(\mu_i - \mu_j)^T \psi(\mu_i - \mu_j) + \frac{1}{2}tr\left\{\Sigma_i^{-1}\Sigma_j + \Sigma_j^{-1}\Sigma_i - 2I\right\} \tag{7.4}$$

where $\psi = \sum_i^{-1} + \sum_j^{-1}$

The distance between two GMMs is computed based on the Mahalanobis distance, squared Mahalanobis distance of a pair of vectors is defined as follows:

$$d^2(x_i, x_{j)} = (x_i - x_j)^T M(x_i - x_j) \tag{7.5}$$

where M is a positive semi-definite matrix. The parameters of Matrix M are estimated using KISSME [12]. i.e $M = \sum_+^{-1} - \sum_-^{-1}$, where $\sum_+$ and $\sum_-$ are feature-difference covariance matrices of positive and negative classes, respectively. Given the mean of two Gaussian distributions, μ_i and μ_j, the positive and negative class covariance matrices are defined as:

$$\sum_+ = \sum_{y_{ij}=+} (\mu_i - \mu_j)(\mu_i - \mu_j)^T \tag{7.6}$$

$$\sum_- = \sum_{y_{ij}=-} (\mu_i - \mu_j)(\mu_i - \mu_j)^T \tag{7.7}$$

where $y_{ij} \in \{+, -\}$ is the ground truth similarity label between pairs of Gaussian distributions (G_i, G_j). Alternatively, matrix M can be estimated using XQDA [18] in similar spirit.

7.6 Experiments and Results

7.6.1 Datasets

We have evaluated our work on two challenging benchmark datasets: *PRID2011* and *iLIDS-VID* and on our own dataset: *CHU Nice dataset*. These datasets were chosen because they provide multiple images per individual (i.e Multi-shot datasets) collected in realistic visual surveillance settings using two cameras.

- *PRID2011* (Hirzer, Beleznai, Roth, & Bischof, 2011): This dataset consists of image frames extracted from two static camera recordings, depicting people walking in different directions. Images from both cameras contain variations in viewpoint, illumination, background and camera characteristics. 475 and 856 person trajectories were recorded via individual cameras, with 245 persons appearing in both views/cameras.
- *iLIDS-VID* (Wang et al., 2014): it contains 300 identities captured in indoor scenes. It is an extended version of iLIDS dataset. It is generally believed that iLIDS-VID is more challenging than *PRID2011* due to extremely heavy occlusion.
- *CHU Nice dataset*: Collected in the hospital of Nice (CHU) in Nice, France. It's related to INRIA Sophia Antipolis. Most of The people recruited for this dataset were elderly people, aged 65 and above, of both genders. It contains 615 videos with 149365 frames. It's also an RGB-D dataset, i.e it provides RGB+Depth images.

7.6.2 Performance Evaluation

Tables 7.1, 7.2 and 7.3 show the recognition rate (%) at different ranks (rank-1, 5, 10, 20) of a baseline method LOMO+XQDA (Liao et al., 2015) and PAM-LOMO+XQDA on PRID2011, iLIDS-VID and CHU Nice datasets, respectively.

From the above experiments, we notice that PAM-LOMO+KISSME achieves good performance on three datasets; it achieves 92.5%, 79.5% and 81.8% rank-1 recognition rates on PRID2011, iLIDS-VID and CHU Nice datasets, respectively. This shows that our adaptation of feature descriptor LOMO and metric learning KISSME to PAM representation is effective.

Figures 7.5, 7.6 and 7.7 show the CMC curves obtained on PRID2011, iLIDS-VID and CHU Nice datasets, respectively, which proved that PAM-LOMO+KISSME (red curve) reached better results on three datasets.

Table 7.1 Top ranked Recognition rates (%) on *PRID2011*

Methods	Rank-1	Rank-5	Rank-10	Rank-20
LOMO+XQDA	–	–	–	–
PAM-HOG+KISSME	55.3	80.7	90.2	95.6
PAM-LOMO+XQDA	–	–	–	–
PAM-LOMO+KISSME	**92.5**	**99.3**	**100.0**	**100.0**

Table 7.2 Top ranked Recognition rates (%) on *iLIDS-VID*

Methods	Rank-1	Rank-5	Rank-10	Rank-20
LOMO+XQDA	53.0	78.5	86.0	93.4
PAM-HOG+KISSME	33.9	60.0	70.2	79.1
PAM-LOMO+XQDA	–	–	–	–
PAM-LOMO+KISSME	**79.5**	**95.1**	**97.6**	**99.1**

Table 7.3 Top ranked Recognition rates (%) on *CHU Nice dataset*

Methods	Rank-1	Rank-5	Rank-10	Rank-20
LOMO+XQDA	30.7	64.6	80.3	90.3
PAM-HOG+KISSME	–	–	–	–
PAM-LOMO+XQDA	38.5	69.2	84.6	100.0
PAM-LOMO+KISSME	**81.8**	**90.9**	**100.0**	**100.0**

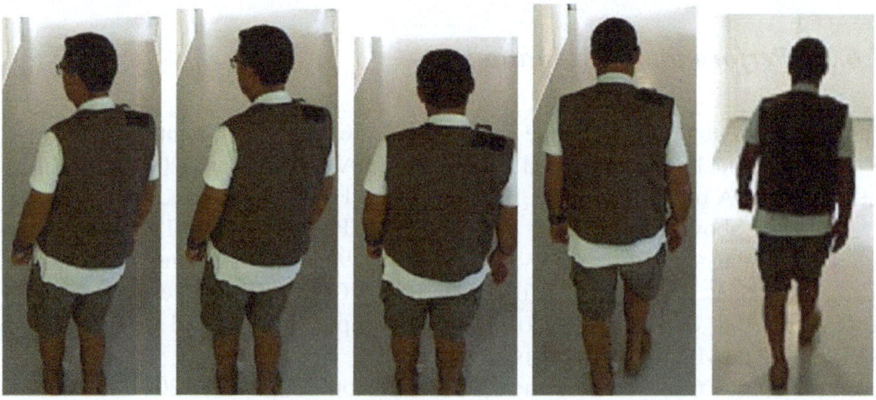

Fig. 7.5 CMC curves obtained on *PRID2011* dataset

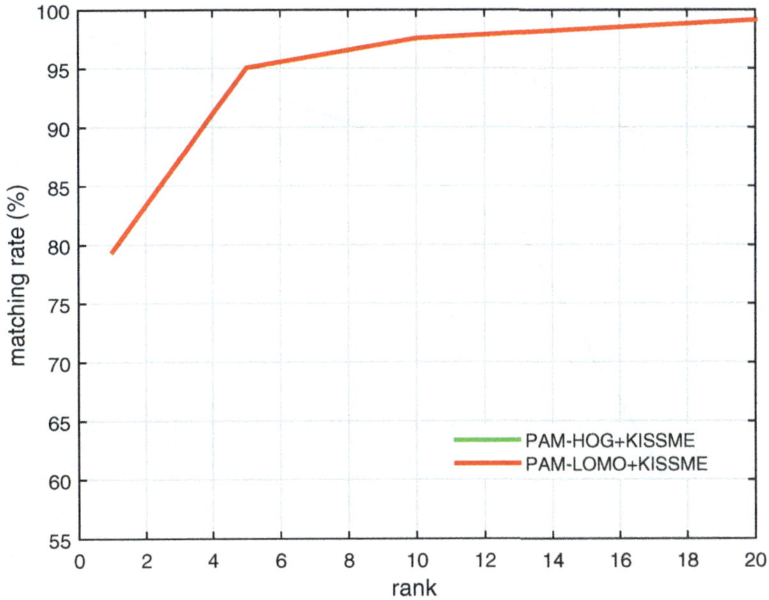

Fig. 7.6 CMC curves obtained on *iLIDS-VID* dataset

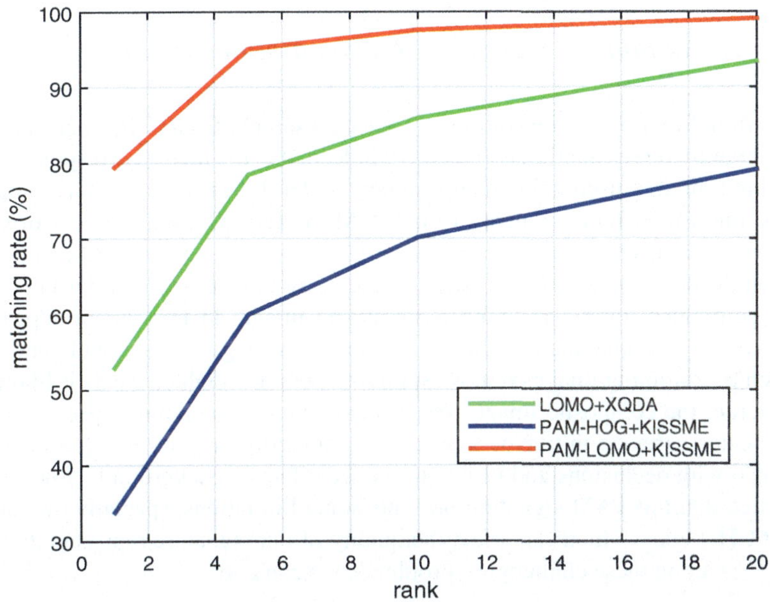

Fig. 7.7 CMC curves obtained on *CHU Nice* dataset

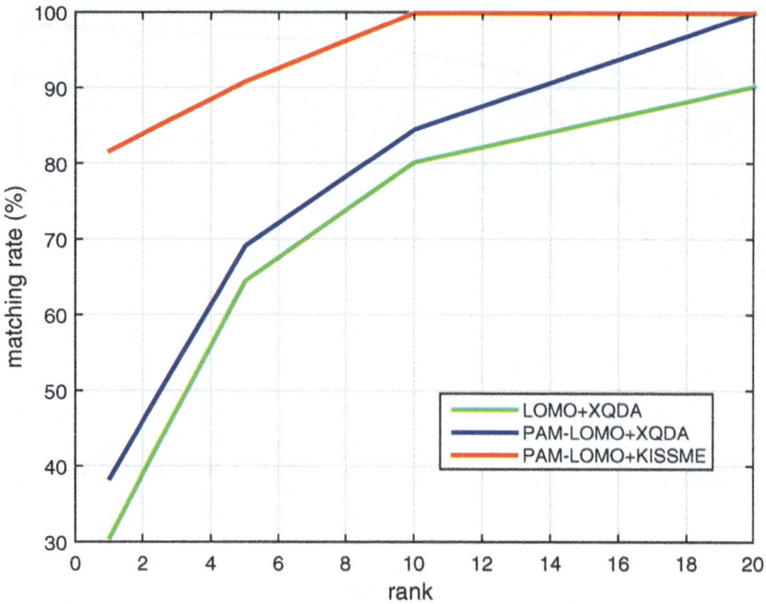

Fig. 7.8 Visualization of selected samples of signature representation from CHU Nice dataset

7.6.3 Evaluation of Signature Representation Quality

As shown in Fig. 7.5, a visualization of a selected sample from CHU Nice dataset of PAM signature representation is presented. Indeed, the first image corresponds to one of the input images used to learn appearance model. It's followed by the composite images, one for each component of the GMM mixture. Optimal number of GMM components for each appearance model varies between persons. GMM components focus on different pose and orientation of the person. Moreover, We visualize each GMM component by constructing a composite image. In fact, given appearance descriptor, we compute the likelihood of an image belonging to a model component and then by summing images of corresponding person weighted by their likelihood we generate the composite image. We can say that our signature representation is able to cater variance in person's pose and orientation as well as illumination, it deals also with occlusions and is able to reduce effect of background. However, we can notice that this PAM signature presents some limitations, specially on our own dataset CHU Nice, which can affect the quality of our signature representation (see Fig. 7.8). Among these challenging problems, we can cite:

- Bad detection
- Number of frames by pose
- Number of GMM components not adequate with the number of person's pose/orientation and depends of the low-level features used.

7.7 Conclusion

Person Re-ID is a challenging task with three aspects: First, ii is important to determine which parts should be segmented and compared. Second, there is a need to generate invariant signatures for comparing the corresponding parts. Third, an appropriate matching function (i.e similarity metric or a ranking function) must be applied to compare these signatures. In most studies, the Re-ID process is designed under the assumption that the appearance of the person is unchanged which doesn't seem reasonable in practise. This chapter presented the person Re-ID process in details beginning by the detection step to the signature computation in order to establish the correspondence. We extracted the tracklets and then compute the signature of persons and represented it based on PAM approach which uses multiple appearance models based on GMM model. Each appearance is described as a probability distribution of some low-level features for a certain part of person's body. Indeed, this improves the appearance descriptors and deals with occlusions and variance in pose/orientation of individuals. The robustness of the quality of this signature representation is verified by extensive experiments. As future work, we are trying to improve the PAM signature representation, by using the skeleton and extracting the pose machines from our dataset, i.e. CHU Nice dataset, which will be soon introduced as a new public dataset in the field of person Re-ID.

References

Ahmed, E., Jones, M., & Marks, T. K. (2015). An improved deep learning architecture for person re-identification. In *Proceedings of the IEEE Conference on Computer Vision and Pattern Recognition* (pp. 3908–3916).

Anh, N. T. L., Khan, F. M., Negin, F., & Bremond, F. (2017). Multi-object tracking using multi-channel part appearance representation. In *2017 14th IEEE International Conference on Advanced Video and Signal Based Surveillance (AVSS)* (pp. 1–6). IEEE.

Bak, S., Kumar, R., & Brémond, F. (2014). Brownian descriptor: A rich meta-feature for appearance matching. In *2014 IEEE Winter Conference on Applications of Computer Vision (WACV)* (pp. 363–370). IEEE.

Bazzani, L., Cristani, M., Perina, A., Farenzena, M., & Murino, V. (2010). Multiple-shot person re-identification by hpe signature. In *2010 20th International Conference on Pattern Recognition (ICPR)* (pp. 1413–1416). IEEE.

Bedagkar-Gala, A., & Shah, S. K. (2014). A survey of approaches and trends in person re-identification. *Image and Vision Computing, 32*(4), 270–286.

Chen, D., Yuan, Z., Hua, G., Zheng, N., & Wang, J. (2015). Similarity learning on an explicit polynomial kernel feature map for person re-identification. In *Proceedings of the IEEE Conference on Computer Vision and Pattern Recognition* (pp. 1565–1573).

Dalal, N., &Triggs, B. (2005). Histograms of oriented gradients for human detection. In *IEEE Computer Society Conference on Computer Vision and Pattern Recognition, 2005. CVPR 2005* (Vol. 1, pp. 886–893). IEEE.

Dikmen, M., Akbas, E., Huang, T. S., & Ahuja, N. (2010). Pedestrian recognition with a learned metric. In *Asian Conference on Computer Vision* (pp. 501–512). Springer.

Farenzena, M., Bazzani, L., Perina, A., Murino, V., & Cristani, M. (2010). Person re-identification by symmetry-driven accumulation of local features. In *2010 IEEE Conference on Computer Vision and Pattern Recognition (CVPR)* (pp. 2360–2367). IEEE.

Gray, D., & Tao, H. (2008). Viewpoint invariant pedestrian recognition with an ensemble of localized features. In *European Conference on Computer Vision* (pp. 262–275). Springer.

Hirzer, M., Beleznai, C., Roth, P. M., & Bischof, H. (2011). Person re-identification by descriptive and discriminative classification. In *Scandinavian Conference on Image Analysis* (pp 91–102). Springer.

Hirzer, M., Roth, P. M., Köstinger, M., & Bischof, H. (2012). Relaxed pairwise learned metric for person re-identification. In *European Conference on Computer Vision* (pp. 780–793). Springer.

Karanam, S., Li, Y., & Radke, R. J. (2015). Person re-identification with discriminatively trained viewpoint invariant dictionaries. In *Proceedings of the IEEE International Conference on Computer Vision* (pp. 4516–4524).

Khan, F. M., & Bremond, F. (2017). Multi-shot person re-identification using part appearance mixture. In *2017 IEEE Winter Conference on Applications of Computer Vision (WACV)* (pp. 605–614). IEEE.

Koestinger, M., Hirzer, M., Wohlhart, P., Roth, P. M., & Bischof, H. (2012). Large scale metric learning from equivalence constraints. In *2012 IEEE Conference on Computer Vision and Pattern Recognition (CVPR)* (pp. 2288–2295). IEEE.

Layne, R., Hospedales, T. M., Gong, S., & Mary, Q. (2012). Person re-identification by attributes. In *Bmvc* (Vol. 2, p. 8).

Li, W., Wu, Y., Mukunoki, M., Kuang, Y., & Minoh, M. (2015). Locality based discriminative measure for multiple-shot human re-identification. *Neurocomputing, 167*, 280–289.

Liao, S., Hu, Y., & Li, S. Z. (2014). Joint dimension reduction and metric learning for person re-identification. arXiv preprint.

Liao, S., Hu, Y., Zhu, X., & Li, S. Z. (2015). Person re-identification by local maximal occurrence representation and metric learning. In *Proceedings of the IEEE Conference on Computer Vision and Pattern Recognition* (pp. 2197–2206).

Liu, C., Gong, S., Loy, C. C., & Lin, X. (2012a). Person re-identification: What features are important? In Fusiello, A., Murino, V., and Cucchiara, R., editors, Computer Vision - ECCV 2012. Workshops and Demonstrations (pp. 391–401). Berlin: Springer.

Liu, K., Ma, B., Zhang, W., & Huang, R. (2015). A spatio-temporal appearance representation for video-based pedestrian re-identification. In *2015 IEEE International Conference on Computer Vision (ICCV)* (pp. 3810–3818). IEEE.

Liu, W., Anguelov, D., Erhan, D., Szegedy, C., Reed, S., Fu, C.-Y., & Berg, A. C. (2016). Ssd: Single shot multibox detector. In *European Conference on Computer Vision* (pp. 21–37). Springer.

Liu, X., Song, M., Zhao, Q., Tao, D., Chen, C., & Bu, J. (2012b). Attribute-restricted latent topic model for person re-identification. *Pattern Recognition, 45*(12), 4204–4213.

Ma, B., Su, Y., & Jurie, F. (2012). Local descriptors encoded by fisher vectors for person re-identification. In *European Conference on Computer Vision* (pp. 413–422). Springer.

Martinel, N., Das, A., Micheloni, C., & Roy-Chowdhury, A. K. (2016). Temporal model adaptation for person re-identification. In *European Conference on Computer Vision* (pp. 858–877). Springer.

Matsukawa, T., Okabe, T., Suzuki, E., & Sato, Y. (2016). Hierarchical gaussian descriptor for person re-identification. In *Proceedings of the IEEE Conference on Computer Vision and Pattern Recognition* (pp. 1363–1372).

Nguyen, T. L. A., Bremond, F., & Trojanova, J. (2016). Multi-object tracking of pedestrian driven by context. In *2016 13th IEEE International Conference on Advanced Video and Signal Based Surveillance (AVSS)* (pp. 23–29). IEEE.

Nguyen, T. L. A., Chau, D. P., & Bremond, F. (2015). Robust global tracker based on an online estimation of tracklet descriptor reliability. In *2015 12th IEEE International Conference on Advanced Video and Signal Based Surveillance (AVSS)* (pp. 1–6). IEEE.

Pala, F., Satta, R., Fumera, G., & Roli, F. (2016). Multimodal person reidentification using rgb-d cameras. *IEEE Transactions on Circuits and Systems for Video Technology, 26*(4), 788–799.

Paul, M., Haque, S. M., & Chakraborty, S. (2013). Human detection in surveillance videos and its applications-a review. *EURASIP Journal on Advances in Signal Processing, 2013*(1), 176.

Prosser, B. J., Zheng, W.-S., Gong, S., Xiang, T., & Mary, Q. (2010). Person re-identification by support vector ranking. In *BMVC* (Vol. 2, p. 6).

Redmon, J., Divvala, S., Girshick, R., & Farhadi, A. (2016). You only look once: Unified, real-time object detection. In *Proceedings of the IEEE Conference on Computer Vision and Pattern Recognition* (pp. 779–788).

Ren, S., He, K., Girshick, R., & Sun, J. (2015). Faster r-cnn: Towards real-time object detection with region proposal networks. In *Advances in Neural Information Processing Systems* (pp. 91–99).

Saghafi, M. A., Hussain, A., Zaman, H. B., & Saad, M. H. M. (2015). Review of person re-identification techniques. arXiv:1511.02319.

Schwartz, W. R., & Davis, L. S. (2009). Learning discriminative appearance-based models using partial least squares. In *2009 XXII Brazilian Symposium on Computer Graphics and Image Processing (SIBGRAPI)* (pp. 322–329). IEEE.

Shen, Y., Lin, W., Yan, J., Xu, M., Wu, J., & Wang, J. (2015). Person re-identification with correspondence structure learning. In *Proceedings of the IEEE International Conference on Computer Vision* (pp. 3200–3208).

Su, C., Yang, F., Zhang, S., Tian, Q., Davis, L. S., & Gao, W. (2015). Multi-task learning with low rank attribute embedding for person re-identification. In *Proceedings of the IEEE International Conference on Computer Vision* (pp. 3739–3747).

Varior, R. R., Haloi, M., & Wang, G. (2016). Gated siamese convolutional neural network architecture for human re-identification. In *European Conference on Computer Vision* (pp. 791–808). Springer.

Vezzani, R., Baltieri, D., & Cucchiara, R. (2013). People reidentification in surveillance and forensics: A survey. *ACM Computing Surveys (CSUR), 46*(2), 29.

Wang, T., Gong, S., Zhu, X., & Wang, S. (2014). Person re-identification by video ranking. In *European Conference on Computer Vision* (pp. 688–703). Springer.

Wang, X., Doretto, G., Sebastian, T., Rittscher, J., & Tu, P. (2007). Shape and appearance context modeling. In *IEEE 11th International Conference on Computer Vision, 2007.* ICCV 2007 (pp. 1–8). IEEE.

Xiong, F., Gou, M., Camps, O., & Sznaier, M. (2014). Person re-identification using kernel-based metric learning methods. In *European Conference on Computer Vision* (pp. 1–16). Springer.

Yan, Y., Ni, B., Song, Z., Ma, C., Yan, Y., & Yang, X. (2016). Person re-identification via recurrent feature aggregation. In *European Conference on Computer Vision* (pp. 701–716). Springer.

You, J., Wu, A., Li, X., & Zheng, W.-S. (2016). Top-push video-based person re-identification. In *Proceedings of the IEEE Conference on Computer Vision and Pattern Recognition* (pp. 1345–1353).

Zeng, M., Wu, Z., Tian, C., Zhang, L., & Hu, L. (2015). Efficient person re-identification by hybrid spatiogram and covariance descriptor. In *2015 IEEE Conference on Computer Vision and Pattern Recognition Workshops (CVPRW)* (pp. 48–56).

Zhao, R., Ouyang, W., & Wang, X. (2013). Person re-identification by salience matching. In *2013 IEEE International Conference on Computer Vision (ICCV)* (pp. 2528–2535). IEEE.

Zhao, R., Ouyang, W., & Wang, X. (2014). Learning mid-level filters for person re-identification. In *Proceedings of the IEEE Conference on Computer Vision and Pattern Recognition* (pp. 144–151).

Zheng, L., Bie, Z., Sun, Y., Wang, J., Su, C., Wang, S., & Tian, Q. (2016). Mars: A video benchmark for large-scale person re-identification.

Chapter 8
A Novel Approach for Speaker Recognition in Degraded Conditions

Sana Boujnah, Xianfang Sun, David Marshall, Paul L. Rosin, and Mohamed Lassaad Ammari

Abstract Human voice is an ideal data source for identifying people in many applications. Because of the increasing need for security in different public places, voice biometrics may be a good solution, as we can easily take voice records. This chapter provides a brief overview of the approaches utilized in recognizing speakers, and then presents a novel approach for recognizing speakers in degraded smart-home conditions. The suggested approach includes a pre-processing phase, a feature extraction phase, and a classification phase, where the feature extraction phase consists of formant extraction to get the spectrum energy maxima of speech audio, dynamic time warping (DTW) to find an optimal alignment between two provided temporal sequences under definite restrictions, and refinement process to improve the results of the DTW system output. The experiments are carried out on a database containing 1,248 samples in order to validate the suggested approach. The latter has good results as regards the state of the art with 94.5% accuracy.

Keywords Biometrics · Speaker recognition · Formant · Dynamic time warping

S. Boujnah (✉) · M. L. Ammari
NOCCS, National Engineering School of Tunis, University of Tunis El Manar, Tunis, Tunisia
e-mail: sana.boujnah@outlook.com

M. L. Ammari
e-mail: mohamed.ammari@eniso.rnu.tn

X. Sun · D. Marshall · P. L. Rosin
HFTC, School of Computer Science, Cardiff University, Wales, UK
e-mail: xianfang.sun@cs.cardiff.ac.uk

D. Marshall
e-mail: Dave.Marshall@cs.cardiff.ac.uk

P. L. Rosin
e-mail: Paul.Rosin@cs.cf.ac.uk

N. Derbel and O. Kanoun (eds.), *Advanced Methods for Human Biometrics*,
Smart Sensors, Measurement and Instrumentation 40,
https://doi.org/10.1007/978-3-030-81982-8_8

8.1 Introduction

Quite recently, considerable attention has been paid to smart living and security like access control and bank-account management through smart phones. It is envisioned to be the standard of living for a large amount of people in the near future. Automatic speaker recognition is based on recognizing people from a spoken phrase via a machine. Its applications are numerous and widespread, e.g. financial, forensic, legal, security, video indexing, monitoring and smart-home applications (Orr & Abowd, 2000). Speaker recognition is a voice processing sub-discipline. It is divided into sub-areas (Perronnin & Dugelay, 2002): verification, and identification. Certainly, speaker recognition is not the most reliable or least intrusive biometric. Nevertheless, when the voice of a person is the only available physical characteristic, biometric identity verification is possible only through speaker recognition. Many information levels are conveyed to the listener by speech signals (Perronnin & Dugelay, 2002). Speech conveys a message via words at the primary level. However, at other levels, speech conveys information about the language being spoken, the gender, the emotion, as well as the speaker identity. The purpose of speech recognition is to discern the words spoken. The automatic speaker recognition systems aims to extract, characterize and recognize the information in the speech signal conveying the speaker identity. A hybrid system consisting of the formant extraction and the Dynamic Time Warping (DTW) method is put forward in this chapter. In addition, a refinement process is introduced to improve the proposed approach. The experiments demonstrate that the new approach is able to produce better results than the state-of-the-art approaches. The rest of the chapter is organized as follows. The well-known state-of-the-art approaches are highlighted in Sect. 8.2. The proposed method is described in Sect. 8.3. Section 8.4 sums up our research results and discussion. The last section deals with the conclusion.

8.2 Related Works

Generally, speaker recognition covers two fundamental tasks: identification and verification. The former determines who is talking from a set of known voices or speakers. The unknown person makes no identity claim, so the system performs a 1:N classification. Normally, the unknown voice comes from a fixed set of known speakers (Perronnin & Dugelay, 2002). Therefore, this task is usually referred to as closed-set identification. Speaker verification (known also as speaker detection or authentication) determines if a person is who they claim to be (a yes/no decision). This is referred to as an open-set task as it is in general presumed that imposters (those falsely claiming to be valid users) are not known to the system. The two tasks for what is called open-set identification can be merged by adding a none-of-the-above option to the closed-set identification task. Ding et al. (2014) presented a comparative study of speaker recognition methods based on Support Vector Machine (SVM),

Gaussian Mixture Model (GMM) and DTW. Indeed, there are several feature extraction techniques. We will consider those extracted from the individual physiological characteristics, for example the frequency of the vibration of the vocal cords. This frequency is known as the fundamental frequency (Jdira et al., 2014), pitch or F0. The existing classification approaches can be organised into these three categories:

- Vector-based approaches: Such as Dynamic Time Warping (DTW), Vector Quantization (VQ) and Support Vector Machine (SVM) (Ding et al., 2014; Limkar et al., 2012).
- Static approaches: e.g. Hidden Markov Model chains (HMM) and Gaussian Mixture Model (GMM) (Yee & Ahmad, 2007).
- Connectionist approaches: Like neural networks (Richardson et al., 2016).

Lozano-Diez et al. (2016) utilized the deep neural network to extract the features for speaker recognition. The work was compared with the standard Mel Frequency Cepstral Coefficients (MFCC) features in an i-vector/Probabilistic Linear Discriminant Analysis (PLDA) speaker recognition system. Alexander et al. (2016) presented a forensic automatic system for speaker recognition. It consists on voice comparison and analysis of the likelihood of speech evidence. This system utilized various algorithms like MFCC, phonetic features (for instance formants), or features of their own choice (e.g. articulation rate and voice quality metrics). Moreover, it includes GMM with [and without] maximal a posteriori adaptation, i-vector extraction with PLDA and cosine distance comparison. Significant corpora available to support speaker recognition research and evaluation were described by Sturim et al. (2016) along with details about the corpora collection and design. PLDA was adopted by Garcia-Romero and McCree (2014) based i-vector speaker recognition systems. Two of the suggested approaches are applicable to a larger class of models (low-rank across-class). Speaker identification was presented in Yusnita et al. (2011) using LPC and formant analysis. Joint factor analysis was explored in Stafylakis et al. (2016) for text-dependent speaker recognition with random digit strings utilizing the RSR2015 (part III) dataset. Hughes et al. (2016) assessed the possible performance enhancement of MFCC-based Automatic Speaker Recognition (ASR) systems with the inclusion of linguistic-phonetic information. Utilizing randomized sets of speakers, testing was run over 20 replications. The estimated quality parameters of speech recording were used in Pop et al. (2013). The authors validated their experiments on the ELSDSR database.

8.3 Proposed Approach

Our principal objective is to provide a suitable and robust speaker identification system in degraded conditions. Speaker recognition consists of three parts: preprocessing, feature extraction and classification. For this, a novel combination of speech processing and classification approaches is presented to recognize speakers in degraded conditions. These steps are illustrated in Fig. 8.1.

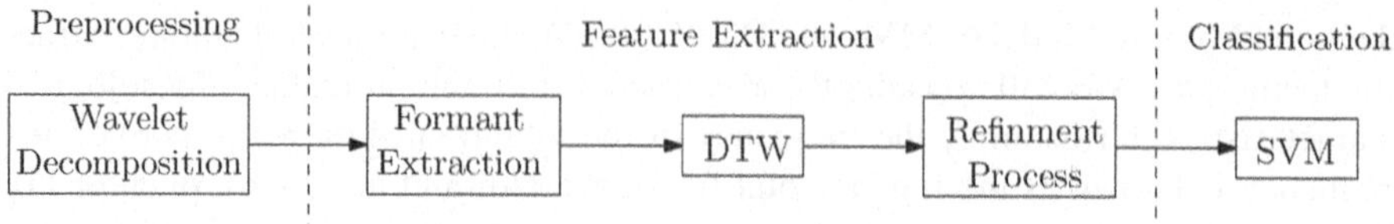

Fig. 8.1 Flowchart of the proposed approach

8.3.1 Pre-processing

Several pre-processing operations are performed to extract the useful data and prepare them for the characterization phase, such as wavelet decomposition (Aggarwal et al., 2011) and finite impulse response filters for denoising raw signals. Stationary wavelet decomposition is chosen for this study in order to denoise raw signals. To ensure a better performance, the silence at the beginning and the end of the signal is automatically removed.

8.3.2 Feature Extraction

It is composed of several steps. The target of this block is to extract the wanted features with the proposed methods and to process the features for classification.

8.3.2.1 Formant Extraction

A speech sound formant is one of the spectrum energy maxima of the speech sound (Hughes et al., 2016). There are various definitions of the word "forming" (vocal tract resonances, poles, etc.). Formants are denoted by the physical notation ϕ (which is measured in Hertz) starting from the first F_1 in low frequencies (Franco-Pedroso & Gonzalez-Rodriguez, 2016). We reserve notation F0 for the fundamental frequency, where its variations over time forms speech intonation. F_1 and F_2, as the first two formants, will not be sufficient to describe the vowel if the language distinguishes, for instance between drawn and rounded anterior vowels as in French (Sathe-Pathak & Panat, 2012). Non diphthong vowels are approximately stationary speech sounds. Therefore, each vowel is characterized by its specific timbre, which is determined as a first approximation by F_1 and F_2. The phonetic discipline correlates the measured values in Hertz of F_1, F_2, the third formant, and the articulations of the phonetic apparatus necessary for the realization of vowels (including lips). As a consequence, F_1 is correlated with the aperture (mouth opening) and F_2 with the previous position (high value of F_2), posterior (lower value of F_2) of the language, as well as with the lips configuration (Hughes et al., 2016). F_3 as a third formant is also an interesting feature, correlated with the lips configuration for the anterior vowels. The first formant is extracted as a feature and will be treated by DTW.

8.3.2.2 Dynamic Time Warping (DTW)

DTW is a widely-known technique for finding an optimal alignment between two given temporal sequences under certain restrictions (Limkar et al., 2012). Intuitively, the sequences are nonlinearly warped to match each other. The DTW algorithm has been used in bioinformatics, computer graphics, audio, and video (Ding et al., 2014). For example, in ASR, we need to consider very variable speech rates. The DTW measures the similarity between two sequences having the ability to vary over time (Yee & Ahmad, 2007); e.g., we can detect the similarities between gaits in videos even though the subject in one or other videos walks slower or faster, or when the subject slows down or accelerates. Generally, DTW (Yee & Ahmad, 2007) is a method that looks for optimal matching between two time series, under certain restrictions. The non-linear transformation of the temporal variable deforms the temporal series, so as to determine a measure of their similarity, independently from certain non-linear time transformations. We mostly use this method of time series alignment in the HMM context. DTW is calculated with the extracted formants. The two utilized sequences are: the testing sequence and all the remaining samples in the database. Then, in order to enhance the results, a refinement process is required.

8.3.2.3 Refinement Process

We opt for improving the results of the DTW system output to ensure a better classification. It is a novel method for enhancing the results. Firstly, we binarize the distance measure of the DTW contained in the output matrix. Next, a threshold is set for each 12 samples of the same person in the binarized matrix.

8.3.3 Classification

There exist several types of classifications (Yee & Ahmad, 2007), but one of the most intuitive and widely used is the supervised classification. The idea of supervised classification is to learn a ranking rule from a set of data whose classification is already known. Once the rule is learned, it is possible to apply it to categorize new unknown data. Based on the SVM, a one-leave-out classification is performed (Ding et al., 2014). SVM is a class of learning algorithms, which is initially defined for discrimination, like predicting a qualitative binary variable. SVM solves classification problems (sorting individuals according to their characteristics) and is important for its good performance (Rückstieß et al., 2011).

8.4 Experimental Results

Comprehensive experiments are carried out for the evaluation of the proposed approach performance compared to a state-of-the-art approach. They are conducted on the Ear and Voice Database in Degraded Conditions (EVDDC). The EVDDC database is acquired from 111 various subjects (50 in the first session and the rest in the second session) with different ages and from varied places of one country (Tunisia) with multiple tones. For every individual, a number of samples, between 12 and 21 voice records and between 12 and 50 images, are taken. The EVDDC contains 40 video records, 2,106 voice records, and 2,742 ear images. It is sequentially numbered for each subject with an integer identification number ("Nperson-Nsample"). It contains samples from 58 males and 53 females. The subjects are from various ages: 2 are more than 60 years, 24 are between 36 and 59 years, 45 are between 15 and 35 years old, and 40 are between 6 and 14 years.

To assess the efficiency of our suggested scheme, classification accuracy is selected. Figure 8.2 depicts the comparative results of these methods and our results. The DTW-based method Fig. 8.2 presents 90.2% accuracy, while the formant-based method (Ding et al., 2014) presents 89.9% accuracy. Clearly, our approach achieves higher a performance with 94.5% accuracy due to the combination of the DTW approach, the formant-extraction method and the refinement process.

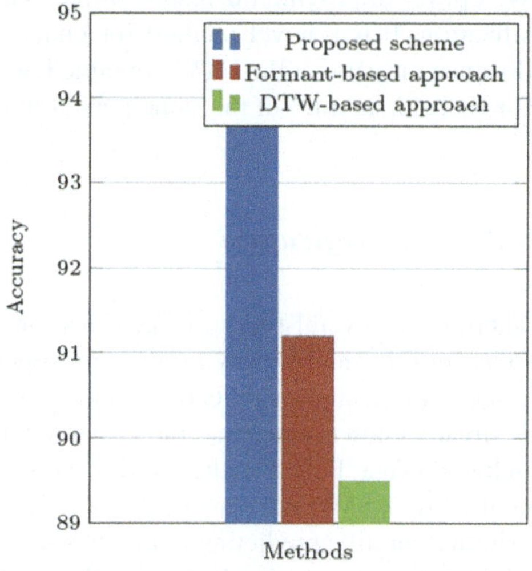

Fig. 8.2 Performances of the proposed scheme and the state-of-the-art methods

8.5 Conclusion

Identity recognition from voice records is an active field of research within the biometric community. Covert distant recording makes voice recognition technology good for security and surveillance applications. Unlike other biometric modalities, with available large datasets, datasets of voice records in degraded conditions are small. We present in this chapter an approach to speaker recognition in degraded smart-home conditions. The results have demonstrated that combining both approaches, DTW and formant extraction, has improved the global performances of the system to 94.5%. A possible application for this approach lies in the development of a multi-model system for person identification in degraded conditions.

References

Aggarwal, R., Singh, J. K., Gupta, V. K., Rathore, S., Tiwari, M., & Khare, A. (2011). Noise reduction of speech signal using wavelet transform with modified universal threshold. *International Journal of Computer Applications, 20*(5), 14–19.

Alexander, A., Forth, O., Atreya, A. A., & Kelly, F. (2016). Vocalise: A forensic automatic speaker recognition system supporting spectral, phonetic, and user-provided features. *Odyssey*

Ding, J., Yen, C.-T., & Ou, D.-C. (2014). A method to integrate GMM, SVM and DTW for speaker recognition. *International Journal of Engineering and Technology Innovation, 4*(1), 38–47.

Franco-Pedroso, J., & Gonzalez-Rodriguez, J. (2016). Linguistically-constrained formant-based i-vectors for automatic speaker recognition. *Speech Communication, 76*, 61–81.

Garcia-Romero, D., & McCree, A. (2014). Supervised domain adaptation for i-vector based speaker recognition. In *IEEE International Conference on Acoustics Speech and Signal Processing (ICASSP)* (pp. 4047–4051).

Hughes, V., Foulkes, P., & Wood, S. (2016). Formant dynamics and durations of um improve the performance of automatic speaker recognition systems. In *16th Australasian Conference on Speech Science and Technology (ASSTA)*, York.

Jdira, M. B., Jemâa, I., & Ouni, K. (2014). Speaker recognition system based on pitch estimation. In *IEEE International Conference on Electrical Sciences and Technologies in Maghreb (CISTEM)* (pp. 1–5).

Limkar, M., Rao, B. R., & Sagvekar, V. (2012). Speaker recognition using VQ and DTW. In *International Conference on Advances in Communication and Computing Technologies* (pp. 18–20).

Lozano-Diez, A., Silnova, A., Matejka, P., Glembek, O., Plchot, O., Pešán, J., Burget, L., & Gonzalez-Rodriguez, J. (2016). Analysis and optimization of bottleneck features for speaker recognition. *Proceedings of Odyssey, 2016*, 352–357.

Orr, R. J., & Abowd, G. D. (2000). The smart floor: A mechanism for natural user identification and tracking. In *CHI'00 Extended Abstracts on Human Factors in Computing Systems* (pp. 275–276). ACM.

Perronnin, F., & Dugelay, J.-L. (2002). Introduction à la biométrie-authentification des individus par traitement audio-vidéo. *Traitement du signal, 19*(4)

Pop, G., Draghicescu, D., & Burileanu, D. (2013). On forensic speaker recognition case pre-assessment. In *Conference on Speech Technology and Human-Computer Dialogue (SpeD)* (pp. 1–8). IEEE.

Richardson, F., Brandstein, M., Melot, J., & Reynolds, D. A. (2016). Speaker recognition using real versus synthetic parallel data for dNN channel compensation. *MIT Lincoln Labortory Lexington United States*, 2796–2800.

Rückstieß, T., Osendorfer, C., & van der Smagt, P. P. (2011). Sequential feature selection for classification. In *Australasian Conference on Artificial Intelligence* (pp. 132–141). Springer.

Sathe-Pathak, B. V., & Panat, A. R. (2012). Extraction of pitch and formants and its analysis to identify 3 different emotional states of a person. *IJCSI International Journal of Computer Science Issues, 9*(4), 296–299.

Stafylakis, T., Alam, M. J., & Kenny, P. (2016). Text-dependent speaker recognition with random digit strings. *IEEE/ACM Transaction on Audio, Speech, and Language Processing, 24*(7), 1194–1203.

Sturim, D. E., Torres-Carrasquillo, P. A., & Campbell, J. P. (2016). Corpora for the evaluation of robust speaker recognition systems. *INTERSPEECH*, 2776–2780.

Yee, L. M., & Ahmad, A. M. (2007). *Comparative study of speaker recognition methods: DTW, GMM and SVM.*

Yusnita, M. A., Paulraj, M. P., Yaacob, S., Bakar, S. A., & Saidatul, A. (2011). Malaysian English accents identification using LPC and formant analysis. In *IEEE International Conference on Control System, Computing and Engineering (ICCSCE)* (pp. 472–476). IEEE.

Chapter 9
Visual Methods for Sign Language Recognition: A Modality-Based Review

Bassem Seddik and Najoua Essoukri Ben Amara

Abstract Sign language visual recognition from continuous multi-modal streams is still one of the most challenging fields. Recent advances in human actions recognition are exploiting the ascension of GPU-based learning from massive data, and are getting closer to human-like performances. They are then prone to creating interactive services for the deaf and hearing-impaired communities. A population that is expected to grow considerably in the years to come. This chapter aims at reviewing the human actions recognition literature with the sign-language visual understanding as a scope. The methods analyzed will be mainly organized according to the different types of unimodal inputs exploited, their relative multi-modal combinations and pipeline steps. In each section, we will detail and compare the related datasets, approaches then distinguish the still open contribution paths suitable for the creation of sign language related services. Special attention will be paid to the approaches and commercial solutions handling facial expressions and continuous signing.

Keywords Sign language recognition · Action recognition · Multi-modal · Continuous signing · Facial expression · Joint · RGB · Depth

9.1 Introduction

Sign language (SL) is a purely visual communication manner relying on a combination of hand gestures, facial expressions and upper-body postures. For nearly 466 million persons world-wide, SL is their first native language. This is particularly the case for those born with total or partial hearing loss, thus with limited or no speech

---

B. Seddik (✉) · N. Essoukri Ben Amara
LATIS - Laboratory of Advanced Technology and Intelligent Systems, Université de Sousse, Ecole Nationale D'Ingénieurs de Sousse, Sousse, Tunisia
e-mail: bassem.seddik@vizmerald.com

N. Essoukri Ben Amara
e-mail: najoua.benamara@eniso.rnu.tn

N. Derbel and O. Kanoun (eds.), *Advanced Methods for Human Biometrics*, Smart Sensors, Measurement and Instrumentation 40,
https://doi.org/10.1007/978-3-030-81982-8_9

capabilities. It is a considerable population that suffers from isolation, inadequate education and very few life services.

This is specially emphasized as most hearing/speaking persons are not able to interact with SL natives. Despite that devices and implants were invented for audition increase, still they are very far from allowing a good hearing. Thus interaction through SL remains a vital need since the youngest babies age (Chestnov et al., 2013).

According to the world health organization, the lack of deaf/hearing-impaired services and employment is causing a global annual loss of 75 billion USD (World Health Organization, 2019). This is an increasingly crucial situation as nearly 900 million persons are expected to show hearing loss by 2050. This explains the importance of considering SL recognition from a technically visual point of view. A situation purely within the computer vision field specialization.

Here are the key-word terminologies that we apprehend in our review: We distinguish different approach families related to the task of human actions recognition. Each one of these is concertized using a specific methodology composed itself of a sequence of consecutive steps leading to the desired result. Within a finer analysis scale, we find that each step could be implemented using a set of techniques according to selected programming algorithms.

In the case where all the steps of a method belonging to a given approach, are implemented from end to end, we consider that we have a fully operational system of some originality.

We will also refer with a capital "D" to the Deaf community as most deaf persons strongly identify to this culture.

The next coming sections are logically organized but also self contained for an easier reading. In Sect. 9.2, we will start by reviewing the general pipeline adopted for human actions recognition. In Sect. 9.3, we will analyze the existing solutions that rely on unimodal inputs. Section 9.4 will then be dedicated to multi-modal approaches. For each of the mentioned sections, we will be pointing the main contributions suitable for SL recognition. An analysis of the commercial solutions actually targeting the Deaf community will be presented in Sect. 9.5. We will rely on it for the proposal of key axis of contribution that could be developed by future techniques.

During our state of the art analysis, we will additionally focus on facial expressions and hand gestures recognition as they strongly contribute to the definition of various SL lexicons (Alon et al., 2009). As shown in Fig. 9.1, hand and face movements are sometimes decisive for distinguishing actions.

9.2 Human Actions Recognition Pipeline

Literature works on SL recognition can be classified according to the different stages of their processing pipeline. While the early approaches were based on static captures (Guo and Lai, 2014; Fasel and Luettin, 2003), the actual trend is towards the recognition from continuous multi-modal streams and large learning populations (Agarwal and Xia, 2014; Sandbach et al., 2012). Thus, several additional processing

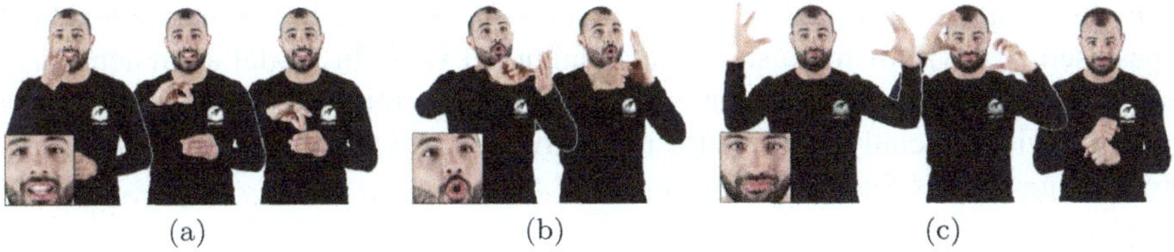

Fig. 9.1 Importance of the face and hand gestures to the sign-language lexicon: **a** One-vote, **b** verify, **c** collect-all

steps became influential to Human Actions Recognition (HAR) in general and SL in particular.

The existing approaches' pipelines could generally be classified according to their: (i) input data, (ii) pre-processing steps, (iii) features types extracted, (iv) data representations, (v) temporal segmentation techniques, (vi) classification methods as well as (vii) post-processing outputs producing the final labels (Cheng et al., 2015). We remind hereafter the main details of these commonly used steps:

(i) **Input data**: They determine the mass and type of the raw input streams within a given method. These streams could be composed of joints, RGB, depth or mixtures of these modalities. We can also opt for a vocabulary related, for example to sign language, daily actions or facial expressions.
(ii) **Preprocessing**: This step implies the usage of any filtering, normalization and data formatting operations in order to improve its readability. Due to multi-modal sources usage, the streams synchronization became also a very important step to ensure their correct subsequent processing.
(iii) **Temporal segmentation**: If we consider recognition from continuous action streams, it is recommended to first segment them into partial sub-actions. This is done by searching the pairs of (beginning, end) temporal positions for each significant sub-gesture relative to the vocabulary.
(iv) **Feature extraction**: This step applies calculations, measures and vectors extraction to form a more consistent meta-information compared to the raw input data. These descriptors can be generated by sophisticated extractors obeying to specific spatio-temporal sampling steps, or more recently, pooled from deep neural-network models (Guo et al., 2016).
(v) **Representation**: In order to further improve the homogeneity and efficiency of the obtained descriptors, additional steps of data representation could be added. They are particularly valuable in the case of learning from large scale populations (e.g. +10K sequences (Guyon et al., 2014)).
(vi) **Classification**: In this step, the extracted information is separated into groups belonging to the different action labels.
(vii) **Post-processing**: In an even more advanced step, there are additional interesting post-processing tools for merging and optimizing the decisions obtained from classifiers through different fusion strategies.

In this chapter, our literature classification will be based on the modalities data-types used. We will distinguish between unimodal vs multi-modal approaches and review their main contributions at each of the aforementioned steps. For each data input, we will especially focus on the respective description, segmentation and post-processing stages techniques applied.

9.3 Unimodal Methods

As the Kinect sensor has been used extensively for acquisition in recent works, synchronized joint, RGB and depth streams became simultaneously available. Despite this, numerous contributions focused solely on a unique modality among them.

We begin this unimodal-dedicated section by presenting the joint-based works. We will then analyze uni-modal literature dedicated to RGB streams and finish with those based on depth. We highlight that we consider "unimodal" the approaches adding hand-made annotation points to RGB available data.

9.3.1 Recognition from Joint Streams

The first joint-based works were based on motion sensors (i.e. Motion Capture (MoCap)) of different types attached to the actor's body. In this configuration, the person (i.e. the signer) is the egocentric carrier of the capture device (Song et al., 2014). Another family of MoCap technologies is based on a remote network of sensors. The use of landmarks (i.e. marker-points) is then generally affordable using infra-red capture devices such as in Han et al. (2010). A review of MoCap-based techniques for HAR is presented in Lara and Labrador (2013).

With the apparition of infra-red sensors similar to the Kinect, marker-less captures become easily achievable by the computer vision community (Han et al., 2017). As a result, various configurations of joints have been proposed based on 15, 20 and 25 positions (among others, see Fig. 9.2) with technologies such as OpenNI, Kinect and MoCap, respectively. A large number of contributions have relied solely on joints to provide very accurate performances. An explanation was provided by Jhuang et al. (2013). The articulation as a high order modality allows (in general) gains of 19–29% of accuracies compared to mono-modal configurations in RGB or depth.

9.3.1.1 Joint-Based Datasets

One the most used datasets is the MSR-Action3D (Wang et al., 2012). It has been first introduced with a depth-based approach, but its joint streams were widely used later as a reference for joint-based methods. And while different training schemes

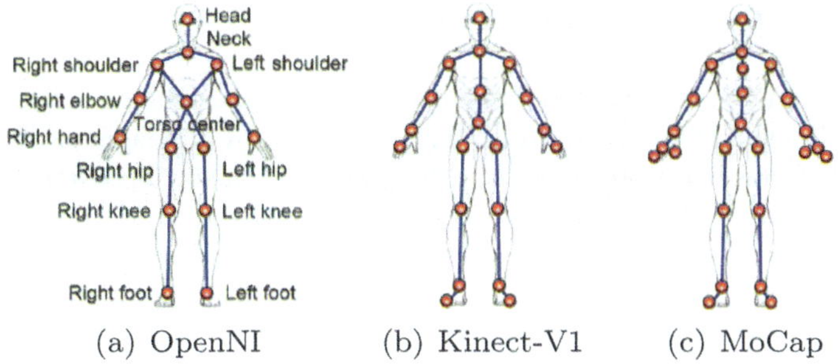

Fig. 9.2 Examples of joint 3D body-landmarks increasing by technology Han et al. (2017)

have been adopted, the most difficult scenarios are those applying Cross-Validation (CV) evaluations.

A similar interesting CV evaluation is also found within the CAD-60 database (Sung et al., 2012). It offers daily-action human movements categorized by rooms. A similar multi-modal dataset is MSR-Daily. A significant information is related to the streams durations. While CAD-60 has a small number of samples, it has the merit of offering streams that span over relatively long periods of time: 50 to 120 s. The larger is the frame samples number, the higher is the likeliness level for intra-class variability. This is also the case within large-scale datasets offering joint streams such as those relative to the Chalearn Gesture Challenges (CGC) (Escalera et al., 2014). At the moment of this survey writing, one of the largest benchmarks offering body joints is the NTU-RGB+D (Shahroudy et al., 2016). Though this dataset offers many other modalities, recent methods are using it to evaluate their joint-only-based performances.

Estimating joint positions from visual data has also contributed to the creation of specialized datasets such as FLIC (Sapp and Taskar, 2013) or MPII (Andriluka et al., 2014). Lately, human pose estimators (i.e. position, rotation and scale in 2D or 3D spaces) such as the OpenPose (Cao et al., 2017) have received interest for the generation of joint data from large-scale RGB datasets. This new trend of estimating pose landmarks from in-the-wild images is prone to further developments (Zhang and Shah, 2015; Belagiannis and Zisserman, 2017; Haque et al., 2016; Güler et al., 2018) as the pose-estimated joint performances are still lower than those sensor-based. This is for instance the case for the state-of-the art approach of Shi et al. (2019) with only 36.9% as recognition rate when using the skeletons generated from the Kinetics RGB dataset (Kay et al., 2017). With the development of synthetic 3D datasets dedicated to human pose estimation such as SURREAL and DensPose (Varol et al., 2017; Güler et al., 2018) promising paths for SL recognition in-the-wild are still expected. A detailed analysis of pose-based HAR is provided by Boualia et al. in (2019). For what follows in this review, we only consider this category of works using pose estimation for the main purpose of actions recognition.

9.3.1.2 Joint-Based Approaches for HAR

The first approaches dedicated for human movements description were derived from the Shape-Context (SC) descriptor initially proposed by Belongie et al. (2002) to describe 2D forms in quantized discrete points. The SC made it possible to generate histograms relative to the distances and angles between the various considered points. A successful extension of the SC descriptor to generate histograms of 3D joints has been proposed in Xia et al. (2012). Figure 9.3 illustrates examples of SC-based approaches in 2D and 3D.

In the representation based on principal joints (EigenJoints), the authors based their contribution on cumulative energies descriptor from joint movements (Yang and Tian, 2014). They applied Principal Component Analysis (PCA) representation and a nearest-neighbor classification. A very close approach has been proposed in Ofli et al. (2014). The authors start by extracting the most discriminative articulations to concatenate their cumulative motion histograms. A related enhancement is proposed in the descriptor ActionLet (Wang et al., 2012). In addition to the movements cumulation and histograms concatenation, the authors have also extracted the temporal pyramid Fourier coefficients to manage the actions' misalignments.

A richer set of cumulative features have been proposed in the MovingPose (Zanfir et al., 2013) representation. The authors propose to compute first-order and second-order motion gradients (i.e. velocities and accelerations) then concatenate N groups of frame contents into unique descriptors per key-frame. The set of descriptors is combined with an SVM classifier to give 91.7% performances on the MSR-Action3D database within its most challenging cross-validation configuration.

The best joint-only-based performance under the CGC-2014 dataset (Escalera et al., 2014) is provided by Evangelidis et al. (2014). They used a selective concatenation of skeletal Quads: Each set of 4 joints is encoded and normalized according to its initial join. Thus, these Quads are designed to analyze the separate movement of arms, feet, torso, etc. In addition, the authors have applied the Fisher-Vectors (FV) representation to gain adaptability to large-scaled data.

More advanced approaches add extra layers of specially designed data representations. For example, specific representation spaces referred as Manifolds hold the

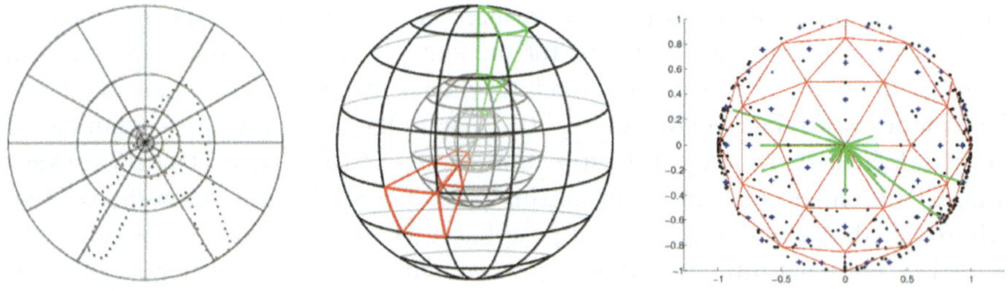

(a) (b) (c)

Fig. 9.3 Examples of SC-based descriptors: **a** in 2D, (**b** and **c**) in 3D contexts best performances reported on the CMU (Han et al., 2010) dataset. Methods such as the Kendall's Manifold (Boulbaba et al., 2015) or the lie-groups (Vemulapalli et al., 2016) have also offered excellent performances. Most of them define specific decision measures in the form of distances adapted to the considered spaces.

In Table 9.1, we present a summary of the progressive contributions proposed to recognize human actions from join-only streams. Following a chronologically ascending order in Table 9.1, we note that the most recent approaches have gradually exploited most of the previous contributions. This is the case for the two most recent approaches presented in Table 9.1. For instance, the AGC-LSTM approach uses an efficient type of Recurrent Neural Network (RNN) known as LSTM (i.e. Long Short-Term Memory) to enhance the convolutional graph while encoding the temporal actions evolution just as in the classic MovingPose descriptors (Zanfir et al., 2013). The top performance is reported on the NTU-RGB+D dataset by the Directed Graphs (Shi et al., 2019). They achieve this as they encapsulate the joint kinetic relations within their deep neural network. From this trend, it is clear that the next performance improvements will be guided by how much of the classic hand-crafted features are simulated using the new deep neural architectures.

9.3.2 *Recognition from RGB Streams*

When analyzing the approaches based on RGB visual data (images or videos), it is noticeable that this modality is the most developed one due to its accessibility. Most contributions related to human actions from RGB streams focused mainly on feature extraction.

While early contributions considered hand-designed features, the recent ones are relying more on variants of the Convolutional Neural Networks (CNN) architecture implemented successfully for the first time by Krizhevsky et al. in (2012) to produce automatically learned descriptors from massive data. We present in the following the RGB most used datasets and their related approaches evolution, both for actions and expressions recognition.

9.3.2.1 RGB-based Datasets for HAR

The first RBG-based human action datasets presented in Table 9.2 focused on using fixed images acquired in the wild. Partially promoted by the VOC challenges (Everingham et al., 2010), the best recognition rates for this task started at 45.7% for the Stanford-40-Actions dataset (Yao et al., 2011) to reach 74.2% for the 89-Action-Dataset (Le et al., 2013). Soon after, a number of video-based datasets started offering labeled action sequences. The KTH (Schuldt, Laptev, & Caputo 2004), Weitzmann (Gorelick et al., 2007) and CAVIAR (Chaquet et al., 2013) datasets have been in used since the 90's. With over a hundred references each (Chaquet et al., 2013), they allow performances close to 100% with nowadays methods.

Table 9.1 Selection of joint-based approaches for HAR and their relative datasets (Presti and Cascia, 2016; Agarwal and Xia, 2014; Lara and Labrador, 2013)

Year	Approach/References	Main contribution	Classifier	Performance per dataset
2019	DGNN (Shi et al., 2019)	2-stream convolutional network that encodes the joint kinematics	Directed CNN	96.1% NTU-RGB+D CV, 36.9% Skeleton-Kinetics
2019	AGC-LSTM (Si et al., 2019)	Description of the spatial an temporal co-occurrence using LSTM networks	LSTM CNN	95.0% NTU-RGB+D CV
2016	LieCurve (Vemulapalli et al., 2016)	Actions representation as points on a lie-curve and rotation quaternions	SVM	93.5% MSR-Action3D CV, 97.1% UTK-Action
2014	Quadruplets (Evangelidis et al., 2014)	Consideration of 4-joint groups and FV representation	SVM	76.8% CGC-2014
2014	Shan and Akella (2014)	Duplication by symmetric joints to improve performance	RF/SVM	84% MSR-Action3D CV, 93.8% CAD-60
2014	Ofli et al. (2014)	Selection of the most informative joints from EigenJoints	χ^2 distance	79.9% Berkeley-MHAD
2014	EigenJoint (Yang and Tian, 2014)	Cumulative movement energies over time then application of PCA	BN	97.8% MSR-Action3D 2/3
2013	MovingPose (Zanfir et al., 2013)	Normalization joint dimensions and concatenate 1st and 2nd order gradients	SVM	91.7% MSR-Action3D CV
2012	H3DJ (Xia et al., 2012)	Description by histograms of distances and orientations (Shape Context)	HMM	96.2% MSR-Action3D 1/3
2012	ActionLet (Wang et al., 2012)	Analysis in Fourier pyramids to temporally partition the action	SVM	85.6% MSR-Daily-Activity
2010	Manifold (Han et al., 2010)	Transformation of inertial data to a surface-based space (manifold)	RF + SVM	98.3% CMU-mocap

1/2, 2/3 and CV: Use of one, two-thirds and cross-validation for learning, respectively

Table 9.2 Main datasets allowing RGB-based approaches for HAR (Betancourt et al., 2015; Cheng et al., 2015; Guo and Lai, 2014; Chaquet et al., 2013)

Year	Dataset	References	Performance (classifier)	Contents	Nb. labels
From video streams					
2019	Kinetics-700	Carreira et al. (2019)	58.7% (LSTM + CNN)	600 seq. of 10s. by action	700 actions
2018	Moments in Time	Monfort et al. (2019)	44.7% (spatiotemporal CNN)	1M seq. of 3s. by action	399 actions
2018	EPIC-Kitchens	Damen et al. (2018)	70%verb; 52%noun; 41%act.	40K seq. 11.5M im. total	125 verb, 331 noun
2017	Kinetics-400	Kay et al. (2017)	61.0% (2-stream CNN)	400 seq. of 10s. by action	400 actions
2015	ActivityNet	Ghanem et al. (2017)	64.8% (CNN)	200 seq. for 648 h.	203 actions
2014	Sports-1M	Karpathy et al. (2014)	63.4% (SVM)	1M seq. total	487 actions
2014	CGC-2014	Peng et al. (2014)	∗ 79.1% (SVM)	14K seq. total	20 actions
2013	CGC-2013	Escalera et al. (2014)	∗ 72.9% (SVM)	13K seq. total	20 actions
2011-12	CGC-2012	Guyon et al. (2013)	∗ 72.9% (SVM)	(One-Shot) Mono exp.	20 actions
2012	UCF-101	Wang et al. (2015)	91.5% (CNN+SVM)	3207 seq. ∼13K im.	24 actions
2011	CAD-60	Sung et al. (2012)	74.7% (HMM)	68 seq. ∼60K im. total	14 actions
2011	HMDB-51	Wang et al. (2015)	65.9% (CNN+SVM)	101 at least per action	51 actions
2010	UCF-50	Ballas et al. (2013)	92.8%	100 to 150 by action	50 actions
2010	Olympic	Wang et al. (2015)	90.4%	50 by action	16 actions
2009	Hollywood-2	Wang et al. (2015)	66.8%	61 to 278 by action	12 actions
2008	UCF-11-youtube	Kuehne et al. (2011)	84.2%	100 per action	11 actions
2008	UCF-sports	Guo and Lai (2014)	93.5%	14 to 35 by action	9 actions
2008	Hollywood	Laptev et al. (2008)	56.8% (SVM)	30 to 140 by action	8 actions
2005	Weizmann	Guo and Lai (2014)	100%	∼100 actions	10 actions
2004	KTH	Guo and Lai (2014)	97.6% (KNN)	789 actions total	6 actions

(continued)

Table 9.2 (continued)

Year	Dataset	References	Performance (classifier)	Contents	Nb. labels
Multi-view video streams					
2010	MuHaVi	Singh et al. (2010)	58.3%	8 cameras and 14 actors	17 actions
2009	i3DPost	Chaquet et al. (2013)	80%	8 cameras and 8 actors	11 actions
2009	HumanEva-2	Sigal et al. (2009)	84.3%	7 cameras and 4 actors	6 actions
2006	IXMAS	Chaquet et al. (2013)	80.4%	5 cameras and 13 actors	13 actions
From fixed action images					
2013	89-action-dataset	Le et al. (2013)	74.2%	2038 im. total	89 actions
2012	VOC-2012	Everingham et al. (2015)	69.1% (SVM)	400 im. min. by action	10 actions
2011	VOC-2011	Everingham et al. (2010)	64.1% (SVM)	200 im. min. by action	10 actions
2011	Stanford-40-Act.	Yao et al. (2011)	45.7%	180 to 300 im. by action	40 actions
2010	VOC-2010	Everingham et al. (2010)	62.2%	50 to 100 im. by action	9 actions
2010	PPMI	Yao and Fei-Fei (2010)	85%	300 im. by action	7 actions
2010	Willow	Delaitre et al. (2010)	62.88%	911 fixed images	7 actions

(∗) Indicates an adaptation of the measure used as a percentage

Known the relative facility of these first video benchmarks (Cheng et al., 2015), the community's concentration shifted towards video sequences with more "confusing" (i.e. real-word) content. We find for instance the Olympic dataset (Niebles et al., 2010), the iterative improvements of UCF (Reddy and Shah, 2013) and Hollywood benchmarks (Marszalek et al., 2009) in addition to the HMDB dataset (Kuehne et al., 2011).

Several other datasets such as i3DPost (Chaquet et al., 2013), MuHaVi Several other datasets such as i3DPost ch9NajouaspsbibspsMuHaVi2010) and HumanEva (Sigal et al., 2009) focused on multi-view and richer visual acquisition contexts (i.e. video streams resolutions, silhouettes masks generation, etc.). In this case, the older contributions reached more than 80% performance, while the most challenging MuHaVi dataset allowed 58.3% accuracy with the RGB-only approach presented in Singh et al. (2010).

On another level, the time-line evolution clearly shows the trend for increasing the learning masses and the action labels: (i) On one hand, after having first used datasets offering hundreds of sample actions (Schuldt et al., 2004), then to the order of 1K learning actions such as UCF (Reddy and Shah, 2013), the 2011–2014 contributions have been dealing with more than 10K videos samples as in CGC (Escalera et al., 2014). (ii) On another hand, the number of action labels has gradually increased from

203 in the ActivityNet dataset (Ghanem et al., 2017) to 399 in the Moments-in-Time one (Monfort et al., 2019) and more recently is reaching 700 within the latest update of the Kinetics dataset (Carreira et al., 2019). According to Carreira et al. in (2019), it is expected to scale up 1K labels and 1M sequences (or to very long videos) in coming years. This is the case for instance with the Sports-1M (Monfort et al., 2019) and the ActivityNet (Ghanem et al., 2017) datasets, respectively, enabling the CNN architectures learning.

For all of these recent benchmarks, the most advanced methods are limited to accuracies between 44.7 and 64.8%. To overcome this limitation, actual evaluation metrics often consider the top-5 correct recognitions. Another promising path is learning the actions from weakly-supervised sequences (Ghadiyaram et al., 2019; Damen et al., 2018). Interestingly, this allows HAR using much larger amounts of noisy social-media sequences and hashtag labels (e.g. 65 M sequences and 10.5 K labels as in Ghadiyaram et al. (2019)). In this case, the labels are rather combinations of verb-noun couples as employed by the EPIC-Kitchens dataset (Damen et al., 2018). This field in particular remains wide open to contributions, specially in first-person related streams (Betancourt et al., 2015).

In Table 9.2, we consider the chronological performances evolution of different RGB-only approaches applied to public fully-labeled benchmarks. Although superior performances exist in the literature, we only maintained those in-line with the evaluation process of the dataset creators (Kuehne et al., 2011).

The literature development of these datasets allowing RGB-only approaches is continued within multi-modal acquisition contexts under the Sect. 9.4.1 and Table 9.5.

9.3.2.2 RGB-based Approaches for HAR

One of the first successful approaches for HAR was the Spatio-Temporal Interest Points (STIP) proposed by Laptev in (2005) and its extension to the 3D temporal volumes (see Fig. 9.4a) in Laptev et al. (2008). In terms of feature vectors, these extractors produce Histograms of Oriented Gradients (HOG) and Histograms of Optical Flows (HOF) (Tabia et al., 2012). Multiple works have relied on the HOG (see Fig. 9.4b) and/or HOF descriptors to produce their low-level local features (Sung et al., 2012; Guyon et al., 2014; Escalera et al., 2014; Wan et al., 2013; Monnier et al., 2014; Hernandez-Vela et al., 2014).

The dense feature extractors shown in Fig. 9.4c came later and added descriptors related to the motion boundary histograms and the motion trajectories (Wang et al., 2013). The improved dense trajectories proposed in Wang and Schmid (2013) were based on a human detector to reduce trajectory features. They enabled one of the best literature performances in conjunction with the Bag of Visual Words (BoVW) sparse representation and the SVM-based classifiers (Guo and Lai, 2014). The combination of BoVW-based descriptors with CNN-generated features brought additional performance improvements (Chandrasekhar et al., 2015). This is partially explained by the capability of BoVW to encapsulate temporal informations complimenting the CNN spatial features.

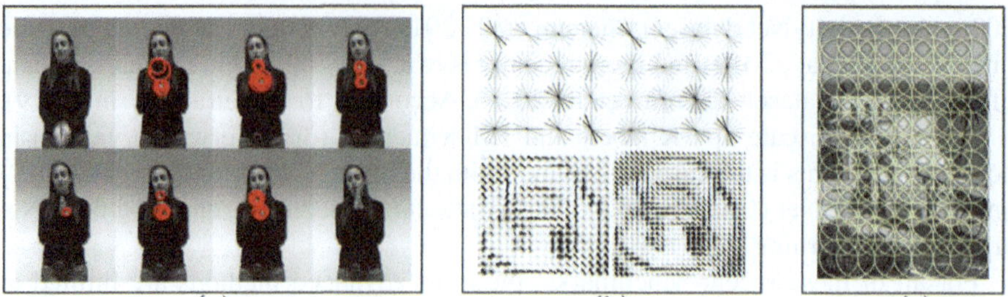

(a) (b) (c)

Fig. 9.4 Example RGB-based descriptors: **a** Spatiotemporal interest points, **b** HOG part-based and **c** Dense grids

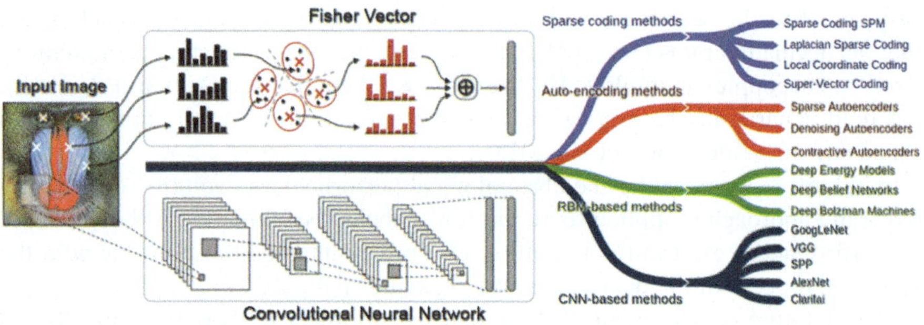

Fig. 9.5 Comparison between the BoVW Fisher vectors and CNN large-scale representations in addition to the relative deep methods families

A comparison between the BoVW Fisher vectors and CNN feature representations from large-scale data is given in Fig. 9.5. According to the study of Guo et al. (2016), the deep learning related approaches could be divided into 4 families: (i) Sparse-coding ones such as Fisher vectors, (ii) BoVW Auto-encoders and (iii) Restricted Boltzmann Machines (RBM) both of them (ii and iii) allowing to reconstruct the input data by automatically extracting (i.e. auto-encoding) the main features, in addition to (iv) CNN ones producing efficient feature representations using different architectures of convolution and pooling layers (Molchanov et al., 2016).

Actually, the main performance improvements are held by the approaches considering convolutional deep architectures to handle the actions temporal data-aspects. This factor is a major performance enabler compared to spatial-only approaches.

This is confirmed by the study of (Ghadiyaram et al., 2019) as learning masses and labels are constantly increasing and leaning towards weakly-supervised conditions.

The incorporation of the temporal evolution of continuous visual streams is affordable using a number of neural architectures: (i) Separately within the RNN or LSTM neural networks as in (Belagiannis and Zisserman, 2017; Si et al., 2019), (ii) using 3D convolutional kernels that exploit volumes of N consecutive frames and pixels (Molchanov et al., 2016), (iii) by applying a cross-time pooling (Pigou et al., 2016) or (iv) by combining the decisions of two-streamed networks, one of them relative to the optical flows (Simonyan and Zisserman, 2014). An example approach that jointly extracts convolutional and spatio-temporal features is proposed in Li et al. (2019). By applying a weight-sharing constraint on the CNN-learned parameters, this method achieves the state-of-the-art performance on the Moments-in-Time challenging dataset (Monfort et al., 2019).

It is also worth mentioning that a number of recent approaches have focused on the use of different learning/testing conditions. Indeed, the number of frames or sequences per action can heavily influence the recognition performances (Ghadiyaram et al., 2019).

This is for instance the case with the CAD-60 dataset (Sung et al., 2012): Since each sequence can span on more than 4,000 frames, the mass of description data reaches more than 15K samples while the videos are limited to 68 sequences.

In other learning scenarios, the use of one-shot learning, as in the CGC-2011/12 competitions (Guyon et al., 2014) creates different performance challenges as the test data is by-far bigger that the learning one.

Another category of approaches focused rather on ego-centric actions recognition (e.g. the EPIC-Kitchens challenge (Damen et al., 2018)). In this case, having a camera attached to the person's head or chest, the recognition of hands motion plays a central role in actions recognition. These particular learning scenarios, as challenging they are, deserve to be further investigated in future works. In Sect. 9.3.3 dealing with depth-only approaches, we pay more attention to hands-dedicated estimation and recognition techniques.

9.3.2.3 RGB-based Datasets for Facial Expressions Recognition

The Facial Expression (FE) is a major information vehicle on human behaviors and emotions. This explains the race for developing many computer vision applications that include: Facial recognition/biometry, gender/age detection, 3D animation generation, FEs and emotions identification (Bettadapura, 2012). Because a very-large population of datasets and contributions exist in relation to the mentioned applications, we focus more on those offering FEs as they are the most useful for SL visual recognition.

Within Table 9.3 a classification of contributions is made chronologically for those having used unimodal images or video sequences. To the difference with body gestures, it is perceivable that many recent facial datasets (e.g. AffectNet (Mollahosseini et al., 2019)) are still exploiting fixed captures. On another level, when considering the labels column in Table 9.3, another classification could be made to 4 families: (i) Those using basic expressions (Lumini et al., 2017; Zhang et al., 2018), (ii) those focusing on Action Units (AU) micro-expressions (Lucey et al., 2010; Benitez-Quiroz et al., 2016), (iii) those considering facial pose estimation (Zhang et al., 2018; Deng et al., 2019) and (iv) the more recent ones dealing with compound emotions under the Valence/Arousal evaluation system (Baveye et al., 2015; Kollias et al.,

Table 9.3 Main RGB-based datasets allowing, among other uses, FEs recognition

Year	Dataset	References	Classifier: Perf.	Content description	Labels
From facial video streams					
2019	Aff-Wild	Kollias et al. (2019)	CNN: 50.4%	298 videos, 1.2M frames in the wild	Valence/Arousal
2018	RAVDESS	Livingstone and Russo (2018)	75% (video only)	26 subjects vocalizing with emotions	8 emotions
2018	Mobiface	Lin et al. (2019)	CNN: 58.6%	80 smart-phone streams, 95k frames	14 attributes
2017	EmotiW	Dhall et al. (2017)	Many solutions: 80.9%	1809 sequences of 5 sec. expressions	6 expressions
2017	3D-MENPO	Deng et al. (2019)	CNN: 79.8% (↔1.68%)	12K frames from many datasets	84 face 3D pts
2017	2D-MENPO	Deng et al. (2019)	// 76.5% (↔0.0024%)	// (semi-frontal set)	68 face 2D pts
2016	DISFA+	Mavadati et al. (2013)	Multiple AUs rates	12 AUs for 27 subjects, 66 landmarks	42 face actions
2015	LIRIS-ACCEDE	Baveye et al. (2015)	Manual: 83.5%/86.2%	160 movies, 9800 ~10s. seq., 73 h.	Valence/Arousal
2015	300-VW	Sagonas et al. (2016)	CNN: 75.2% (↔0.06)	300 sequences, 218K faces	68/51 face pts
2013	CASME-II	Wen-Jing Yan et al. (2013)	SVM: 63.4%	3000 sequences, 247 micro-exp.	6 expressions
2013	AM-FED	McDuff et al. (2013)	Multiple AUs rates	242 webcams, 5268 subjects, 168k im.	22 pts, 6 AUs
2012	SEMAINE	McKeown et al. (2012)	75% (↔0.07)	150 subjects, 959 5mn. discussions	cognitive states
2011	AFEW	Dhall et al. (2011)	59.2%	1156 seq., 113K im., gender	pose/6 exp.
2010	CK+	Lucey et al. (2010)	SVM+AAM: 94.5%	593 sequences, 23 AUs	6 expressions
2005	MMI-DB	Pantic et al. (2005)	SVM: 92.3%	848 AU videos and 740 images	79 emotions
Year	Dataset	References	Classifier: Perf.	Content description	Labels
From fixed face images					
2017	AffectNet	Mollahosseini et al. (2019)	CNN: 55.5%	1M faces, 8 emotions	Valence/Arousal
2017	RAF-DB	Li and Deng (2019)	CNN: 74.2% and 44.6%	30K faces x 40 compound labels	7/11 expressions
2016	EmotionNet	Benitez-Quiroz et al. (2016)	CNN: 75.5%	950K images, 12 AU rates	12 AUs
2015	ExpW	Zhang et al. (2018)	CNN: 70.2% (subset)	92K faces in the wild	7 expressions
2013	FER-2013	Carrier and Courville (2013)	71.2%	36K facial images	7 expressions

(continued)

Table 9.3 (continued)

Year	Dataset	References	Classifier: Perf.	Content description	Labels
2013	CASME	Wang et al. (2014)	60.82%	195 micro expressions	35 emotions
2012	HELEN	Le et al. (2012)	Euclidean min. dist.	2330 high-resolution images	194 pts
2011	SFEW	Dhall et al. (2011)	CNN: 48.6%	700 facial images	6 expressions
2011	LFPW	Belhumeur et al. (2013)	SVM+Bayesian: 97%	1432 celebrity faces	29 pts
2008	Multi-PIE	Lu et al. (2013)	SVM: 92.3%	337 Subjects, 15 views, 19 illum.	5 expressions
2004	IMM	Nordstrøm et al. (2004)	AAM: ~100%	240 face images, 40 persons	58 pts, 3 exp.
1999	JAFFE	Lyons et al. (1999)	Multiple: ~100%	213 Japanese female images	7 expressions
1998	AR	Martinez and Benavente (1998)	99%	126 subjects, 4000 images, 2 sessions	22 pts
1997	YALE	Georghiades et al. (2001)	99.3%	165 images, 15 subjects	5 expressions
1994	ORL	Roychowdhury and Emmons (2015)	83%	400 images, 40 subjects, occlusion	4 expressions
1993	FERET	Lumini et al. (2017)	97%	14K images, 1199 subjects, 2 sessions	2 expressions

(//) same configuration as above, (↔) The performance is relative to the tolerance Margin-error considered

2019). In addition to these FE recognition families, the active communities will be also described in what follows.

Image-based expressions analysis: The first image-based datasets offered increasing numbers of expressions. They started with 2 expressions within the first benchmarking FERET dataset (Lumini et al., 2017). It was the first multi-session dataset with 14K captures and 1199 subjects. Within the next *AT&T* dataset, known as ORL, 4 expressions could be found in addition to occlusion with glasses (Roychowdhury and Emmons, 2015). Later, 5 different expressions were offered in YALE dataset (Georghiades et al., 2001), to finally reach all 7 basic expressions (see next Sect. 9.3.2.4) within the Japanese female faces dataset, i.e. JAFFE (Lyons et al., 1999).

Many next coming image datasets offered richer learning sets. The Multi-PIE extended its previous CMU-PIE version to become the first multi-view (15 cameras) benchmark with 5 expressions and additional 19 illuminations (Gross et al., 2008). The Static Facial Expressions in the Wild (SFEW) dataset, extracted from its dynamic version AFEW, was the first to offer challenging in-the-wild expressions (Dhall et al.,

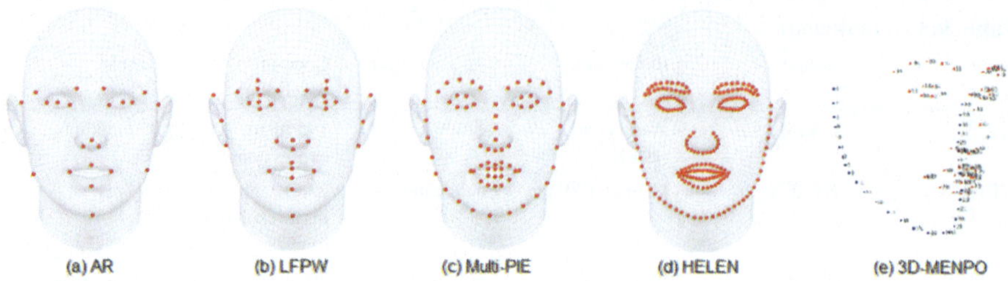

Fig. 9.6 Examples of facial landmarks: (**a**, **b**, **c** and **d**) 2D (Sagonas et al., 2016) and (**e**) 3D datasets (Deng et al., 2019)

2011). The training population continued to grow in the FER-2013 (Carrier and Courville, 2013 and ExpW (Zhang et al., 2018) datasets, to reach 36 K then 92 K faces, respectively.

Video-based emotions: Starting from 2005, the first video datasets made their appearance with MMI-DB (Pantic et al., 2005) and the extended Cohn-Canade (CK+) dataset offering 523 sequences and 23 AUs (Lucey et al., 2010). At this level, the dynamic analysis of AUs and emotions started receiving dedicated attention. This was the case with AFEW (Dhall et al., 2011) with 1156 in-the-wild sequences, then with the AM-FED dataset with 6 AUs relative to spontaneous expressions from webcam recordings for a total of 168k frames (McDuff et al., 2013). The extended CASME-II dataset offered on its turn 3000 videos, 247 micro-expressions and focused on high temporal resolutions (200fps) (Wen-Jing Yan et al., 2013). DISFA+ is one of the latest improvements with stereo-videos particularly useful for analyzing the finest temporal AU changes within posed and spontaneous expressions (Mavadati et al., 2013). Nonetheless, the biggest AU dataset with 950K in-the-wild captures is held by EmotioNet (Benitez-Quiroz et al., 2016).

Pose estimation: From another point of view, other datasets allowed expressions analysis using pose estimation. Indeed, the facial pose vehicles information on both individual and group emotions (Zhang et al., 2018). For this purpose, increasingly richer benchmarks have offered different configurations of facial landmark points. As illustrated in Fig. 9.6, multiple 2D facial landmarks have been proposed (Sagonas et al., 2016). The recent advancements are even considering 3D landmarks extraction from RGB-only faces (Deng et al., 2019).

Starting with 22 points within the AR dataset (Martinez and Benavente, 1998), the facial landmark annotations have increased to 29 points within the LFPW dataset (Belhumeur et al., 2013). Within the IMM image-dataset, 58 landmark points have been given (Nordstrøm et al., 2004). Then within 300-W and 300-VW dataset both 51 and 68 landmarks have been offered (Sagonas et al., 2016). And while the highest number of annotations is held by the Helen dataset with 194 points (Le et al., 2012), these positions could be interpolated from fewer numbers. Within the recent 2D-MENPO video dataset, 68 points were considered from 12K frame samples. For the interesting 3D-MENPO dataset, the 84 landmark annotations were used for the first time in 3D from RGB-only faces (Deng et al., 2019).

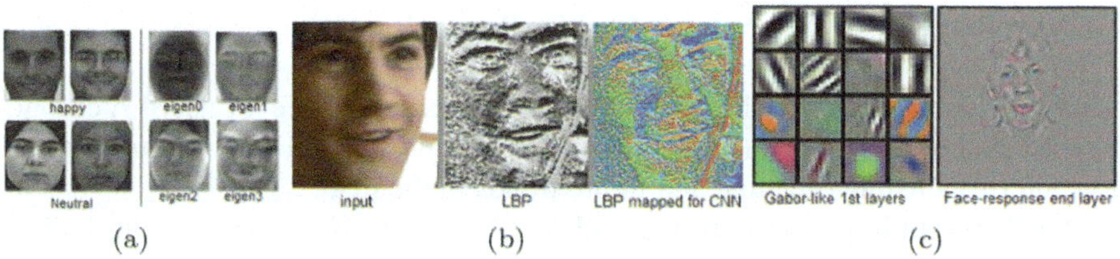

Fig. 9.7 Examples of commonly used facial features: **a** PCA-based Eigenfaces, **b** Local Binary Patterns and their adaption to CNNs and **c** visualization of facial neural responses at different CNN layers

Compound emotions (Affect) analysis: If we continue in scale increase, we find that recent datasets were targeting the 1M frames in size and that they were "forced" to consider compound labeling (e.g. disgusted-smile, sad-smile, etc.). The first attempt with 160 open movies and 73 h has been offered by the LIRIS-ACCEDE dataset (Baveye et al., 2015). Next, camc the RAF-DB dataset with an interesting set of 40 labels per FE. The selection of those with-consensus allowed the definition of 11 compound labels (Li and Deng, 2019). On the top positions of Table 9.3, we find that the AffectNet dataset (Mollahosseini et al., 2019) is actually the biggest image-based benchmark with 1M faces relative to 8 emotions. The biggest number of videos with 298 sequences and over 1.2M frames is actually held by the Aff-Wild dataset (Kollias et al., 2019). While LIRIS-ACCEDE and RAF-DB used crow-sourcing for labeling (Baveye et al., 2015; Li and Deng, 2019), the AffectNet and Aff-Wild (Mollahosseini et al., 2019; Kollias et al., 2019) where annotated by 2 and 6 experts, respectively.

Active communities: Multiple recent competitions created by specialized communities exist and are open for new contributions. One interesting competition is the one organized by the MediaEval community. It has proposed multiple benchmarks using, among others, the LIRIS-ACCEDE large data collection (Baveye et al., 2015). The iBUG research group, on its side, contributed to the definition of multiple benchmarks in recent years (Chrysos et al., 2018). Examples include: MMI-DB with 848 videos (Pantic et al., 2005), SEMAINE with 959 human-agent discussion sequences (McKeown et al., 2012), the 300-W and 300-VW pose estimation datasets (Sagonas et al., 2016). Active recent competitions include: (i) EmotiW offering near 2K sequences for group level emotions recognition (Dhall et al., 2017), (ii) the Aff-Wild challenge with 1.2M frames of random emotional states (Kollias et al., 2019), and (iii) 3D-MENPO for facial pose estimation in real-world coordinates (Deng et al., 2019).

Within Table 9.3, we give a comparative ordering for the RGB-based datasets allowing FEs approaches. While many competitions have proposed iterative content improvements, we summarize by keeping the most representative recent versions (e.g. 300-VW instead of 300-W (Sagonas et al., 2016)) and indicate the references relative to the best respective performances. This list is continued under multi-modal acquisitions in Sect. 9.4.4 and Table 9.6.

9.3.2.4 RGB-based Approaches for FEs Recognition

Our RGB-based FE approaches analysis will be pursued at three levels: (i) A chronological cover of FEs recognition evolution, (ii) a focus on CNN-based feature extraction and (iii) a listing of actual and future trends in the field.

Development in expressions recognition: One of the fundamental contributions is the Facial Action Coding System (FACS) of Ekman et al. (1997) as it defines the standard coding of facial movements. Many approaches use the basic FACS system composed of 7 expressions: '*the neutral state*', '*surprise*', '*anger*', '*fear*', '*disgust*', '*sadness*' and '*happiness*'. The more advanced approaches consider the finer facial motions described within FACS using 44 basic AUs (for reminder: Action Units) relative to the muscles involved in the expression, in addition to other AUs for face and eye movements. For example, AU1 and 2 are relative to inner and outer brows raising, AU26 is for the jaw drop and AU45 is for eye blink (Ekman and Rosenberg, 1997).

For expression recognition, the first approaches exploited appearance-based projection methods such as PCA (see Fig. 9.7a) and LDA (i.e. Linear Discriminant Analysis). The faces' eigenvectors are computed, then all learning images are projected within a reduced space to calculate their weights. The obtained parameters are used as face descriptors. Multiple variants of these representations were tested on the first benchmarks (e.g. FERET and Multi-PIE datasets (Lu et al., 2013)). A representative solution using them in combination with neural networks is found in Agarwal et al. (2010).

A next improvement has been brought by the Active Shape Models (ASM) and Active Appearance Models (AAM) developed by Cootes et al. (2001). Using a statistical model with controllable parameters, they have combined: (i) Geometrical deformable shapes with (ii) localized appearance representations such as PCA.

By modulating predefined deformable models, the ASMs and AAMs were first used for the assisted generation of 2D and 3D textured faces (Blanz and Vetter, 1999). Later, they were behind multiple solutions aiming at facial pose estimation (Cootes et al., 2001; Nordstrøm et al., 2004).

At this point, literature started considering the task of facial detection from non-cropped faces. Using 2D Haar features and a cascaded learning model, a successful facial detector was proposed by Viola and Jones (2001). While detectors' robustness continued to improve using HOG and CNN features (Li and Deng, 2018), they are still impacting the performance of recent approaches (Sagonas et al., 2016). Challenging situations include small facial resolutions, occlusion, extreme poses and illumination variations. Additional preprocessing techniques such as facial colors normalization and pose alignment are actually applied to overcome these problems (Corneanu et al., 2016).

A hole set of approaches have also focused on the definition of more efficient facial descriptors. The Local Binary Patterns (LBP) have been extensively applied as they are illumination invariant and save the micro-texture variations of pixels neighborhood. Coupled with SVM classifiers, they have led to the best 63.4% accuracy on CASME-II (Wen-Jing Yan et al., 2013). A recent combination of different LPB setups (i.e. orientations, thresholds) as an RGB image is shown in Fig. 9.7b. These maps served as an optimized input for multiple CNN learning paths leading to a 40% performance gain on the EmotiW 2015 challenge (Levi and Hassner, 2015). Other orientation-and-scale invariant descriptors such as HOG and SIFT have also been evaluated separately or in combination with LBPs on datasets such as SFEW, AFEW and RAF-DB (Dhall et al., 2011; Li and Deng, 2019).

From Gabor filters to CNN-based features: Gabor filters figure also among the most used facial feature extractors (Riche et al., 2013; Corneanu et al., 2016). When used in banks of different orientations and scales, they can encode both spatial and frequency dispositions of pixels intensities at different resolutions. Interestingly, it has been proved by Krizhevsky et al. in (2012) that the CNN first layers within the AlexNet architecture behave in Gabor-like filters when learned from a massive datasets such as ImageNet. This has been confirmed by the visualization toolkit elaborated by Yosinski et al. using generative neural networks (Nguyen et al., 2019). As illustrated in Fig. 9.7c, the left-images synthesized to produce high activations for the first layer neurons, resemble greatly to Gabor filter banks. Accordingly, CNN layers gradually learn abstract combinations of these filters until reaching facial or emotion-specific specialized neurons, producing activations similar to the right-image synthesized in Fig. 9.7c (Nguyen et al., 2019).

In Alekseev and Bobe (2019), a tentative for the replacement of CNN first layers by auto-generated Gabor filters has been introduced in the GaborNet model. The performance evaluation on the AffectNet FEs dataset has reported similar performances with a much faster model convergence. More recently within the FACS3D-Net CNN model, a combination of 3D-CNN spatial features with LSTM temporal modeling has been successfully applied for AUs dynamic recognition (Yang et al., 2019). Another special CNN architecture known as the Hourglass model has been particularly successful lately both for face and body points estimation from RGB (Deng et al., 2019). Its symmetrical architecture resembles visually to a sand-hourglass: It is composed of a first down-sampling using pooling to a very-low resolution, then of an up-sampling by pixel-wise production. Different combinations of this architecture have successfully applied stacking or cascading techniques to predict 2D poses then extrapolate 3D real-wold positions out of them (Guo et al., 2018).

Actual and coming orientations: In addition to movies and Internet in-the-wild frames, actual acquisitions are also considering smart-phone streams (Lin et al., 2019) and the finest facial motions such as mouth vocalization (Livingstone and Russo, 2018; Kollias et al., 2019). And beside individual's emotions, even interpersonal relations (e.g. friendliness, dominance, etc.) could be predicted using additional attributes such as gender, age and automatic pose tracking (Zhang et al., 2018; Deng et al., 2019). For what concerns facial pose tracking, as already initiated for the body (Güler et al., 2018), the next coming challenges will also consider dense points estimation in 3D world coordinates (Deng et al., 2019).

New proper annotation methods will be required for the latter tasks. This is also valid for the circular expression representation system controlled by the Valence and Arousal measures (a detailed analysis is provided in Mollahosseini et al. (2019)). As the use of compound-labels is shifting to continuous evaluation systems (Kollias et al.,

2019; Noroozi et al., 2018), labels locality preservation methods such the one used in Li and Deng (2019) could bring partial solutions. But the best will be to provide more annotators and labels consensus in future contributions. A proof of this difficulty is found with the AffectNet dataset where the 2 expert annotators agreed only on 60.7% of the labels due to emotions variations. And while crowd-sourcing could provide more consensus, non-expert annotators might mis-understand the fine-details of AU labels as within the LIRIS-ACCEDE dataset (Baveye et al., 2015). Thus, apart from data collection, proper procedures orienting the annotators to better labeling are still expected to improve.

Finally, the application of previous FEs recognition approaches into concrete robotic platforms is among the most promising future exploitations. Initial tentatives for facial mimicking and autism therapy could be found within the SEER (Todo, 2018) and ENIGMA (Marinoiu et al., 2018) projects, respectively.

9.3.3 Recognition from Depth Streams

The problem of occlusion represents a major challenge for correct HAR. It is present in real-word scenarios, and specially those involving inter-person/object interactions (Wang et al., 2018). It causes considerable noise within the joint modality and hardens the body-parts segmentation within RGB streams. Besides using costly multi-view acquisitions, the depth modality represents an optimal solution for occlusion handling. This is logically explained by the extra Z dimensionality offered by depth as it simulates the human stereo vision system. A related analysis is proposed in what follows.

9.3.3.1 Depth-Based Datasets for HAR

For acquisition, there is a number of databases offering, among others, the depth modality using the Infra-Red (IR) sensors of Kinect V_1 and V_2 (Oreifej and Liu, 2013; Rahmani et al., 2014). Many depth-only methods exploited the MSR-Action3D database (Wang et al., 2012) within its 1/3, 2/3 or cross-validation configurations. Other works exploited the large-scale capabilities of the CAD and CGC benchmarks (Guyon et al., 2013; Cippitelli et al., 2016) reaching 12K actions sequences.

Another easier way to reach large scale depth datasets is their synthetic 3D generation as employed within Microsoft's MSR dataset (Shotton et al., 2013). Different IR sensors were also used such as the Asus Xtion in Haque et al. (2016) or Intel's RealSense depth cameras in Yuan et al. (2017).

On another level, after a number of datasets dedicated for actions description (Vieira et al., 2012; Rahmani et al., 2014; Cippitelli et al., 2016), we find that the recent trend is rather for pose estimation applications (Haque et al., 2016; Yuan et al., 2017; Garcia-Hernando et al., 2018). Indeed, the need for better virtual-reality experiences has accelerated the production of datasets focusing on hand gestures in

(a) (b) (c)

Fig. 9.8 Example depth-related approaches for: **a** Depth generation from videos, **b** motion-energy features extraction and **c** hand-joints pose estimation

frontal and egocentric views (Doosti, 2019). Examples include the MSR-Action-pairs (Oreifej and Liu, 2013) with 12 action labels and 360 sequences related to gaming scenarios. More recent pose estimation datasets include: (i) The BigHand2.2M with 2.2M depth images of hands and 21 joint position labels, (ii) the F-PHAB offering similar joints with 1.2K sequences, 105K frames and additional 45 action labels, and (iii) HO-3D offering 16 joint positions and 3 object models for the challenging task of coupled hands and objects pose estimation. A combination of these datasets is offered within the one-million-hands challenge (Camgoz et al., 2017).

Another distinguishable category of datasets are those rather focused on varying the acquisition view-angles. One example includes the UWA3D-multiview (Rahmani et al., 2014) with 300 sequences and 30 action labels. Another more recent example is the ITOP dataset (Haque et al., 2016) applied for pose estimation with 100K front/top captures and 14 joint labels.

Finally, we also highlight a totally different set of databases such as the Mannequin dataset (Li et al., 2019). With 2000 RGB sequences containing fixed human actors, the goal is to reconstruct the depth maps from the intrinsic motion. While not directly related to action recognition or pose estimation, these datasets applied to depth creation from stereo-vision are full of promise when applied to moving actors (see Fig. 9.8a) (Valentin et al., 2018). Within Table 9.4, we highlight a selection of depth-only methods for actions recognition, pose estimation and their relative datasets.

9.3.3.2 Depth-Based Approaches for HAR

At the depth modality's level, although the apparition of the Kinect sensor in late 2010 enabled more related work (Firman, 2016), the number of research contributions found is still smaller than those using RGB or joint modalities. One reason might be the limited accessibility to this modality and the need to compensate for the noise generated by the infra-red sensors. A number of methods have also changed the data format from depth images to sets of 3D point clouds or 3D meshes after triangulation (Blackburn et al., 2010). Overall, the depth-based approaches could be divided in two families: (i) Those using the raw depth stream for feature extraction and (ii) those exploiting the displacement information for pose estimation. Both cases can enable HAR afterwards.

Depth-based actions description: In terms of efficiency, we highlight that using the depth streams, a number of methods surpassed their equivalent in RGB (Zhang et al., 2019). This is the case for example for the approach proposed in Guyon et al. (2013) within the CGC-2012 one-shot challenge. The best performance of 90.1% obtained from the depth data was far above the 72.9% accuracy obtained using the RGB streams. As features, a number of methods have focused on calculating energy-maps of displacements (see Fig. 9.8b) at different frame instants from one action sequence (Vieira et al., 2012; Guyon et al., 2013). Other features were also proposed using the 3D HOG descriptors (Yang et al., 2012) or the histograms of discrete normals (Oreifej and Liu, 2013). Similar features could also extracted from the reconstructed 3D point clouds: (i) A PCA-based representation was proposed in Rahmani et al. (2014), (ii) an analysis of the 3D point clouds arrangements within well-defined networks has been also proposed in Cippitelli et al. (2016).

More recently, many approaches have migrated to CNN-based feature extractors: (i) A solution for the recognition of finger spelling using the basic CNN architecture AlexNet of Krizhevsky et al. (2012) has been proposed in Kang et al. (2015), (ii) a more recent application of the 3D-CNN architecture (Guo et al., 2016) to depth maps has also been adopted by Duan et al., as an improvement to previous BoVW methods in Duan et al (2018).

Depth-based pose estimation: One advantage offered by the depth modality is the relatively-easy extraction of 3D joint skeletal positions compared to RGB. It was used under Microsoft's Kinect pose estimator proposed by Shotton et al. (2013). The authors relied on depth displacement differences and random decision forests to implement their real-time solution. Knowing that their algorithm was optimized for speed rather than accuracy, their resulting joint positions were relatively noisy and required additional post-processing (Shotton et al., 2013). A related improvement was proposed by Ganapathi et al. for the incorporation of skeletal and temporal constraints. Using the 3D point clouds generated from depth, they have used a probabilistic model for the body and the iterative-closest-point algorithm to improve the joints tracking (Ganapathi et al., 2012).

Within the contributions dedicated to hands pose estimation, a similar extension could be found: A kinematic model defining the fingers pose relationships has recently been tuned within the MANO model (Hampali et al., 2019) composed of 21 joints (see Fig. 9.8c). A successful incorporation of the kinematic constraints within the learning process of a deep CNN neural model has been also proposed in the HOLI-CNN model (Ye et al., 2016). To do so, the authors apply a hybrid approach composed of: (i) A cascaded CNN description for the partial joint poses, and (ii) a particle swarm optimization that integrates the kinematic constraints within the regression model.

The MANO and HOLI-CNN models were successfully applied to hands pose estimation on the Ho-3D and BigHand2.2M datasets. Within a 5mm error range, they have approximated 60 and 90% accuracies, respectively (Hampali et al., 2019; Yuan et al., 2017). Under the more challenging egocentric views, we find that the HOLI-

Table 9.4 Main references and datasets allowing depth-based description/pose-estimation (Chen et al., 2013; Agarwal and Xia, 2014; Ye et al., 2013)

Year	References	Perf.	Content description	Labels	Dataset
Spatial/temporal description from depth					
2018	Garcia-Hernando et al. (2018)	78.7%	1.2K hand seq., 105K frames	45 acts.	F-PHAB
2016	Cippitelli et al. (2016)	93.9%	68 seq. total, 4 actors	14	CAD-60
2014	Tabia et al. (2012)	84.9%	~300 seq. total	30	UWA3D-multiview
2014	Rahmani et al. (2014)	96.2%	336 seq. total	12	MSR-Gesture-3D
2013	Oreifej and Liu (2013)	98.2%	360 seq. total	12	MSR-Action-pairs
2012	Guyon et al. (2013)	$*$ 90.1%	(One-Shot)	8 to 12	CGC-2012
2012	Vieira et al. (2012) ($\frac{1}{3}$)	98.3%	567 seq. total	20	MSR-Action3D
2012	Vieira et al. (2012) ($\frac{2}{3}$)	96.8%	//	//	//
2012	Yang et al. (2012) (CV)	91.6%	//	//	//
Pose estimation from depth					
2019	Li et al. (2019) (↔20px)	41%	2000 frozen actors sequences	21 pts.	Mannequin
2018	Hampali et al. (2019) (↔5mm)	~60%	15K depth im., 8 pers, 3 obj.	16	HO-3D
2018	Garcia-Hernando et al. (2018) (↔14mm)	69.2%	1.2K hand seq., 105K frames	21	F-PHAB
2017	Yuan et al. (2017) (↔5mm)	90%	2.2M depth hand im., 10 pers.	21	BigHand2.2M
2017	Yuan et al. (2017) (↔20mm)	~61%	290K egocentric depth images	21	//
2016	Haque et al. (2016) (↔10cm)	74.1%	100K front/top depth images	14	ITOP
2012	Ganapathi et al. (2012)	84%	9K depth images, 24 seq.	8	EVAL
2011	Shotton et al. (2013)	77%	900K synthetic/real depth im.	16	MSR

($*$) Adaptation as a percentage, (//) same condition as above

CNN estimator has been combined with LSTM networks for the 69.2% action recognition performance obtained on the F-PHAB dataset (Garcia-Hernando et al., 2018). But, when using the egocentric subsample of the more challenging BigHand2.2M dataset, only 61% performance was obtained within the relatively wide tolerance of 20 mm (Yuan et al., 2017). This proves that more improvements are still expected.

9.3.4 Unimodal Temporal Segmentation Approaches

When using spontaneous action sequences, the actions flow consecutively with minor resting momentums. The lexicon of recognizable gestures becomes then unlimited and correlated. The Temporal Segmentation (TS) step allows the decomposition of these continuous streams into manageable segments of more distinguishable actions. The TS results then in multiple $(start, end)$ frame couples delimiting the action positions. An example of the TS segments obtained for two moving hands is illustrated in Fig. 9.9.

In the following, we present a classification of TS related approaches into 4 method families:

(i) **Heuristic**: Using predefined thresholds, these methods allow Change-Detection (CD) (Alippi et al., 2013) from 1D signals (i.e. one computed feature). They are then suitable for real-time decisions over time-series of data. Peng et al. (2014) search for the most frequented body position with RGB frames, then extract the action segments having motions farther than a given T threshold. Liang et al. (2014) use rather the joint modality to implement a similar scheme and decide on the dominant hand in motion. The CD approach presented by Seddik et al. (2015) uses the joint inputs to segment both the streams and the learning population into right, left and bi-handed actions (see Fig. 9.9), thus allowing classifiers specialization in later steps.

(ii) **Dynamic programming**: The commonly used techniques include the Dynamic Time Warping (DTW) (Sempena et al., 2011) and Hidden Markov Models (HMM) (Alon et al., 2009). In both case, the temporal slices are obtained by search for low-cost paths within generated weight matrices (Guyon et al.,

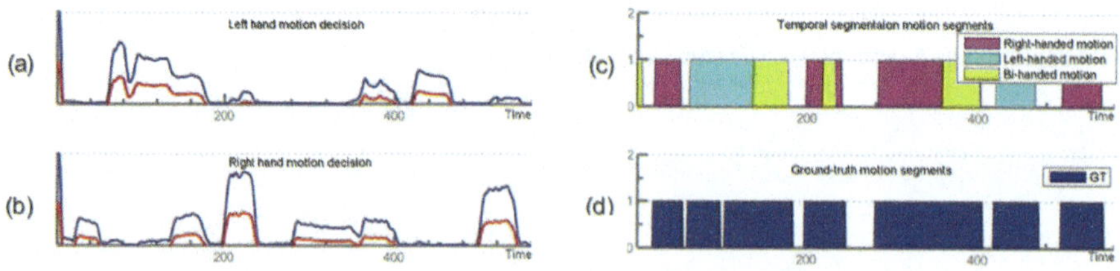

Fig. 9.9 Example temporal segmentation results: **a** and **b** Left and right hands motion curves, **c** TS results and **d** the ground truth (Seddik et al., 2015)

2014). The HMMs allowed top ranked performances in the Chalearn competitions requiring TS (Escalera et al., 2013). Though DTW is a quite old technique (Müller, 2007), many approaches are still successfully using it (Wan et al., 2013). A slightly improved DTW version using Gaussian Mixture Models (GMM) has been proposed in Hernandez-Vela et al. (2014). A more recent DTW combination with deep CNN architectures is proposed by Junfu et al. in (2019) for the purpose of continuous SL recognition.

(iii) **Sliding windows**: This method is used on a wider scale and consists of defining a number of frames (i.e. a window of *N* seconds) used to segment and analyze the stream's features. This window is then slided with a predefined translation factor to browse the full stream (Laguna et al., 2011; Escalera et al., 2013). Many approaches extract motion energy features (Shao et al., 2012) that encapsulate the temporal change differences. They can be extracted at different resolutions, from discrete grids of RGB, depth or directly from skeletal-joints positions (Poppe, 2010).

(iv) **Learning-based**: More advanced approaches are considering kernel-based or neural-networks-based solutions for the binary classification of frames into motion/non-motion (Gong et al., 2012. These classifiers could be optimized for accelerated executions (Chen and Koskela, 2013) or for more robust segmentations as proposed by Neverova et al. in (2016). More advanced approaches are combining RNN and CNN techniques to apply both the temporal segmentation and recognition within the neural architecture as proposed by Pigou et al. in (2016). This last configuration allowed the state-of-the-art performance on the challenging CGC-2014 benchmark.

9.4 Multi-modal Methods

Several recent approaches and databases are now profiting from multi-modal data availability. Many works benefited from 2D and 3D data affordable combinations. We focus here additionally on the approaches oriented towards static 3D actions/faces as-well-as dynamic ones. We summarize in what follows the details of these two research-work families.

9.4.1 Multi-modal Datasets for HAR

While the initial contributions combined the inertial motion sensors streams with RGB images (Sigal et al., 2009; De la Torre et al., 2008), the following advancements involved adding the actors' silhouettes masks (Singh et al., 2010) before giving place to a dominance of combined RGB, depths, and joint streams mainly coming from the Kinect sensor. This is the case for instance for the CGC and CAD benchmarks-based approaches (Guyon et al., 2014; Escalera et al., 2014; Sung et al., 2012; Koppula et al., 2013).

The most recent advances are considering learning on from large-scale data masses. This is for instance behind CGC-2011's dataset reformatting into the latest versions of ConGD and IsoGD 2016 datasets (Wan et al., 2016) offering continuous and isolated action streams, respectively. This is also the case with one of the most recent and largest multi-camera dataset using Kinect-V2: NTU-RGB+D (Shahroudy et al., 2016). In the case of multi-view datasets, different viewing angles are combined to achieve higher performances. This is the case of Berkeley-MHAD, where 100% of the actions are recognized (Ofli et al., 2013) in a multi-modal and multi-views configurations.

We summarize in Table 9.5, the multi-modal contents for the most recent datasets. We also distinguish the solutions dealing with SL recognition and those offering continuous streams and therefore requiring a temporal segmentation stage before action recognition.

9.4.2 Multi-modal Fusion Approaches

The combination of different modalities allows gaining their complementary informations for better performances (Vrigkas et al., 2015). This implies the use of different fusion strategies. The most used ones are detailed hereafter:

(i) **Feature-concatenation fusion**: The concatenation of low-level descriptors can be applied to combine multiple features coming from a unique modality (Wang and Schmid, 2013). It can also be applied with different modalities such as RGB and depth to reduce feature variability (Wan et al., 2013). Additionally, the selective extraction of the RGB and depth features to concatenate has been also applied using the joints as sparse position indicators (Escalera et al., 2014; Koppula et al., 2013). In this special case where user-mask or joint positions are used for feature selection, high performance improvements have been reported (Jhuang et al., 2013). This transition from low-level to mid and high level descriptors is illustrated in Fig. 9.10. Finally, full concatenation of the RGB, depth and joint descriptors has been also proven interesting (Seddik et al., 2017).

(ii) **Score fusion**: At this level, the fusion combines the scores produced by different unimodal classifiers. They can be generated using probabilistic or discriminative decision models. In data complementarity is available, the fusion offers better performances than the separate models (Vrigkas et al., 2015). Weighted fusion schemes have been applied from of multiple SVM classifiers' outputs in (Cippitelli et al., 2016; Seddik et al., 2017). They have also been directly implemented within deep neural architectures in (Neverova et al., 2016; Pigou et al., 2016).

(iii) **Representation fusion**: Peng et al. proved (Peng et al., 2014) that the fusion of different BoVW representations improves considerably HAR. The fusion process can be then improved using concatenation or pooling techniques to

Table 9.5 Description of multi-modal datasets Yu et al. (2016)

Year	Dataset	References	C	D	J	I	M	Samples	Actions	Sign.	TS
Mono-angle view											
2019	ASL-100	Jing et al. (2019)	✓	✓	✓	–	–	15 subjects, 4K seq.	100 act.	✓	–
2016	ConGD and IsoGD	Wan et al. (2016)	✓	✓	–	–	–	21 subjects, 48K seq.	249 act.	✓	✓
2015	UTD-MHAD	Chen et al. (2015)	✓	✓	✓	✓	–	8 subjects, 861 seq.	27 act.	✓	–
2014	CGC-2014	Escalera et al. (2014)	✓	✓	✓	–	✓	40 subjects, 14K seq.	20 act.	✓	✓
2013	CGC-2013	Escalera et al. (2013)	✓	✓	✓	–	✓	20 subjects, 13K seq.	20 act.	✓	✓
2013	CAD-120	Koppula et al. (2013)	✓	✓	✓	–	–	4 subjects, 120 seq.	10 act.	–	✓
2013	Berkeley-MHAD	Ofli et al. (2013)	✓	✓	✓	–	–	8 subjects, 861 seq.	27 act.	–	–
2013	J-HMDB	Jhuang et al. (2013)	✓	–	✓	–	✓	∞ subjects, 5K seq.	51 act.	✓	–
2015	BBC extended	Pfister (2015)	✓	–	✓	–	–	40 subjects, 92 seq.	∞ act.	✓	✓
2012	CAD-60	Sung et al. (2012)	✓	✓	✓	–	–	4 subjects, 68 seq.	14 act.	–	–
2011-12	CGC-2012	Guyon et al. (2013)	✓	✓	–	–	–	One-Shot	8 to 12 act.	–	✓
2012	MSR-Daily	Wang et al. (2012)	✓	✓	✓	–	–	10 subjects, 320 seq.	16 act.	–	–
2010	MSR-Action3D	Li et al. (2010)	–	✓	✓	–	–	10 subjects, 557 seq.	8 act.	–	–
2010	BBC	Buehler et al. (2010)	✓	–	✓	–	–	9 subjects, 20 seq.	∞ act.	✓	✓
Multi-cameras or multi-views											
2016	NTU-RGB+D	Shahroudy et al. (2016)	✓	✓	✓	-	✓	40 subjects, 57K seq.	3× 60 act.	✓	–
2015	UWA3D-multiview-2	Rahmani et al. (2014)	✓	✓	✓	–	–	–	5× 30 act.	✓	–
2015	Berkeley-MHAD	Ofli et al. (2013)	✓	✓	✓	✓	–	12 subjects, 647 seq.	4× 11 act.	–	–
2014	Human3.6M	Ionescu et al. (2014)	✓	✓	✓	–	–	–	–	–	–
2010	MuHaVi	Singh et al. (2010)	✓	–	–	–	✓	14 subjects, 136 seq.	8× 17 act.	–	–
2009	Human-Eva-2	Sigal et al. (2009)	✓	–	–	✓	–	4 subjects, 56K im.	6 act.	–	–
2008	CMU-MMAC	De la Torre et al. (2008)	✓	–	–	✓	–	–	5 act.	–	–

'C': Color, 'D': Depth, 'J': Joints, 'M': actor's Mask, 'I': Inertial sensor. 'Sign.': Relativeness to SL, 'TS': Temporal Segmentation usage

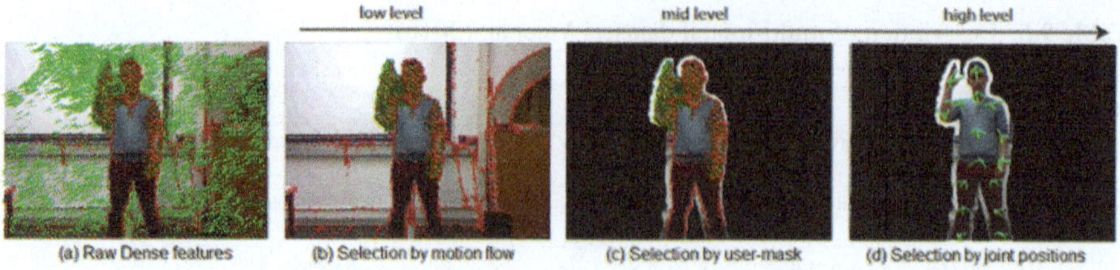

Fig. 9.10 Selective concatenation of **a** raw dense descriptors using **b** motion-flow, **c** user-mask and **d** joint positions to produce high-level features

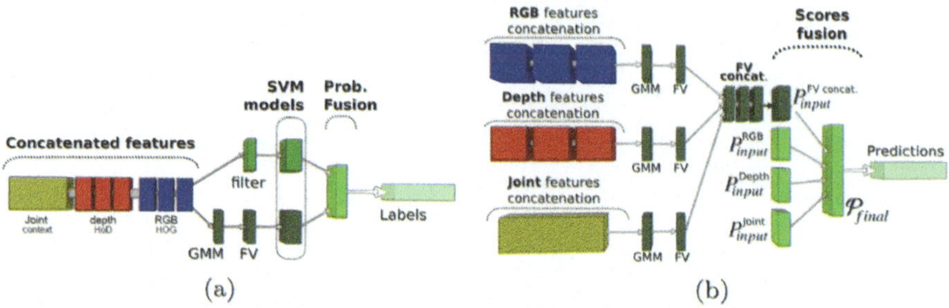

Fig. 9.11 Examples of hybrid fusion schemes: Different concatenation and score fusions are applied in (**a**) and (**b**). Additional representation fusion is applied with FVs in (**b**) Seddik et al. (2017)

combine the different representations. Within the CNN architectures, the fusion at the representation level is also considered a key part of the neural pooling process specially when using multi-modal inputs (Pigou et al., 2016; Wu et al., 2016).

(iv) **Hybrid fusion**: The hybrid-fusion methods apply combinations of previous fusions as illustrated in Fig. 9.11. Using a hybrid super-vector approach, advanced performances were obtained by Peng et al. (2016) on multiple datasets by combining the decisions produced by multiple BoVW architectures. Similarly, in Seddik et al. (2017) the authors proposed a multi-layered fusion scheme of Kinect modalities for the recognition of SL actions from continuous streams. Their solution starts by concatenating different modality-specific features, learns specialized SVM classifiers at local and global temporal scales, then optimizes their scores merging iteratively.

9.4.3 Multi-modal Datasets for 3D FEs Recognition

Before getting into multi-modal SL recognition, one special set of useful techniques are those exploiting 3D multi-modal facial data. These datasets can offer 3D meshes, point clouds, surface normals, IR maps in addition to RGB and landmark points.

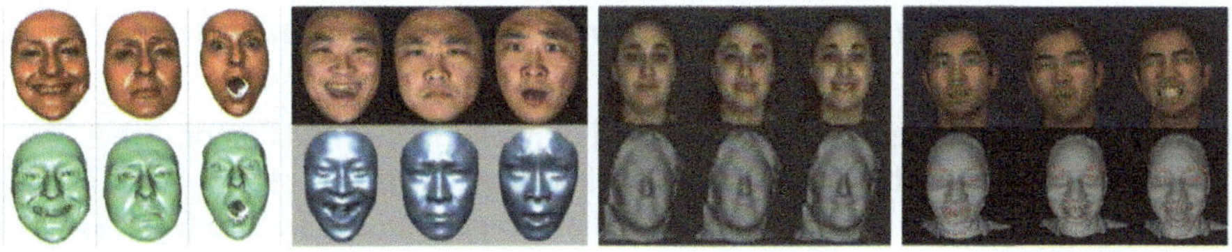

(a) Bosphorus (b) BU-3DFE (c) BU-4DFE (d) BP4D

Fig. 9.12 Examples of 3D FE datasets: **a**, **b** Static and **c**, **d** dynamic

The acquisitions are then naturally less vulnerable to the frequent face alignment and lighting problems found with RGB-only methods (Li and Deng, 2018). They could be acquired using high resolution 3D scanners as for the Biwi-3D and the 3D-dynamic-DB datasets (Messer et al., 2003; Alashkar et al., 2014) or with multiple cameras and consumer depth sensors as within the Biwi-kinect (Fanelli et al., 2013) and the 4DFAB (Cheng et al., 2018) datasets.

A classification into static 3D or dynamic 4D facial datasets is offered in Table 9.6. Many successful static datasets such as BU-3DFE (Yin et al., 2006) and Bospherus (Savran et al., 2008) focused on fixed captures at peaks of expressions or AUs. Soon after, dynamic analysis of 3D FEs started gaining interest using, for instance, the extended versions of BU-4DFE (Yin et al., 2008) and BP4D+ (Zhang et al., 2016) datasets. An illustration of these benchmarks is given in Fig. 9.12.

In relation to pose estimation, many datasets also included 2D and/or 3D landmarks. This was the case with 6 points offered in 2D coordinates within Eurecom-Kinect dataset allowing a good separability by genre (Min et al., 2014). Other recent dynamic 4D datasets such as B4PD (Zhang et al., 2014) incorporated 41 landmark annotations on top of both RGB images and 3D dynamic models.

The largest set of static acquisitions is actually found within the MeIn3D dataset (Booth et al., 2018) with 9663 subjects and a total of 12K 3D scans. It has served for the definition of the biggest morphable model till now. More recently, a 1.8M scale suitable for training deep-learning architectures has been provided for the first time under the 4DFAB multi-modal 3D dataset. Including 180 subjects, it holds spontaneous and posed expressions with applications related to both biometric and behavioral analysis.

Other interesting sets of 3D data include those considering IR thermal information as they are un-vulnerable to ambient lighting (Li and Deng, 2018). Example datasets include the Natural Visible and Infrared facial Expression database (NVIE (Wang et al., 2010)), the Kotani Thermal Facial Emotion (KTFE (Nguyen et al., 2014)) and the richer 4D-dynamic BP4D+ dataset with additional physiological sensors measuring blood pressure, heart and respiration rates (Zhang et al., 2016).

9.4.4 Multi-modal Approaches for 3D FEs Recognition

We distinguish hereafter between static and dynamic multi-modal approaches.

Static 3D approaches for FEs: One of the first works to exploit the RGB+D for FEs recognition was proposed in Tsalakanidou et al. (2003). The features were extracted using PCA analysis from RGB and depth from the XM2VTS database, then nearest-neighbor classification was applied. The decision of belonging to a given class (e.g. happiness, fear, etc.) was obtained by a score fusion process multiplying 2 euclidean distances obtained from each modality (Turk and Pentland, 1991).

After face cropping and alignment, a large number of contributions have also considered the landmark positions to compute distance measures of facial variations (Sandbach et al., 2012). These landmarks could be projected on RGB and depth maps from 2D coordinates, or directly exploited in 3D coordinates.

Frequency-based feature extractors such as Gabor filters have been successfully applied on top of these positions for localized expressions description (Savran et al., 2008). Other baseline methods proposed extended versions of the LBP descriptors with the BU-3DFE and Eurecom Kinect datasets (Yin et al., 2006; Min et al., 2014). An extension of RGB Multi-LBP maps (Levi and Hassner, 2015), to a richer set of LPB maps deduced from 2D+3D combinations has been presented in (Fu et al., 2019). By applying tensor low-rank modeling, both dimensionality reduction and classification were made with 82.9 and 75.9% accuracies on BU-3DFE and Bospherus datasets, respectively.

A different set of methods relied rather on automatic discretization into facial curves or interest points (Sandbach et al., 2012). This was the case with the SIFT-based features extractor proposed by Berretti et al. in (2011). A wider comparison of different 3D facial datasets and approaches using BoVW techniques such as SIFT is found in Ouarda et al. (2015). A related contribution optimizing the space of possible gradients between 3D facial poses to generate a 3D facial morphing model is also presented in Li et al. (2010).

Logically, more recent approaches have been considering deep neural architectures. Using accurate facial parts localization for the brows, eyes, nose and mouth, a fusion of 4 dedicated CNN paths has been applied in Jan et al. (2018) leading to 88.5% in expressions recognition accuracy on BU-3DFE. A fusion at the representation level of the CNN features produced from RGB and depth paths has been presented with an accuracy of 89.3% on the same dataset in Oyedotun et al. (2017). Due to noise in eyes regions, it has been shown that fusion of 2D and 3D facial CNN features improves performance. A fusion of CNN paths from depth, RGB, curvature and the 3 X/Y/Z normal maps has been proven efficient on subsets of the BU-3DFE and Bosphrus datasets in Li et al. (2017). More recently, data focused methods have also been successfully applied. Trimech et al. 2017 reached 94.9% accuracy by doubling the BU-3DFE dataset's size using data-augmentation with the coherent-point-drift algorithm and deep neural networks for learning and classification.

Dynamic 4D approaches for FEs: Classic methods rely on description of all data frames within a sequence, then on per-frame classification for a decision on the

Table 9.6 Static and dynamic 3D FEs datasets Sandbach et al. (2012); Corneanu et al. (2016)

Year	Dataset	References	Nb. Faces	Content description
Approaches based on dynamic RGB+4D models				
2018	4DFAB	Cheng et al. (2018)	180	1.8M 3D meshes, posed and spontaneous videos
2016	BP4D+	Zhang et al. (2016)	140	10 emotions, 2D, 3D, thermal, physiological sensors
2016	BAUM-1	Zhalehpour et al. (2017)	31	1502 emotions videos, frontal stereo + mono lateral
2014	BP4D	Zhang et al. (2014)	41	Dynamic FACS emotions, 3D and 2D landmarks
2014	3D-dynamic-DB	Alashkar et al. (2014)	58	Dynamic 3D faces, 15 fps, 4K vertex, pose free
2012	Hi4D-ADSIP	Matuszewski et al. (2012)	80	3360 high-resolution sequences, 14 expressions
2011	D3DFACS	Cosker et al. (2011)	10	100 3D frames per subject, 38 AUs
2008	BU-4DFE	Yin et al. (2008)	101	60K 3D frames, 606 sequences, 35K vertex par model
Approaches based on static RGB+3D models				
2017	MeIn3D	Booth et al. (2018)	9663	12K 3D facial scans
2016	FERG-DB	Aneja et al. (2016)	6	55767 annotations on 3D cartoon emotion faces
2014	KTFE	Nguyen et al. (2014)	26	Thermal+ RGB posed and spontaneous expressions
2013	Eurecom-kinect	Min et al. (2014)	52	Fixed RGB-D Kinect content with 6 annotated pts
2013	Biwi-kinect	Fanelli et al. (2013)	20	Head pose and kinect RGB-D
2010	NVIE	Wang et al. (2010)	100	Thermal+ RGB 6 expressions
2010	Biwi-3D	Fanelli et al. (2010)	20	Emotions and RGB-D
2010	Texas-3D	Gupta et al. (2010)	105	RGB-D; Facial recog. with 25 annotated pts
2008	York-3D	Heseltine et al. (2008)	350	3D content with textures
2008	Bosphorus	Savran et al. (2008)	105	3D model with textures and 24 annotated pts

(continued)

Table 9.6 (continued)

Year	Dataset	References	Nb. Faces	Content description
2008	ETH	Breitenstein et al. (2008)	20	Head pose and depth
2007	CASIA	Zhong et al. (2007)	123	4624 3D scans, ~37 poses/illum./expression
2007	Gavdb	Sandbach et al. (2012)	61	427 3D faces, 3 expressions
2006	BU-3DFE	Yin et al. (2006)	100	6 3D facial exp., 100 subjects, 83 annotated pts
2003	XM2VTS	Messer et al. (2003)	295	speaking+rotating RGB sequences + a 3D model

whole sequence label based on majority votes. BoVW-based methods offer additional robustness in this case as they are able to encode the expression using the most representative codebooks. Related to BoVWs, the FisherFaces representation allowed 82.3% performance for the dynamic expressions recognition task on top of the Hi4D-ADSIP dataset (Matuszewski et al., 2012).

On top of the BP4D+ dynamic dataset, different other baseline methods have been evaluated separately on each modality. For instance, the use of Gabor features and ASM refinement led to 91.6% 3D landmarks tracking (with 10% tolerance error) on top of the thermal data of 60 random subjects. Using another similar random subset, 91% was obtained using SIFT features and SVM classifiers for expressions recognition.

In addition to expressions description, many representation-based approaches (PCA in particular) are extensively used for 3D datasets analysis and the generation of 3D parametric models suitable for dynamic 3D animations. By extracting the K prominent components representing the data, it is then possible to control the dynamic creation of synthetic 3D faces with their relative textures as in Booth et al. (2018) or facial-animation blendshape transitions as in Cheng et al. (2018).

A representative solution for dynamic FEs recognition and 3D avatars animation using Kinect streams has been proposed by Weise et al. in (2011). Their assisted learning algorithm prompts users to record 19 different FEs then the RGB and depth informations are combined within a Gaussian mixture model. Using expectation-maximization, they were able to predict associations between the on-line streams and those recoded with robustness towards depth noise. A similar solution for 3D avatar animation using the 7 basic FACS emotions has been proposed later in Seddik et al. (2013). It relied on the PCA representations with the advantage of automatically learning the expressions to avoid user input. A more recent tentative to model 3D avatar's FEs using CNNs was proposed in Aneja et al. (2016). Though it used fixed 3D captures, it could easily be applied to dynamic 3D animations generation.

Recently, recurrent neural network of type LSTM have been encountering success for encoding the dynamic evolution of temporally spaced features. As features, the

PCA representations of the deformations with regard to the neutral state have been used in Cheng et al. (2018). They have led to a 70.1% expressions recognition rate on the 4DFAB large-scale dataset.

Beyond this chapter's scope, it is also interesting to consider voice combination with visual data for dynamic 3D FEs recognition. BAUM-1 is representative of a challenging set of recent benchmarks (e.g. RAVDESS in 2D (Livingstone and Russo, 2018)) for understanding dynamic vocalization. Using multi-path CNN fusions, a combination of sound and 3D visual data, the best performance obtained is 54.6% Zhalehpour et al. (2017). Thus, improvements are still possible.

9.5 Main Contributions Related to SL Recognition

Sign Language research works can be divided in two categories: (i) The first family of works is dedicated to the exhaustive enumeration of all gestures related to a given SL vocabulary. In this context we find a wide range of vocabulary datasets offering limited instances by action (von Agris et al., 2008). They are therefore rather oriented to the task of the data indexation. Their exploitation is then more often limited to web-based searches using keywords and meta-data (Athitsos et al., 2008). (ii) The second category of works concerns those oriented for the task of action recognition. In this context, it is worth mentioning that technologies based on markers such as colored gloves or sensor-based hand wearables (e.g. cybernetic glove) (Mohandes, 2013) allow very good performances. Nevertheless, considering visual and markerless gestures recognition, the task becomes more delicate. In this precise context related to computer vision, we find that the vocabularies become much smaller and that the number of samples per action are greedily increasing with the most recent datasets (Wan et al., 2016), thus allowing for deep-learning-based implementations.

9.5.1 SL Datasets

In this context, we distinguish two types of datasets: Those designed mainly as an actions' dictionary for a given SL, and those rather suitable to visual actions recognition. We detail both types hereafter.

9.5.1.1 SL Dictionary Indexation Datasets

Although a large number SL databases are already established, many are not suitable for computer vision recognition tasks. This is mainly due to uncontrolled acquisition conditions including inadequate labeling, reduced resolutions, variable view-angles, moving cameras, occlusion and noise (Ong and Ranganath, 2005). Many dataset creation efforts were rather targeting the categorization and indexation of the different sign gestures. Consequently, their common exploitations were related to viewers-accessibly and SL education purposes. Some limited contributions tried to extract video descriptive features, but rather focused on searching videos by content similarities (Athitsos et al., 2008).

Some more advanced efforts have involved the generation of long signing phrases from predefined gesture lexicons (von Agris et al., 2008). When exploiting computer avatars for this task, many interesting applications have been proposed:

(i) Under the ELAN project, a realistic 3D avatar performing American Sign Language (ASL) has been iteratively improved. Its last version offers realistic multi-person interactions and expressions (McDonald et al., 2016).
(ii) Other recent solutions are exploiting stylized 3D avatars for SL synthesis and interaction. Successful examples include the Mimix and Turjuman applications related to ASL and Arabic SL (ArSL) (Darawsheh and Al-Kilani, 2019).
(iii) The Hand-Talk solution is also offering a coverage for Brazilian SL (BrSL) (Tenório, 2019). As in Darawsheh and Al-Kilani (2019), it recently extended its avatar to perform as a Web-based SL translator.
(iv) Another contribution to the indexation of Tunisian Sign Language (TnSL) is presented by Bouzid and Jemni (2016). It combines 3D avatars with the written SL notation (Parkhurst and Parkhurst, 2010; Balti, 2013).

Interesting future extensions of the signing-capable avatars include their exploitation in social interactions and virtual-reality experiences as in Pan et al. (2016).

9.5.1.2 SL Visual-Recognition Dedicated Datasets

The analysis of Table 9.7 shows that the databases dedicated to visual SL recognition have evolved from fixed images (Menasy, 2001) and MoCap recordings (Gao et al., 2004) to include richer sets of isolated video sequences (von Agris et al., 2008), long and continuous signing phrases (Escalera et al., 2014) gradually towards deep-learning suitable scales (Koller et al., 2015; Wang et al., 2016). Some datasets followed multiple enrichments for theirs contents and labels. This is the case for CGC (Wan et al., 2016) and specially the RWTH-PHOENIX benchmark as one of the most cited for visual SL recognition. Actually, the richest SL-related content is held by the KETI dataset with near 15 K video sequences acquired by stereo cameras and performed by 14 signers (12 of them are native). It offers labels for 524 signs and 127 landmarks points for body, hands and face. Figure 9.13 illustrates example contents for the listed datasets.

From another point of view, we can distinguish different SL communities having been particularly active in recent years. A large number of advanced works have concerned the British Sign Language (BSL) (Pfister, 2015) from continuous or pre-segmented acquisitions relative to television streams (Kadir et al., 2004). Similarly, many other works were dedicated to the ASL (Athitsos et al., 2008; Koller et al., 2015). The Dicta-Sign project is one of the representative research actions facilitating Web communication with native SL speakers (Efthimiou et al., 2012). Promoted by

Table 9.7 Datasets and communities dedicated to the visual recognition of SL (Alon et al., 2009; von Agris et al., 2008; Ong et al., 2014)

Year	Research community: Dataset	References	Nbr. Sequences	Nbr. Signs
2019	**Korean sign language: KETI ⋆**	Ko et al. (2019)	**14,672 stereo seq.**	**524 by 14 signers + 124 face/body pts**
2019	Tunisian sign language: Icharati-Sawti	International Foundation for Electoral Systems (2019)	200 electoral phrases	∞ by 7 signers
2018	South African sign language: Real-SASL	for the Deaf (2001)	Multiple seq. per lexicon	800 on-line submissions
2016	Arabic sign language: ArSL	Aliyu et al. (2016)	400 seq.	20
2015	British sign language: BBC-Extended	Pfister (2015)	92 long seq.	∞
2015	**German sign language: RWTH-PHOENIX-2014 ⋆**	Koller et al. (2015)	**7000 seq. total**	**1081 by 9 signers + ~41 face/body pts**
2015	French sign language: FSL	Zbakh et al. (2015)	2500 seq.	500
2015	Chinese sign language: Large-vocab.-SL	Wang et al. (2016)	3000 seq. total	1000
2015	Chinese sign language: Daily-SL	Wang et al. (2016)	1850 seq. total	370
2014	Italian sign language: CGC-2014	Escalera et al. (2014)	14K seq. total	20
2013	Italian sign language: CGC-2013	Escalera et al. (2013)	13K seq. total	20
2012	Greec sign language: GSL	Cooper et al. (2012)	4920 seq.	984
2011	Turkish sign language: TuSL	Altun and Albayrak (2011)	16 seq.	29
2011	Multiples sign vocabularies: CGC-2012	Guyon et al. (2014)	(One-Shot)	249
2010	British sign language: BBC	Buehler et al. (2010)	20 long seq.	100
2008	American sign language: ASLLVD	Athitsos et al. (2008)	~3000 seq. total	1200
2008	British sign language: BSL-corpus	von Agris et al. (2008)	8100 seq. total	263
2008	German sign language: SIGNUM	von Agris et al. (2008)	19K seq., 780 phrases	455 by 25 signers
2004	British sign language: BSL	Kadir et al. (2004)	1640 seq. total	164
2004	Chinese sign language: CSL	Gao et al. (2004)	61K Mocaps	5113
2001	Arabic sign language: Menasy	Menasy (2001)	~1600 images	~1600

∞ : Continuous and uncontrolled signs. (⋆) The 2 most trending datasets

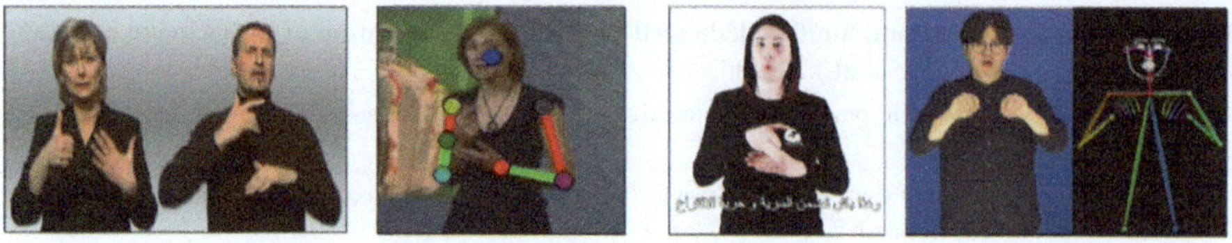

(a) RWTH-PHOENIX (b) BBC-Extended (c) Icharati-Sawti (d) KETI

Fig. 9.13 Examples of SL dataset contents in chronological order. Notice the availability of different modalities in addition to RGB

the Chalearn gesture challenges, many emerging works have exploited Italian SL (ISL) actions (Escalera et al., 2014).

Different other datasets have concerned other SL lexicons (von Agris et al., 2008). Example dictionaries include the ArSL (Mohandes et al., 2014), German SL (GeSL), (Koller et al., 2015), Chinese SL (CSL) (Wang et al., 2016), French SL (FSL) (Zbakh et al., 2015), Greek SL (GSL) (Cooper et al., 2012), Turkish SL (TSL) (Auephanwiriyakul et al., 2013), etc. Our approaches analysis in next section will consider their relative contributions.

One interesting recent dataset for Tunisian SL has been published under the name Icharati-Sawti (2019). It targets 200 electoral concepts and has the specificity of offering long explanation phrases in Tunisian SL with text and voice transcripts. Interesting paths for contribution include: (i) visual signs recognition and (ii) the automatic learning of associations between text/audio annotations and visual streams, as proposed by Bowden et al. in (2009a).

We highlight in this context that the normalization of SL dictionaries through the creation of exhaustive visual datasets remains an open task for many languages (Pietro and Nicole, 2016). This is specially the case as many SLs are actually mixtures of other lexicons (e.g. Tunisian SL contains among others ISL, FSL and ArSL (Balti, 2013)). Additional influencing factors include: (i) Vocabulary dissimilarities between younger and older aged signers, then (ii) geographic lexicon variations within a same country (Pietro and Nicole, 2016). One solution that could help solving this problem is the call to crowd-sourcing for both data collection and labeling. This is the case with the RealSASL South African dataset (for the Deaf, 2001). On-line video submissions are accepted, then a validation procedure is applied afterwards by expert votes. They have been able to collect 800 sign video sequences from an initial dictionary composed of 6500 language lexicons (for the Deaf, 2001). Applying this method is an interesting path for many other SL datasets.

9.5.2 SL Visual-Recognition Based Works

Despite notable efforts by some research groups, the visual recognition of SLs remains less developed than the other branches of computer vision. Since the first steps engaged in this field under non-constrained (i.e. in the wild) and constrained experimental environments (von Agris et al., 2008; Starner and Pentland, 1995), this task encountered multiple challenges mainly related to body-parts segmentation and tracking. This is why the majority of the initial works focused on the use of colored and cybernetic gloves to circumvent members tracking problems.

Later, the apparition of multi-modal/depth-capable sensors (e.g. Kinect), in addition to the advances in RGB streams description (Nguyen et al., 2019) and in human pose estimation (Zhang and Shah, 2015), allowed the development of interesting recent contributions. In what follows, we distinguish two categories of visual approaches: (i) The research-based contributions and (ii) the commercial solutions.

9.5.2.1 Research-Works Related to SL Recognition

As listed in Table 9.7, the SL visual recognition works can be grouped into specialized communities. We describe hereafter their main contributions.

Works of Zisserman et al.: Several contributions have been proposed by Zisserman and his team since 2004 for BSL recognition. A good number of them have focused on mono-modal recognition from television streams dedicated to the Deaf community (Kadir et al., 2004; Buehler et al., 2008; Pfister, 2012). These first contributions were kept unimodal and focused on body-members segmentation. Logically, the authors concentrated their next efforts in a number of recent works on the actor's pose estimation (generating the joints positions) using hands/arms images labeled with ground-truth positions. Thus, they benefited from RGB and joints modalities during actions recognition (Pfister et al., 2013, 2014). Their latest advances followed the trend of large scale learning. This allowed their latest arms pose-estimator to reach an accuracy of 95.6% within a 6-pixels reach from ground-truth (Charles et al., 2016. They have used for this task a personalized CNN model that adapts to the occlusions found in each sequence.

Works of Bowden et al.: Working on their turn on multiple lexicons, Bowden et al. published many contributions related both to SL and HAR. In order to overcome the lack of large-scale databases dedicated to SL recognition, Cooper and Bowden proposed in Cooper and Bowden (2009b) an approach capable of learning within the special conditions of reduced dataset sizes or ground truth absence. They have used pseudo-Haar volume filters to capture the sign actions from arbitrary temporal volumes. An extension of this approach to partially labeled sequences has been proposed in Cooper and Bowden (2009a). It exploited the textual annotations offered within SL television newscasts. In their recent contributions, Bowden et al. incorporated CNN architectures for the recognition of localized actions relative to different body parts: The mouth in Koller et al. (2015), then the hands in Cihan Camgoz et al. (2017) and more recently for full SL translation (Cihan Camgoz et al., 2018).

For mouth shapes learning, they have forwarded the CNN soft-max outputs to an HMM model inspired from speech recognition models to reach 55.7% as precision on a subset of 168K frames coming from the RWTH-PHOENIX dataset (Koller et al., 2015). For hand-shape recognition, they have been able to propose the 80.3% top-1 and 93.9% top-5 state-of-art SubUNets architecture on the one-million-hands dataset (Camgoz et al., 2017). Their approach relied on: (i) The CaffeNet CNN architecture for spatial description, (ii) on Bidirectional LSTMs (BLTSM) for temporal modeling and (iii) on a 12GB hardware GPU configuration that maximizes the frames loaded into GPU memory. One key contribution of BLSTMs is that they offer a better knowledge on the whole sequence by fusion of two opposite direction LSTM temporal layers.

Lower performances were obtained for continuous SL translation using SubUNets. For this purpose, they have proposed an improved approach that, as within Natural Language Processing (NLP) methods, accounts for grammar and words ordering for SL decoding. The evaluation process of their hybrid CNN-RNN-HMM model has reported 45.5 and 19.3% performance under the ROUGE and BLEU-4 respective scores on a large specifically-annotated subset of RWTH-PHOENIX (Cihan Camgoz et al., 2018). As there are many equally valid translations within a given language, the ROUGE and BLEU-1,2,3,4 measures are used within the newly emerging field of machine translation to describe the performance at different phrase scales (Cihan Camgoz et al., 2018).

Chalearn challenge related Works: The Chalearn community has been very active in recent years for the creation of competition benchmarks related to SL actions, facial emotions and human behaviors (Escalera, 2019). In its first CGC-2011 benchmark, the related competition proposed a large number of gestures belonging to multiple vocabularies (e.g. Taekwondo referees' actions, basketball, Indian SL, numbers, signs of divers, baby gestures, etc.) (Guyon et al., 2014). The dataset offered limited numbers of gestures (around 10 for each vocabulary) made by non-native actors for one-shot learning goals. Thus, the research teams were forced to use 1 training sequence per action.

In CGC-2013 and CGC-2014 datasets, the competitions proposed the first large scale human-actions database with 12K video sequences. The data inputs have been enriched to become multi-modal (from the Kinect sensor). The performance evaluation requested also the temporal segmentation from continuous streams (Escalera et al., 2013, 2014). Later, the larger 2016 ConJD and IsoGD benchmarks combined formatted content from previous CGC datasets. In Duan et al (2018) an IsoGD related contribution reached 67.3% accuracy by combining the decisions of 3 different temporal and spatial CNNs for both the RGB and depth modalities. A better performance of 76.1% has been reported in Jing et al. (2019) using 5 CNN paths adding the hands and face extracted streams. Many other challenges, specially related to facial recognition applications are still actually active within the Charlean community (Escalera, 2019).

Other recent works: The development of new solutions/devices assisting Deaf or hard of hearing people for SL communication is already receiving the attention from dedicated conferences such as M-enabling (Hazan-Tessler et al., 2019). They allow SL promotion and the creation of services adapted to the concerned community.

At the spatial features level, having a set of joint positions for the body, face and/or hands, an open path for contribution is the extraction of high-level features from landmarks (i.e. joints) and their combination with the convolutional features

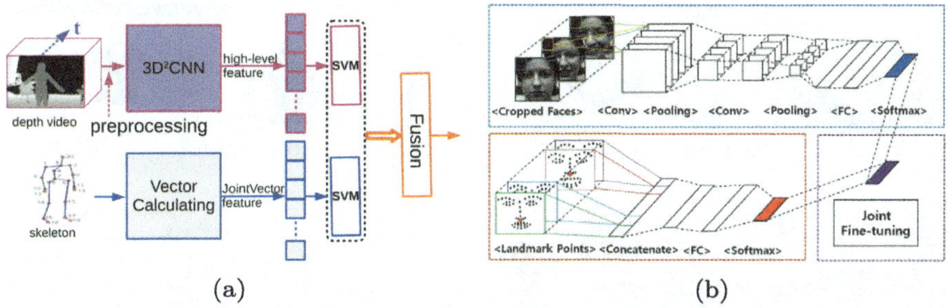

Fig. 9.14 Example approaches merging skeletal (high-level) and spatial (low-level) modalities within deep architectures: **a** Body-related and **b** face-related

given by CNNs. This is particularly the case for Liu et al. (2016) merging the depth and skeleton body-relative modalities. A similar contribution is proposed by Jung et al. by combining the RGB and landmark facial modalities (Jung et al., 2015). Figure 9.14 illustrates their respective contributions.

Many other recent research efforts have focused on modeling the temporal aspects of sign actions both for jointly their automatic description and segmentation. This could be made using a sequential pattern of interval descriptors and a decision-tree classifier as in Ong et al. (2014), or calling to newly specialized recurrent networks as in Molchanov et al. (2016). A solution avoiding temporal segmentation in the case of continuous streams is proposed in Huang et al. (2018).

The use of 2D/3D avatars for communication is additionally on of the very in-demand solutions for SL communication. Efthimiou et al. have proposed a recognition based on HOG/HOF descriptors in the first communication direction (Deaf → speaking person) and through the use of an avatar character in the reverse direction (speaking → Deaf person). The recent advances as in European Union of the Deaf et al. (2019) are offering increasingly realistic interactive characters.

Finally, it is a fact that the combination of different modalities is a major performance improver. As recent methods are enabling the estimation of in-the-wild depth maps for the classically challenging non-rigid human bodies (Li et al., 2019), there is place for major improvements in the actual human pose estimators (Cao et al., 2017; Güler et al., 2018). As interesting are their relative results, the methods having exploited them are already indicating the need for next generation more efficient pose estimators (Ko et al., 2019). In this case, the progress made in the generation of depth maps can bring the required multi-modal boost in SL visual recognition performances.

9.5.2.2 Commercial Solutions Related to SL Recognition

The first successful SL recognition solutions focused on the use of sensors attached to the actor's upper body. The use of colored gloves or marker-based motion sensors on

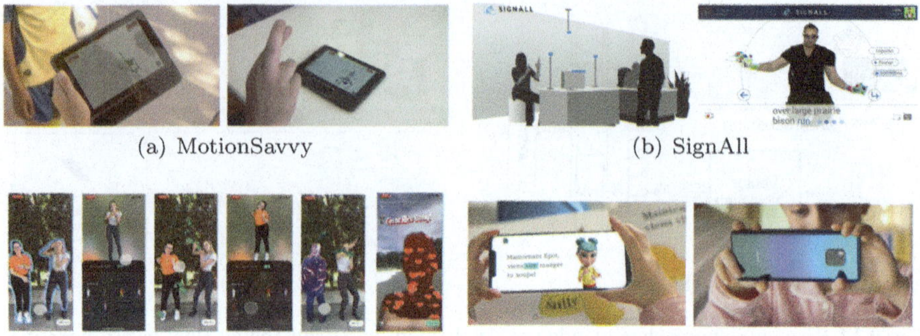

(a) MotionSavvy (b) SignAll

(c) OCTI (d) StorySign

Fig. 9.15 Examples of commercial solutions using gesture recognition

the face, body and fingers allowed effective and accurate motion tracking (Mohandes, 2013; European Union of the Deaf et al., 2019). They relied on multiple camera views with short and separate sequences of selected vocabularies. As such approaches require heavy hardware configurations, their applicability to real-world scenarios was kept limited (von Agris et al., 2008).

The main challenges encountered by commercial solutions are relative to the amount of sensors to exploit and the recognition effectiveness outcome. We present hereafter 4 active commercial solutions exploiting visual informations for the creation of Deaf-dedicated services:

(i) There is currently a solution called *MotionSavvy* (Opalka and Miller-Jacobson, 2018) targeting the American ASL users. It is based on the Leap Motion sensor packaged within a tablet interface (see Fig. 9.15a). It allows both signs to text/voice translation (first direction) and voice to text recognition (second direction). Although taking advantage of the infra-red-generated depth stream of the sensor, it does not consider FEs. It also limits the actor's movements within a small space where both hands are accessible to the sensor. Therefore, it remains prone to future improvements.

(ii) Another commercial solution based on the Kinect sensor is currently distributed under the *SignAll* label (Robotka et al., 2018). It offers significantly more freedom for the actor, takes into account FEs and hand gestures through colored light-weight hand gloves. Though its first iterations suffered from relatively limited performances, it holds an active development team. Shown in Fig. 9.15b, its latest version exploits the improved Kinect-v2 in conjunction with multiple cameras, sound sensors and natural language processing for sentences generations. This product allows a Deaf agent to communicate through a terminal with speaking persons using ASL and Hungarian SL. Thus additional contributions are possible for other SL lexicons.

(iii) For what concerns RGB-only-based action recognition, the early solutions functioned within constrained environments. In this case, they required fixed and easily removable backgrounds. Interestingly, new deep-neural based solutions such as the *OCTI* application (Martin, 2018) (see Fig. 9.15c) are starting to appear on last smart-phone generations to allow background removal and basic actions recognition for consumer visual affects. This new trend of smart-phone applications is expected to continue improving in coming years specially with the recent advances on body, face and hands poses estimation (Zhang and Shah, 2015; Belagiannis and Zisserman, 2017; Haque et al., 2016; Güler et al., 2018).

(iv) One of the recent contributions joining the efforts of European, British and Australian Deaf organizations resulted in the *StorySign* mobile application. It has been released in early 2019 in collaboration with Huawei AI (European Union of the Deaf et al., 2019) (see Fig. 9.15d) to facilitate SL based education. Its originality consists in the use of state-of-the-art FEs recognition and motion capture technologies to allow high quality facial, hand and body performances by the 3D friendly avatar. As the amount of stories per SL are still limited within the *StorySign* solution, this specific task of content creation and formating still remains wide open for new contributions.

9.6 Conclusion and Discussion

We have presented in the previous sections a review of HAR methods with consideration to their specific exploitation for visual SL recognition. We have followed a classification based on their input data-modalities. For each of the joint, RGB and depth input streams, we have listed the most used dataset contents and approach details. We have also accorded attention to the temporal segmentation task, in addition to the static and dynamic FEs recognition due to their importance in SL interpretation. Next, we detailed the new trend of multi-modal datasets and their relative contributions.

In each section, we have been able to: (i) Discuss the solutions advantages, (ii) report the current best performances and (iii) highlight the most promising improvement paths. We finally payed special interest to the different working groups and commercial solutions targeting the Deaf and hearing-impaired community.

We summarize hereafter the main contribution paths that we recommend at the levels of (i) datasets, (ii) approaches and (iii) commercial solutions for the goal of SL recognition.

9.6.1 Datasets Level

From previous dataset listings, appears the need for multi-modal, large-scaled and SL-specific benchmarks. To our best knowledge, the nearest multi-modal dataset to this goal is the KETI dataset proposed by Ko et al. in (2019) focusing on Korean SL and offering stereo RGB + landmarks streams. Though it relies on the OpenPose

(Cao et al., 2017) estimator to generate its 127 landmarks, these positions allow only 55.3% accuracy on the validation set for a limited selection of 105 sentences. Its creators already place as perspective the call to more robust pose estimators and the enrichment to larger-scaled recordings (Ko et al., 2019). The nearest multi-modal and large-scaled (13K sequences) dataset is CGC-2014. But it has the limitation of offering only 20 signs. Though many SL-dedicated RGB datasets might be suitable for large-scaled learning with appropriate data-augmentation techniques (Molchanov et al., 2016) (e.g. RWTH-PHOENIX dataset with 7000 sequences and 1081 sign labels), they remain either unimodal, or lack robust landmark annotations. Thus they remain open to contributions applying enriched formatting and labeling. This explains, for instance, the different extensions applied to the RWTH-PHOENIX dataset (Koller et al., 2015).

The design of any new SL dataset is preferred to be multi-modal, large-scaled and offering body, face and hands landmarks. Knowing that SL lexicons have thousands of gestures and follow different grammar-rules than spoken languages, the annotations should target the largest affordable lexicon coverage, both in isolated and continuous spontaneous streams. Also, as SL lexicons are constantly changing, it is preferable to record native SL signers after consensus on the dictionary to adopt. This was for instance the case for the relatively-small Tunisian electoral SL dataset proposed in International Foundation for Electoral Systems (2019). Due to the logistic complexity of this task, specially if multi-view conditions were to considered as in Shahroudy et al. (2016), this task remains a challenge to overcome by coming contributions.

9.6.2 Approaches Level

After a relative prevalence for hand-crafted features in combination with BoVW representations, the capability of deep CNNs to learn automatic descriptive features from massive data offered a major performance leap in computer vision (Krizhevsky et al., 2012). From this perspective, a majority of community efforts have been oriented towards the collection of large-scaled datasets targeting actually the 1M samples and 1K labels (Kollias et al., 2019; Mollahosseini et al., 2019; Monfort et al., 2019; Karpathy et al., 2014). This is an additional argument encouraging the creation of large SL-dedicated benchmarks as they will automatically lead to better performances using the actual CNN architectures. Next, we categorize the future promising improvements at the approach's level.

Streams preprocessing: Many intuitive contributions have successfully considered the preprocessing of input streams into more CNN-suitable formats to improve their accuracies. Example manipulations include the extension to a more-dimensional formatting (e.g. convolution along orthogonal views of volumetric videos (Li et al., 2019)) or the de-correlation of input modalities to a more easily-processable formatting (e.g. LBP descriptors for faces (Li and Deng, 2018)).

Multiple neural paths fusion: Considering that the incorporation of multi-modal fusion paths is a performance improver (Neverova et al., 2016; Seddik et al., 2017), many recent contributions have combined two-streamed (and more) convolutional network paths for the description of both temporal and spatial features (Wang et al., 2018). The two neural streams could fully or partially be dedicated to the description of action's evolution in time. This is the case with the dedication of one neural path for the motion flows in Simonyan and Zisserman (2014). The CNN paths could also be relative to different spatial modalities as employed by Duan et al. (2018) for the fusion of two 3D convolutional networks (i.e. from space-time volumes of pixels) one for RGB and the other for depth. Other joint-based contributions have also considered merging two LSTM paths: One for the joints description and the other for extracting the kinematic relations in-between isolated joint groups (Si et al., 2019). This can be seen as a deep neural equivalent to the Joint-Quadruplets (Evangelidis et al., 2014) and the Moving-Pose (Zanfir et al., 2013) hand-crafted features.

Better neural description architectures: The last comparison reveals the actual orientation towards the creation of deep neural networks automating previously existing descriptors. Interestingly, even a spatiotemporal CNN-based equivalent to the widely used STIP descriptor of Laptev et al. (2008) has been recently claimed in Li et al. (2019). Some specific features such as motion trajectories, optical flows, joint shape contexts, etc. remained for a while absent within the CNN produced features. This resulted in a first set of contributions combining BoVW and CNN decisions for better performances (Wang et al., 2015; Liu et al., 2016). Within next approaches, the temporal descriptors started to be modeled within the deep neural architectures using for instance RNN models and their optimized LSTM implementations. This allowed for the successful handling of both temporal segmentation and sign actions recognition tasks in Pigou et al. (2016) and Neverova et al. (2016). And though the spatial and temporal behaviors of videos have received literature attention, there is still place for incorporating the contextual and structural information offered for instance by shape-context features (Mansouri et al., 2015) within the deep neural architectures (Wang et al., 2018).

Adaptation to challenging training conditions: The capability of actual CNN methods to deal with small, weakly-supervised or even previously unseen data within the learning population (Guyon et al., 2014; Damen et al., 2018; Ghadiyaram et al., 2019) is also an interesting challenge. As transfer-learning has been an attractive research topic in recent years (e.g. through SVM classifiers specialization (Mâamatou et al., 2016)), more related contributions are to be expected within future CNN/RNN architectures. Next generation contributions will have to automatically adapt their recognition schemes to the modality-types used as in Neverova et al. (2016), and also, to the scenes' contexts as in Charles et al. (2016).

Actions prediction in real-time: Finally, just as humans do, the optimal SL recognition solution would inevitably function in real time with the highest reliability. To do so, humans are capable of expecting the next coming behaviors by analyzing the actual attitudes of the target individuals. One promising path is the incorporation of language models as an equivalent component to the visual features extracted. Also, updating existing natural language processing methods to the SL grammar could help these predictions. This dual challenge of actions prevision and on-line multi-modal streams handling has been considered by some promising contributions:

(i) In Liang et al. (2019), the authors predicted both the interactions labels and motion trajectories happening between pedestrians and cars. They employed CNN to segment the scene contents and describe the surroundings of the persons found, then forwarded this to LSTM networks to encode the final scene-person features. Both classification and regression layers were used for robust predictions.

(ii) In Ke et al. (2019), Ke et al. obtained state-of-the-art actions prediction within short and long ranges on the EPIC-Kitchen dataset. They applied conditional neural connections skipping to keep only the prominent temporal features within a one-shot leaning scheme.

The adaptation of this technology to continuous SL recognition would enable understanding multi-persons signing interactions. The creation of such a service that transcripts in real-time SL discussions is awaiting for affiliated contributions.

9.6.3 Commercial Solutions Level

From the previous per-modality analysis, turns out that access to the depth stream is a performance enabler for real-time solutions. The access to better sensors with miniaturized sizes and wider ranges is expected in coming years (Yuan et al., 2017). Interestingly, as last-generation smart-phones are already equipped with IR depth sensors for facial authentication, pushing further their exploitation for FEs and SL recognition is very promising. With the recent mobile computation capabilities, there is also place for generating depth maps from the video's intrinsic motion as applied in (Valentin et al., 2018; Martin, 2018) for augmented reality experiences. Once the milestone of depth generation achieved, multi-modal SL implementations will be unlocked.

On another level, from the previously listed commercial solutions, appears the need for extending existing products (mostly focused on ASL and BSL (Robotka et al., 2018; Opalka and Miller-Jacobson, 2018)) to cover a wider range of SL lexicons. It is also possible to extend the actual robotic platforms (Todo, 2018; Marinoiu et al., 2018) to consider SL interactions. The versatility of SL communities (see Table 9.7) requires adapting the actual artificial intelligence models to learn multiple SL lexicons and provide more scalable services.

The Deaf and hearing-impaired communities have also numerous demands for dedicated interactive contents including: Educational scenarios as within the StorySign initiative (European Union of the Deaf et al., 2019), communication as aimed by the in-progress SignAll solution (Robotka et al., 2018) or even entertainment as within the OCTI mobile application (Martin, 2018). Still, to mention a few to our best knowledge, no proper SL automatic transcription services exist for TV broadcasts, movies or outdoor facilities. The concerned population's scale and economic valuation are by far bigger than the actual numeric services (World Health Organization, 2019).

References

Agarwal, M., Jain, N., Kumar, M. M., & Agrawal, H. (2010). Face recognition using eigen faces and artificial neural network. *International Journal of Computer Theory and Engineering, 2*(4), 624–629.

Aggarwal, J., & Xia, L. (2014). Human activity recognition from 3D data: a review. *Pattern Recognition Letters, 48*, 70–80. Celebrating the life and work of Maria Petrou.

Alashkar, T., Ben Amor, B., Daoudi, M., & Berretti, S. (2014). A 3D dynamic database for unconstrained face recognition. In *5th International Conferences and Exhibition on 3D Body Scanning Technologies*, Lugano, Switzerland.

Alekseev, A., & Bobe, A. (2019). Gabornet: Gabor filters with learnable parameters in deep convolutional neural networks. arXiv:1904.13204

Alippi, C., Boracchi, G., & Roveri, M. (2013). Just-in-time classifiers for recurrent concepts. *IEEE Transactions on Neural Networks and Learning Systems, 24*(4), 620–634.

Aliyu, S., Mohandes, M., Deriche, M., & Badran, S. (2016). Arabie sign language recognition using the microsoft kinect. In *IEEE Multi-Conferences on Systems Signals Devices* (pp. 301–306).

Alon, J., Athitsos, V., Yuan, Q., & Sclaroff, S. (2009). A unified framework for gesture recognition and spatiotemporal gesture segmentation. *IPAMI, 31*(9), 1685–1699.

Alon, J., Athitsos, V., Yuan, Q., & Sclaroff, S. (2009). A unified framework for gesture recognition and spatiotemporal gesture segmentation. *IEEE Transactions on Pattern Analysis and Machine Intelligence, 31*(9), 1685–1699.

Altun, O., & Albayrak, S. (2011). Turkish fingerspelling recognition system using generalized Hough transform, interest regions, and local descriptors. *Pattern Recognition Letters, 32*(13), 1626–1632.

Andriluka, M., Pishchulin, L., Gehler, P., & Schiele, B. (2014). 2D human pose estimation: New benchmark and state of the art analysis. In *The IEEE Conferences on Computer Vision and Pattern Recognition (CVPR)* (pp 3686–3693).

Aneja, D., Colburn, A., Faigin, G., Shapiro, L., & Mones, B. (2016). Modeling stylized character expressions via deep learning. In *Asian Conference on Computer Vision* (pp. 136–153). Springer.

Athitsos, V., Neidle, C., Sclaroff, S., Nash, J., Stefan, A., Yuan, Q., & Thangali, A. (2008). The American sign language Lexicon video dataset. In *2008 IEEE Computer Society Conferences on Computer Vision and Pattern Recognition Workshops* (pp. 1–8).

Auephanwiriyakul, S., Phitakwinai, S., Suttapak, W., Chanda, P., & Theera-Umpon, N. (2013). Thai sign language translation using scale invariant feature transform and hidden Markov models. *Pattern Recognition Letters, 34*(11), 1291–1298.

Ballas, N., Yang, Y., Lan, Z. Z., Delezoide, B., Prêteux, F., & Hauptmann, A. (2013). Space-time robust representation for action recognition. In *2013 IEEE International Conference on Computer Vision*, (pp. 2704–2711).

Balti, M. A. (2013). *120 configurations de la main*. The Sign Writing Press. http://www.signwriting.org/tunisia/.

Baveye, Y., Dellandré, E., Chamaret, C., & Chen, L. (2015). Deep learning versus kernel methods: Performance for emotion prediction in videos. In *2015 International Conference on Affective Computing and Intelligent Interaction (ACII)* (pp. 77–83).

Belagiannis, V., & Zisserman, A. (2017). *Recurrent human pose estimation* (pp. 468–475).

Belhumeur, P. N., Jacobs, D. W., Kriegman, D. J., & Kumar, N. (2013). Localizing parts of faces using a consensus of exemplars. *IEEE Transactions on Pattern Analysis and Machine Intelligence, 35*(12), 2930–2940.

Belongie, S., Malik, J., & Puzicha, J. (2002). Shape matching and object recognition using shape contexts. *IEEE Transactions Pattern Analysis and Machicne Intelligence, 24*(4), 509–522.

Benitez-Quiroz, C. F., Srinivasan, R., & Martinez, A. M. (2016). Emotionet: An accurate, real-time algorithm for the automatic annotation of a million facial expressions in the wild. In *2016 IEEE Conferences on Computer Vision and Pattern Recognition (CVPR)* (pp. 5562–5570).

Berretti, S., Ben Amor, B., Daoudi, M., & del Bimbo, A. (2011). 3D facial expression recognition using sift descriptors of automatically detected keypoints. *The Visual Computer, 27*(11), 1021–1036.

Betancourt, A., Morerio, P., Regazzoni, C. S., & Rauterberg, M. (2015). The evolution of first person vision methods: A survey. *IEEE Transactions on Circuits and Systems for Video Technology, 25*(5), 744–760.

Bettadapura, V. (2012). *Face expression recognition and analysis: The state of the art*. arXiv:1203.6722.

Blackburn, J., Kubacki, D., & Stratton, J. (2010). *Incremental 3D model generation using depth cameras*. Technical report, University of Illinois, Urbana.

Blanz, V., & Vetter, T. (1999). A morphable model for the synthesis of 3D faces. In *26th Annual Conferences on Computer Graphics and Interactive Techniques, SIGGRAPH '99*, New York, NY, USA (pp. 187–194). ACM Press/Addison-Wesley Publishing Co.

Booth, J., Roussos, A., Ponniah, A., Dunaway, D., & Zafeiriou, S. (2018). Large scale 3D morphable models. *International Journal of Computer Vision, 126*(2), 233–254.

Boualia, S. N., & Essoukri Ben Amara, N. (2019). Pose-based human activity recognition: a review. In *15th International Wireless Communications Mobile Computing Conference (IWCMC)* (pp. 1468–1475).

Boulbaba, B. A., Su, J., & Anuj, S. (2015). Action recognition using rate-invariant analysis of skeletal shape trajectories. In *IEEE Transactions on Pattern Analysis and Machine Intelligence* (pp 1–14).

Bouzid, Y., Khenissi, M. A., Essalmi, F., & Jemni, M. (2016). Using educational games for sign language learning—A signwriting learning game: Case study. *Educational Technology & Society, 19*(1), 129–141.

Breitenstein, M. D., Kuettel, D., Weise, T., van Gool, L., & Pfister, H. (2008). Real-time face pose estimation from single range images. In *IEEE Conferences on Computer Vision and Pattern Recognition* (pp. 1–8).

Buehler, P., Everingham, M., Huttenlocher, D. P., & Zisserman, A. (2008). Long term arm and hand tracking for continuous sign language TV broadcasts. *British Machine Vision Conferences*.

Buehler, P., Everingham, M., & Zisserman, A. (2010). Employing signed TV broadcasts for automated learning of British sign language. In *Workshop on the Representation and Processing of Sign Languages: Corpora and Sign Language Technologies* (pp. 22–23).

Camgoz, N. C., Hadfield, S., Koller, O., & Bowden, R. (2017). Subunets: End-to-end hand shape and continuous sign language recognition. In *IEEE International Conference on Computer Vision (ICCV)*.

Cao, Z., Simon, T., Wei, S.-E., & Sheikh, Y. (2017). Realtime multi-person 2D pose estimation using part affinity fields. In *The IEEE Conference on Computer Vision and Pattern Recognition (CVPR)* (pp. 7291–7299).

Carreira, J., Noland, E., Hillier, C., & Zisserman, A. (2019). A short note on the kinetics-700 human action dataset.

Carrier, P. L., & Courville, A. (2013). Challenges in representation learning: A report on three machine learning contests. In *Neural Information Processing*, Berlin, Heidelberg (pp. 117–124). Berlin, Heidelberg: Springer.

Chandrasekhar, V., Lin, J., Morère, O., Goh, H., & Veillard, A. (2015). *A practical guide to CNNS and fisher vectors for image instance retrieval*. arXiv:1508.02496.

Chaquet, J. M., Carmona, E. J., & Fernández-Caballero, A. (2013). A survey of video datasets for human action and activity recognition. *Computer Vision and Image Understanding, 117*(6), 633–659.

Charles, J., Pfister, T., Magee, D., Hogg, D., & Zisserman, A. (2016). Personalizing human video pose estimation. In *IEEE Conference on Computer Vision and Pattern Recognition* (pp. 3063–3072).

Chen, C., Jafari, R., & Kehtarnavaz, N. (2015). UTD-MHAD: A multimodal dataset for human action recognition utilizing a depth camera and a wearable inertial sensor. In *IEEE International Conference on Image Processing (ICIP)* (pp. 168–172).

Chen, L., Wei, H., & Ferryman, J. (2013). A survey of human motion analysis using depth imagery. *pattern recognition letters, 34*(15), 1995–2006. Smart Approaches for Human Action Recognition.

Chen, X., & Koskela, M. (2013). Classification of RGB-d and motion capture sequences using extreme learning machine. In J.-K. Kämäräinen, & M. Koskela (Eds.), *Image Analysis*, Berlin, Heidelberg (pp. 640–651). Berlin, Heidelberg: Springer.

Cheng, G., Wan, Y., Saudagar, A. N., Namuduri, K., & Buckles, B. P. (2015). Advances in human action recognition: A survey. arXiv:1501.05964.

Cheng, S., Kotsia, I., Pantic, M., & Zafeiriou, S. (2018). 4dfab: A large scale 4D database for facial expression analysis and biometric applications. In *2018 IEEE/CVF Conference on Computer Vision and Pattern Recognition* (pp. 5117–5126).

Chestnov, O., Mendis, S., Chadha, S., Mariotti, S. P., & Kocur, I. (2013). *Multi-country assessment of national capacity to provide hearing care*. Geneva, Switzerland: WHO Document Production Services.

Chrysos, G. G., Antonakos, E., Snape, P., Asthana, A., & Zafeiriou, S. (2018). A comprehensive performance evaluation of deformable face tracking "in-the-wild". *International Journal of Computer Vision, 126*(2), 198–232.

Cihan Camgoz, N., Hadfield, S., Koller, O., & Bowden, R. (2017). Subunets: End-to-end hand shape and continuous sign language recognition. In *The IEEE International Conference on Computer Vision (ICCV)* (pp 3056–3065).

Cihan Camgoz, N., Hadfield, S., Koller, O., Ney, H., & Bowden, R. (2018). Neural sign language translation. In *The IEEE Conference on Computer Vision and Pattern Recognition (CVPR)* (pp. 7784–7793).

Cippitelli, E., Gasparrini, S., Gambi, E., & Spinsante, S. (2016). A human activity recognition system using skeleton data from RGBD sensors. *Computational Intelligence and Neuroscience, 2016*, 14.

Cooper, H., & Bowden, R. (2009a). Learning signs from subtitles: A weakly supervised approach to sign language recognition. In: *IEEE Conference on Computer Vision and Pattern Recognition, 2009. CVPR 2009* (pp 2568–2574).

Cooper, H, & Bowden, R. (2009b). Sign language recognition: Working with limited corpora. In *International Conference on Universal Access in Human-Computer Interaction* (vol. 1, pp. 472–481). San Diego, CA, USA: Addressing Diversity.

Cooper, H., Ong, E.-J., Pugeault, N., & Bowden, R. (2012). Sign language recognition using subunits. *Journal of Machine Learning Research, 13*(1), 2205–2231.

Cootes, T. F., Edwards, G. J., & Taylor, C. J. (2001). Active appearance models. *IEEE Transactions on Pattern Analysis and Machine Intelligence, 23*(6), 681–685.

Corneanu, C. A., Simon, M. O., Cohn, J. F., & Guerrero, S. E. (2016). Survey on RGB, 3D, thermal, and multimodal approaches for facial expression recognition: History, trends, and affect-related applications. *IEEE Transactions on Pattern Analysis and Machine Intelligence, 38*(8), 1548–1568.

Cosker, D., Krumhuber, E., & Hilton, A. (2011). A facs valid 3D dynamic action unit database with applications to 3D dynamic morphable facial modeling. In *2011 International Conference on Computer Vision* (pp 2296–2303).

Damen, D., Doughty, H., Farinella, G. M., Fidler, S., Furnari, A., Kazakos, E., Moltisanti, D., Munro, J., Perrett, T., Price, W., & Wray, M. (2018). Scaling egocentric vision: The epic-kitchens dataset. In *European Conference on Computer Vision (ECCV)* (pp 720–736).

Darawsheh, M., & Al-Kilani, M. (2019). *Mimix, turjuman*. https://handtalk.me/.

De la Torre, F., Hodgins, J., Bargteil, A., Martin, X., Macey, J., Collado, A., & Beltran, P. (2008). *Guide to the Carnegie Mellon University multimodal activity (CMU-MMAC) database* (p. 15213). Technical report, Pittsburgh, PA.

Delaitre, V., Laptev, I., & Sivic, J. (2010). Recognizing human actions in still images: a study of bag-of-features and part-based representations. In *British Machine Vision Conference* (pp. 97.1–97.11). BMVA Press. https://doi.org/10.5244/C.24.97.

Deng, J., Roussos, A., Chrysos, G., Ververas, E., Kotsia, I., Shen, J., & Zafeiriou, S. (2019). The Menpo benchmark for multi-pose 2D and 3D facial landmark localisation and tracking (vol. 127, pp. 599–624).

Dhall, A., Goecke, R., Ghosh, S., Joshi, J., Hoey, J., & Gedeon, T. (2017). From individual to group-level emotion recognition: Emotiw 5.0. In *19th ACM International Conference on Multimodal Interaction, ICMI '17* (pp. 524–528). New York, NY, USA: ACM.

Dhall, A., Goecke, R., Lucey, S., & Gedeon, T. (2011). Static facial expression analysis in tough conditions: Data, evaluation protocol and benchmark. In *2011 IEEE International Conference on Computer Vision Workshops (ICCV Workshops)* (pp. 2106–2112).

Doosti, B. (2019). *Hand pose estimation: A survey*. arXiv:1903.01013.

Duan, J., Wan, J., Zhou, S., Guo, X., & Li, S. Z. (2018). A unified framework for multi-modal isolated gesture recognition. *ACM Transactions on Multimedia Computing, Communications, and Applications, 14*(1s), 21:1–21:16.

Efthimiou, E., Fotinea, S.-E., Hanke, T., Glauert, J., Bowden, R., Braffort, A., Collet, C., Maragos, P., & Lefebvre-Albaret, F. (2012). The dicta-sign wiki: Enabling web communication for the deaf. In: *13th International Conference on Computers Helping People with Special Needs-Volume Part II, ICCHP'12*, Berlin, Heidelberg (pp. 205–212). Springer.

Ekman, P., & Rosenberg, E. L. (1997). *What the face reveals: Basic and applied studies of spontaneous expression using the facial action coding system (FACS)*. Series in Affective Science Edition. Oxford, New York: Oxford University Press.

Escalera, S. (2019). *Chalearn gesture challenge (CGC)*. http://gesture.chalearn.org.

Escalera, S., Baró, X., González, J., Bautista, M., Madadi, M., Reyes, M., Ponce-López, V., Escalante, H., Shotton, J., & Guyon, I. (2014). Chalearn looking at people challenge 2014: Dataset and results. *ECCV Workshops* (pp. 459–473).

Escalera, S., Gonzàlez, J., Baró, X., Reyes, M., Lopes, O., Guyon, I., Athitsos, V., & Escalante, H. (2013). Multi-modal gesture recognition challenge 2013: Dataset and results. *ICMI* (pp. 445–452).

European Union of the Deaf, Deaf Australia Inc, and the British Deaf Association (2019). Huawei storysign. https://consumer.huawei.com/en/campaign/storysign/.

Evangelidis, G., Singh, G., & Horaud, R. P. (2014). Continuous gesture recognition from articulated poses. In *European Conference on Computer Vision Workshops*, Zurich, Switzerland (pp. 595–607).

Everingham, M., Eslami, S. M. A., Van Gool, L., Williams, C. K. I., Winn, J., & Zisserman, A. (2015). The pascal visual object classes challenge: A retrospective. *International Journal of Computer Vision, 111*(1), 98–136.

Everingham, M., Van Gool, L., Williams, C. K. I., Winn, J., & Zisserman, A. (2010). The pascal visual object classes (VOC) challenge. *International Journal of Computer Vision, 88*(2), 303–338.

Fanelli, G., Dantone, M., Gall, J., Fossati, A., & Van Gool, L. (2013). Random forests for real time 3D face analysis. *International Journal of Computer Vision, 101*(3), 437–458.

Fanelli, G., Gall, J., Romsdorfer, H., Weise, T., & Gool, L. V. (2010). A 3-D audio-visual corpus of affective communication. *IEEE Transactions on Multimedia, 12*(6), 591–598.

Fasel, B., & Luettin, J. (2003). Automatic facial expression analysis: A survey. *Pattern Recognition, 36*(1), 259–275.

Felzenszwalb, P. F., Girshick, R. B., McAllester, D., & Ramanan, D. (2010). Object detection with discriminatively trained part-based models. *IEEE Transactions Pattern Analysis and Machine Intelligence, 32*(9), 1627–1645.

Firman, M. (2016). RGBD datasets: Past, present and future. *CVPR Workshop on Large Scale 3D Data: Acquisition, Modelling and Analysis* (pp. 19–31).

Deaf, D. I. (2001). *For the real South African sign language*. https://www.realsasl.com.

Fu, Y., Ruan, Q., Luo, Z., Jin, Y., An, G., & Wan, J. (2019). Ferlrtc: 2D + 3D facial expression recognition via low-rank tensor completion. *Signal Processing, 161*, 74–88.

Ganapathi, V., Plagemann, C., Koller, D., & Thrun, S. (2012). Real-time human pose tracking from range data. In A. Fitzgibbon, S. Lazebnik, P. Perona, Y. Sato, & C. Schmid (Eds.), *Computer vision–ECCV 2012*, Berlin, Heidelberg (pp. 738–751). Berlin Heidelberg: Springer.

Gao, W., Fang, G., Zhao, D., & Chen, Y. (2004). A Chinese sign language recognition system based on SOFM/SRN/HMM. *Pattern Recognition, 37*(12), 2389–2402.

Garcia-Hernando, G., Yuan, S., Baek, S., & Kim, T.-K. (2018). First-person hand action benchmark with RGB-D videos and 3D hand pose annotations. In *Proceedings of Computer Vision and Pattern Recognition (CVPR)*.

Georghiades, A., Belhumeur, P., & Kriegman, D. (2001). From few to many: Illumination cone models for face recognition under variable lighting and pose. *IEEE Transactions Pattern Analysis and Machine Intelligence, 23*(6), 643–660.

Ghadiyaram, D., Tran, D., & Mahajan, D. (2019). *Large-scale weakly-supervised pre-training for video action recognition* (pp. 12046–12055).

Ghanem, B., Niebles, J. C., Snoek, C., Heilbron, F. C., Alwassel, H., Krishna, R., Escorcia, V., Hata, K., & Buch, S. (2017). *Activitynet challenge 2017 summary*. arXiv:1710.08011.

Gong, D., Medioni, G., Zhu, S., & Zhao, X. (2012). Kernelized temporal cut for online temporal segmentation and recognition. In: *12th European Conference on Computer Vision—Volume Part III, ECCV'12*, Berlin, Heidelberg, (pp. 229–243). Springer.

Gorelick, L., Blank, M., Shechtman, E., Irani, M., & Basri, R. (2007). Actions as space-time shapes. *Transactions on Pattern Analysis and Machine Intelligence, 29*(12), 2247–2253.

Gross, R., Matthews, I., Cohn, J., Kanade, T., & Baker, S. (2008). Multi-pie. In *2008 8th IEEE International Conference on Automatic Face Gesture Recognition* (pp. 1–8).

Güler, R. A., Neverova, N., & Kokkinos, I. (2018). *Densepose: Dense human pose estimation in the wild* (pp. 7297–7306).

Guo, G., & Lai, A. (2014). A survey on still image based human action recognition. *Pattern Recognition, 47*(10), 3343–3361.

Guo, J., Deng, J., Xue, N., & Zafeiriou, S. (2018). *Stacked dense u-nets with dual transformers for robust face alignment* (p. 44).

Guo, Y., Liu, Y., Oerlemans, A., Lao, S., Wu, S., & Lew, M. S. (2016). Deep learning for visual understanding: A review. *Neurocomputing, 187*, 27–48. Recent Developments on Deep Big Vision.

Gupta, S., Markey, M. K., & Bovik, A. C. (2010). Anthropometric 3D face recognition. *International Journal of Computer Vision, 90*(3), 331–349.

Guyon, I., Athitsos, V., Jangyodsuk, P., & Escalante, H. (2014). The Chalearn gesture dataset (CGD 2011). *Machine Vision and Applications, 25*(8), 1929–1951.

Guyon, I., Athitsos, V., Jangyodsuk, P., Escalante, H., & Hamner, B. (2013). Results and analysis of the Chalearn gesture challenge 2012. Lecture notes in computer scienceIn X. Jiang, O. Bellon, D. Goldgof, & T. Oishi (Eds.), *Advances in depth image analysis and applications* (Vol. 7854, pp. 186–204). Berlin, Heidelberg: Springer.

Hampali, S., Oberweger, M., Rad, M., & Lepetit, V. (2019). *HO-3D: A multi-user, multi-object dataset for joint 3D hand-object pose estimation*. arXiv:1907.01481.

Han, F., Reily, B., Hoff, W., & Zhang, H. (2017). Space-time representation of people based on 3D skeletal data: A review. *Computer Vision and Image Understanding, 158*, 85–105.

Han, L., Wu, X., Liang, W., Hou, G., & Jia, Y. (2010). Discriminative human action recognition in the learned hierarchical manifold space. *Image and Vision Computing, 28*(5), 836–849.

Haque, A., Peng, B., Luo, Z., Alahi, A., Yeung, S., & Fei-Fei, L. (2016). Towards viewpoint invariant 3D human pose estimation. In B. Leibe, J. Matas, N. Sebe, & M. Welling (Eds.), *Computer vision—ECCV 2016* (pp. 160–177). Cham: Springer International Publishing.

Hazan-Tessler, P., Krause, K., Gessner, R., & Bianchi, F. C. (2019). *M-enabling summit*. www.m-enabling.com.

Hernandez-Vela, A., Bautista, M. A., Perez-Sala, X., Ponce-Lopez, V., Escalera, S., Baro, X., Pujol, O., & Angulo, C. (2014). Probability-based dynamic time warping and bag-of-visual-and-depth-

words for human gesture recognition in RGB-D. *Pattern Recognition Letters, 50*, 112–121. Depth Image Analysis.
Heseltine, T., Pears, N., & Austin, J. (2008). Three-dimensional face recognition using combinations of surface feature map subspace components. *Image and Vision Computing, 26*(3), 382–396. 15th Annual British Machine Vision Conference.
Huang, J., Zhou, W., Zhang, Q., Li, H., & Li, W. (2018). *Video-based sign language recognition without temporal segmentation*. arXiv:1801.10111.
International Foundation for Electoral Systems. (2019). Icharati sawti. http://www.ifes-esll.com.
Ionescu, C., Papava, D., Olaru, V., & Sminchisescu, C. (2014). Human 3.6 m: Large scale datasets and predictive methods for 3D human sensing in natural environments. *IEEE Transactions on Pattern Analysis and Machine Intelligence, 36*(7), 1325–1339.
Jan, A., Ding, H., Meng, H., Chen, L., & Li, H. (2018). Accurate facial parts localization and deep learning for 3D facial expression recognition. In *2018 13th IEEE International Conference on Automatic Face Gesture Recognition (FG 2018)* (pp 466–472).
Jhuang, H., Gall, J., Zuffi, S., Schmid, C., & Black, M. J. (2013). Towards understanding action recognition. *ICCV* (pp. 3192–3199).
Jing, L., Vahdani, E., Huenerfauth, M., & Tian, Y. (2019). Recognizing American sign language manual signs from RGB-D videos. arXiv:1906.02851.
Jung, H., Lee, S., Yim, J., Park, S., & Kim, J. (2015). Joint fine-tuning in deep neural networks for facial expression recognition. In *International Conference on Computer Vision* (pp 2983–2991).
Kadir, T., Bowden, R., Ong, E. J., & Zisserman, A. (2004). Minimal training, large lexicon, unconstrained sign language recognition. In *British Machine Vision Conference*.
Kang, B., Tripathi, S., & Nguyen, T. Q. (2015). Real-time sign language fingerspelling recognition using convolutional neural networks from depth map. In *3rd IAPR Asian Conference on Pattern Recognition, ACPR 2015*, Kuala Lumpur, Malaysia, November 3–6, 2015 (pp. 136–140).
Karpathy, A., Toderici, G., Shetty, S., Leung, T., Sukthankar, R., & Fei-Fei, L. (2014). Large-scale video classification with convolutional neural networks. In *IEEE Conference on Computer Vision and Pattern Recognition (CVPR)* (pp. 1725–1732).
Kay, W., Carreira, J., Simonyan, K., Zhang, B., Hillier, C., Vijayanarasimhan, S., Viola, F., Green, T., Back, T., Natsev, P., Suleyman, M., & Zisserman, A. (2017). *The kinetics human action video dataset*. arXiv:1705.06950.
Ke, Q., Fritz, M., & Schiele, B. (2019). Time-conditioned action anticipation in one shot. In *The IEEE Conference on Computer Vision and Pattern Recognition (CVPR)* (pp. 9925–9934).
Klaser, A., Marszalek, M., & Schmid, C. (2008). A spatio-temporal descriptor based on 3d-gradients. In M. Everingham, C. Needham & R. Fraile (Eds.), *BMVC* (Vol. 275, pp. 1–10). British Machine Vision Association.
Ko, S.-K., Kim, C. J., Jung, H., & Cho, C. (2019). Neural sign language translation based on human keypoint estimation. *Applied Sciences, 9*(13).
Koller, O., Forster, J., & Ney, H. (2015a). Continuous sign language recognition: Towards large vocabulary statistical recognition systems handling multiple signers. *Computer Vision and Image Understanding, 141*, 108–125.
Koller, O., Ney, H., & Bowden, R. (2015b). Deep learning of mouth shapes for sign language. In *Third workshop on assistive computer vision and robotics, ICCV*, Santiago, Chile (pp. 477–483).
Kollias, D., Tzirakis, P., Nicolaou, M. A., Papaioannou, A., Zhao, G., Schuller, B., Kotsia, I., & Zafeiriou, S. (2019). Deep affect prediction in-the-wild: Aff-wild database and challenge, deep architectures, and beyond. *International Journal of Computer Vision, 127*(6), 907–929.
Koppula, H. S., Gupta, R., & Saxena, A. (2013). Learning human activities and object affordances from RGB-D videos. *International Journal of Robotics Research, 32*(8), 951–970.
Krizhevsky, A., Sutskever, I., & Hinton, G. E. (2012). Imagenet classification with deep convolutional neural networks. In F. Pereira, C. J. C. Burges, L. Bottou, & K. Q. Weinberger (Eds.), *Advances in neural information processing systems 25* (pp. 1097–1105). Curran Associates Inc.

Kuehne, H., Jhuang, H., Garrote, E., Poggio, T., & Serre, T. (2011). Hmdb: A large video database for human motion recognition. In *2011 International Conference on Computer Vision* (pp. 2556–2563).

Laguna, J. O., Olaya, A. G., & Borrajo, D. (2011). A dynamic sliding window approach for activity recognition. In *Proceedings of the 19th International Conference on User Modeling, Adaption, and Personalization, UMAP'11*, Berlin, Heidelberg (pp. 219–230). Springer.

Laptev, I. (2005). On space-time interest points. *International Journal of Computer Vision, 64*(2–3), 107–123.

Laptev, I., Marszalek, M., Schmid, C., & Rozenfeld, B. (2008). Learning realistic human actions from movies. In *CVPR* (pp. 1–8).

Lara, O. D., & Labrador, M. A. (2013). A survey on human activity recognition using wearable sensors. *IEEE Communications Surveys Tutorials, 15*(3), 1192–1209.

Le, D. T., Bernardi, R., & Uijlings, J. (2013). Exploiting language models to recognize unseen actions. In *Proceedings of the 3rd ACM Conference on International Conference on Multimedia Retrieval, ICMR '13*, New York, NY, USA (pp 231–238). ACM.

Le, V., Brandt, J., Lin, Z., Bourdev, L., & Huang, T. S. (2012). Interactive facial feature localization. In A. Fitzgibbon, S. Lazebnik, P. Perona, Y. Sato & C. Schmid (Eds.), *Computer vision—ECCV 2012*, Berlin, Heidelberg (pp. 679–692). Berlin, Heidelberg: Springer.

Levi, G., & Hassner, T. (2015). Emotion recognition in the wild via convolutional neural networks and mapped binary patterns. In: *Proceedings of the 2015 ACM on International Conference on Multimodal Interaction, ICMI '15*, New York, NY, USA (pp. 503–510). ACM.

Li, C., Zhong, Q., Xie, D., & Pu, S. (2019a). Collaborative spatiotemporal feature learning for video action recognition. In *The IEEE Conference on Computer Vision and Pattern Recognition (CVPR)* (pp. 7872–7881).

Li, H., Sun, J., Xu, Z., & Chen, L. (2017). Multimodal 2D + 3D facial expression recognition with deep fusion convolutional neural network. *IEEE Transactions on Multimedia, 19*(12), 2816–2831.

Li, H., Weise, T., & Pauly, M. (2010). Example-based facial rigging. *ACM Transactions on Graphics, 29*(4), 32:1–32:6.

Li, S. and Deng, W. (2018). Deep facial expression recognition: A survey. arXiv:1804.08348.

Li, S., & Deng, W. (2019). Reliable crowdsourcing and deep locality-preserving learning for unconstrained facial expression recognition. *IEEE Transactions on Image Processing, 28*(1), 356–370.

Li, W., Zhang, Z., & Liu, Z. (2010). Action recognition based on a bag of 3D points. In *2010 IEEE Computer Society Conference on Computer Vision and Pattern Recognition—Workshops* (pp. 9–14).

Li, Z., Dekel, T., Cole, F., Tucker, R., Snavely, N., Liu, C., & Freeman, W. T. (2019b). *Learning the depths of moving people by watching frozen people* (pp. 4521–4530).

Liang, B., & Zheng, L. (2014). Multi-modal gesture recognition using skeletal joints and motion trail model. In *ECCV Workshops* (pp. 623–638).

Liang, J., Jiang, L., Niebles, J. C., Hauptmann, A. G., & Fei-Fei, L. (2019). Peeking into the future: Predicting future person activities and locations in videos. In *The IEEE Conference on Computer Vision and Pattern Recognition (CVPR)*, (pp. 5725–5734).

Lin, Y., Cheng, S., Shen, J., & Pantic, M. (2019). Mobiface: A novel dataset for mobile face tracking in the wild. In *14th IEEE International Conference on Automatic Face & Gesture Recognition* (pp. 1–8).

Liu, Z., Zhang, C., & Tian, Y. (2016). 3D-based deep convolutional neural network for action recognition with depth sequences. *Image and Vision Computing, 55*, 93–100. Handcrafted versus Learned Representations for Human Action Recognition.

Livingstone, S. R., & Russo, F. A. (2018). The Ryerson audio-visual database of emotional speech and song (Ravdess): A dynamic, multimodal set of facial and vocal expressions in north American English. *PLOS ONE, 13*(5), 1–35.

Lu, C., Zhao, D., & Tang, X. (2013). Face recognition using face patch networks. In *2013 IEEE International Conference on Computer Vision* (pp. 3288–3295).

Lucey, P., Cohn, J. F., Kanade, T., Saragih, J., Ambadar, Z., & Matthews, I. (2010). The extended Cohn-Kanade dataset (ck+): A complete dataset for action unit and emotion-specified expression. In *CVPR Workshops* (pp. 94–101).

Lumini, A., Nanni, L., & Brahnam, S. (2017). Ensemble of texture descriptors and classifiers for face recognition. *Applied Computing and Informatics, 13*(1), 79–91.

Lyons, M. J., Budynek, J., & Akamatsu, S. (1999). Automatic classification of single facial images. *IEEE Transactions on Pattern Analysis and Machine Intelligence, 21*(12), 1357–1362.

Mâamatou, H., Chateau, T., Gazzah, S., Goyat, Y., & Essoukri Ben Amara, N. (2016). Transductive transfer learning to specialize a generic classifier towards a specific scene. In *VISIGRAPP* (pp. 411–422).

Mansouri, J., Seddik, B., Gazzah, S., & Chateau, T. (2015). Coarse localization using space-time and semantic-context representations of geo-referenced video sequences. In *IPTA* (pp. 355–359).

Marinoiu, E., Zanfir, M., Olaru, V., & Sminchisescu, C. (2018). 3D human sensing, action and emotion recognition in robot assisted therapy of children with autism. In *The IEEE Conference on Computer Vision and Pattern Recognition (CVPR)* (pp. 2158–2167).

Marszalek, M., Laptev, I., & Schmid, C. (2009). Actions in context. In *2009 IEEE Conference on Computer Vision and Pattern Recognition* (pp. 2929–2936).

Martin, D. (2018). Octi. https://www.octi.tv/.

Martinez, A., & Benavente, R. (1998). The ar face database. *Computer vision center technical report* (p. 24).

Matuszewski, B. J., Quan, W., Shark, L.-K., McLoughlin, A. S., Lightbody, C. E., Emsley, H. C., & Watkins, C. L. (2012). Hi4D-adsip 3-D dynamic facial articulation database. *Image and Vision Computing, 30*(10), 713–727. 3D Facial Behaviour Analysis and Understanding.

Mavadati, S. M., Mahoor, M. H., Bartlett, K., Trinh, P., & Cohn, J. F. (2013). Disfa: A spontaneous facial action intensity database. *TAC, 4*(2), 151–160.

McDonald, J., Wolfe, R., Schnepp, J., Hochgesang, J., Jamrozik, D. G., Stumbo, M., Berke, L., Bialek, M., & Thomas, F. (2016). An automated technique for real-time production of lifelike animations of American sign language. *Universal Access in the Information Society, 15*(4), 551–566.

McDuff, D., Kaliouby, R. E., Senechal, T., Amr, M., Cohn, J., & Picard, R. (2013). Affectiva mit facial expression dataset (am-fed) naturalistic and spontaneous facial expressions collected "in the wild". In *IEEE Conference on Computer Vision and Pattern Recognition Workshops* (pp. 881–888).

McKeown, G., Valstar, M., Cowie, R., Pantic, M., & Schroder, M. (2012). The semaine database: Annotated multimodal records of emotionally colored conversations between a person and a limited agent. *IEEE Transactions on Affective Computing, 3*(1), 5–17.

Menasy, N. M. H. (2001). *Arabic lexicon dictionary*. http://www.menasy.com/.

Messer, K., Kittler, J., Sadeghi, M., Marcel, S., Marcel, C., Bengio, S., Cardinaux, F., Sanderson, C., Czyz, J., Vandendorpe, L., Srisuk, S., Petrou, M., Kurutach, W., Kadyrov, A., Paredes, R., Kepenekci, B., Tek, F., Akar, G., Deravi, F., & Mavity, N. (2003). *Face verification competition on the xm2vts database* (Vol. 2688, pp. 964–974).

Min, R., Kose, N., & Dugelay, J.-L. (2014). Kinectfacedb: A kinect database for face recognition. *IEEE Transactions on Systems, Man, and Cybernetics: Systems, 44*(11), 1534–1548.

Mohandes, M., Deriche, M., & Liu, J. (2014). Image-based and sensor-based approaches to Arabic sign language recognition. *IEEE Transactions on Human-Machine Systems, 44*(4), 551–557.

Mohandes, M. A. (2013). Recognition of two-handed Arabic signs using the Cyberglove. *Arabian Journal for Science and Engineering, 38*(3), 669–677.

Molchanov, P., Yang, X., Gupta, S., Kim, K., Tyree, S., & Kautz, J. (2016). Online detection and classification of dynamic hand gestures with recurrent 3D convolutional neural networks. In *2016 IEEE Conference on Computer Vision and Pattern Recognition (CVPR)* (pp. 4207–4215).

Mollahosseini, A., Hasani, B., & Mahoor, M. H. (2019). Affectnet: A database for facial expression, valence, and arousal computing in the wild. *IEEE Transactions on Affective Computing, 10*(1), 18–31.

Monfort, M., Andonian, A., Zhou, B., Ramakrishnan, K., Bargal, S. A., Yan, T. et al. (2019). Moments in time dataset: one million videos for event understanding. *IEEE Transactions on Pattern Analysis and Machine Intelligence*, 1–8.

Monnier, C., German, S., & Ost, A. (2014). A multi-scale boosted detector for efficient and robust gesture recognition. In *ECCV Workshops* (pp. 491–502).

Müller, M. (2007). *Information retrieval for music and motion*. Secaucus, NJ, USA: Springer New York Inc.

Neverova, N., Wolf, C., Taylor, G., & Nebout, F. (2016). Moddrop: Adaptive multi-modal gesture recognition. *IEEE Transactions on Pattern Analysis and Machine Intelligence, 38*(8), 1692–1706.

Nguyen, A., Yosinski, J., & Clune, J. (2019). *Understanding neural networks via feature visualization: A survey* (pp. 55–76).

Nguyen, H., Kotani, K., Chen, F., & Le, B. (2014). A thermal facial emotion database and its analysis. In R. Klette, M. Rivera & S. Satoh (Eds.), *Image and video technology*, Berlin, Heidelberg (pp. 397–408). Springer, Berlin Heidelberg.

Niebles, J. C., Chen, C.-W., & Fei-Fei, L. (2010). Modeling temporal structure of decomposable motion segments for activity classification. In *Proceedings of the 11th European Conference on Computer Vision: Part II, ECCV'10*, Berlin, Heidelberg (pp. 392–405). Springer.

Nordstrøm, M. M., Larsen, M., Sierakowski, J., & Stegmann, M. B. (2004). The IMM face database—an annotated dataset of 240 face images. Technical report, Informatics and Mathematical Modelling, Technical University of Denmark, DTU.

Noroozi, F., Kaminska, D., Corneanu, C., Sapinski, T., Escalera, S., & Anbarjafari, G. (2018). Survey on emotional body gesture recognition. *IEEE Transactions on Affective Computing*.

Ofli, F., Chaudhry, R., Kurillo, G., Vidal, R., & Bajcsy, R. (2013). Berkeley MHAD: A comprehensive multimodal human action database. In *2013 IEEE Workshop on Applications of Computer Vision (WACV)*, (pp. 53–60).

Ofli, F., Chaudhry, R., Kurillo, G., Vidal, R., & Bajcsy, R. (2014). Sequence of the most informative joints (SMIJ): A new representation for human skeletal action recognition. *The Journal of Visual Communication and Image Representation, 25*(1), 24–38.

Ong, E. J., Pugeault, N., & Bowden, R. (2014). Sign spotting using hierarchical sequential patterns with temporal intervals. In *2014 IEEE Conference on Computer Vision and Pattern Recognition*, (pp. 1931–1938).

Ong, S. C., & Ranganath, S. (2005). Automatic sign language analysis: A survey and the future beyond lexical meaning. *IEEE Transactions on Pattern Analysis and Machine Intelligence, 27*(6), 873–891.

Opalka, A., & Miller-Jacobson, B. (2018). Motionsavvy. https://www.motionsavvy.com/.

Oreifej, O., & Liu, Z. (2013). Hon4d: Histogram of oriented 4D normals for activity recognition from depth sequences. In *CVPR* (pp. 716–723).

Ouarda, W., Trichili, H., Alimi, A. M., & Solaiman, B. (2015). Bag of face recognition systems based on holistic approaches. In *2015 15th International Conference on Intelligent Systems Design and Applications (ISDA)* (pp. 201–206).

Oyedotun, O. K., Demisse, G., Shabayek, A. E. R., Aouada, D., & Ottersten, B. (2017). Facial expression recognition via joint deep learning of RGB-depth map latent representations. In *2017 IEEE International Conference on Computer Vision Workshops (ICCVW)* (pp. 3161–3168).

Pan, X., Slater, M., Beacco, A., Navarro, X., Bellido Rivas, A. I., et al. (2016). The responses of medical general practitioners to unreasonable patient demand for antibiotics—A study of medical ethics using immersive virtual reality. *PLOS ONE, 11*(2), 1–15.

Pantic, M., Valstar, M., Rademaker, R., & Maat, L. (2005). Web-based database for facial expression analysis. In *2005 IEEE International Conference on Multimedia and Expo* (pp. 317–321).

Parkhurst, S., & Parkhurst, D. (2010). *A corss-Linguistic guide to signWriting: A phonetic approach*.

Peng, X., Wang, L., & Cai, Z. (2014). Action and gesture temporal spotting with super vector representation. In *ECCV Workshops* (pp. 518–527).

Peng, X., Wang, L., Wang, X., & Qiao, Y. (2016). Bag of visual words and fusion methods for action recognition: Comprehensive study and good practice. *Computer Vision and Image Understanding*, 1–17.

Pfister, T. (2015). *Advancing human pose and gesture recognition*. Ph.D. thesis, University of Oxford.

Pfister, T., Charles, J., Everingham, M., & Zisserman, A. (2012). Automatic and efficient long term arm and hand tracking for continuous sign language TV broadcasts. In *British Machine Vision Conference* (pp. 1105–1114).

Pfister, T., Charles, J., & Zisserman, A. (2013). Large-scale learning of sign language by watching TV (using co-occurrences). In *British Machine Vision Conference*.

Pfister, T., Simonyan, K., Charles, J., & Zisserman, A. (2014). Deep convolutional neural networks for efficient pose estimation in gesture videos. In *Asian Conference on Computer Vision*, (pp. 538–552).

Pietro, C., & Nicole, V. (2016). Teaching reading and writing with the intramorphic method to deaf children. *Journal of Applied Psycholinguistics, 16*(2), 85–99.

Pigou, L., van den Oord, A., Dieleman, S., Van Herreweghe, M., & Dambre, J. (2016). Beyond temporal pooling: Recurrence and temporal convolutions for gesture recognition in video. *International Journal of Computer Vision*, 1–10.

Poppe, R. (2010). A survey on vision-based human action recognition. *Image and Vision Computing, 28*(6), 976–990.

Presti, L. L., & Cascia, M. L. (2016). 3d skeleton-based human action classification: A survey. *Pattern Recognition, 53*, 130–147.

Pu, J., Zhou, W., & Li, H. (2019). Iterative alignment network for continuous sign language recognition. In *The IEEE Conference on Computer Vision and Pattern Recognition (CVPR)* (pp. 4165–4174).

Rahmani, H., Mahmood, A., Q Huynh, D., & Mian, A. (2014). *Hopc: Histogram of oriented principal components of 3D point clouds for action recognition* (pp. 742–757).

Reddy, K. K., & Shah, M. (2013). Recognizing 50 human action categories of web videos. *Machine Vision and Applications, 24*(5), 971–981.

Riche, N., Mancas, M., Duvinage, M., Mibulumukini, M., Gosselin, B., & Dutoit, T. (2013). Rare 2012: A multi-scale rarity-based saliency detection with its comparative statistical analysis. *Signal Processing: Image Communication, 28*(6), 642–658.

Robotka, Z., Rovnyai, J., Gerlis, S., Retek, D., Pintér, M., & Croasmun, D. (2018). Signall media kit. www.signall.us.

Roychowdhury, S., & Emmons, M. (2015). *A survey of the trends in facial and expression recognition databases and methods*. arXiv:1511.02407.

Sagonas, C., Antonakos, E., Tzimiropoulos, G., Zafeiriou, S., & Pantic, M. (2016). 300 faces in-the-wild challenge. *Image Vision Computer, 47*(C), 3–18.

Sandbach, G., Zafeiriou, S., Pantic, M., & Yin, L. (2012). Static and dynamic 3d facial expression recognition: A comprehensive survey. *Image and Vision Computing, 30*(10), 683–697. 3D Facial Behaviour Analysis and Understanding.

Sapp, B., & Taskar, B. (2013). Modec: Multimodal decomposable models for human pose estimation. In *The IEEE Conference on Computer Vision and Pattern Recognition (CVPR)* (pp. 3674–3681).

Savran, A., Alyüz, N., Dibeklioğlu, H., Çeliktutan, O., Gökberk, B., Sankur, B. et al. (2008). *Bosphorus database for 3D face analysis* (pp. 47–56). Berlin, Heidelberg: Springer.

Schuldt, C., Laptev, I., & Caputo, B. (2004). Recognizing human actions: A local SVM approach. In *Proceedings of the 17th International Conference on Pattern Recognition (ICPR'04)* (Vol. 3, pp. 32–36).

Seddik, B., Gazzah, S., & Essoukri Ben Amara, N. (2015). Modalities combination for Italian sign language extraction and recognition. In V. Murino & E. Puppo (Eds.), *ICIAP (2), Volume 9280 of Lecture Notes in Computer Science* (pp. 710–721). Springer.

Seddik, B., Gazzah, S., & Essoukri Ben Amara, N. (2017). Human-action recognition using a multi-layered fusion scheme of kinect modalities. *IET Computer Vision, 11*(7), 530–540.

Seddik, B., Mâamatou, H., Gazzah, S., Chateau, T., & Essoukri Ben Amara, N. (2013). Unsupervised facial expressions recognition and avatar reconstruction from kinect. In: *2013 10th International Multi-Conference on Systems, Signals Devices (SSD)* (pp. 1–6).

Sempena, S., Maulidevi, N. U., & Aryan, P. R. (2011). Human action recognition using dynamic time warping. In *2011 International Conference on Electrical Engineering and Informatics (ICEEI)* (pp. 1–5).

Shahroudy, A., Liu, J., Ng, T., & Wang, G. (2016). NTU RGB+D: A large scale dataset for 3D human activity analysis. arXiv:1604.02808.

Shan, J., & Akella, S. (2014). 3D human action segmentation and recognition using pose kinetic energy. In *ARSO* (pp. 69–75).

Shao, L., Ji, L., Liu, Y., & Zhang, J. (2012). Human action segmentation and recognition via motion and shape analysis. *Pattern Recognition Letters, 33*(4), 438–445. Intelligent Multimedia Interactivity.

Shi, L., Zhang, Y., Cheng, J., & Lu, H. (2019). Skeleton-based action recognition with directed graph neural networks. In *The IEEE Conference on Computer Vision and Pattern Recognition (CVPR)* (pp. 7912–7921).

Shotton, J., Girshick, R., Fitzgibbon, A., Sharp, T., Cook, M., Finocchio, M., Moore, R., Kohli, P., Criminisi, A., Kipman, A., & Blake, A. (2013). Efficient human pose estimation from single depth images. *IEEE Transactions on Pattern Analysis and Machine Intelligence, 35*(12), 2821–2840.

Si, C., Chen, W., Wang, W., Wang, L., & Tan, T. (2019). An attention enhanced graph convolutional LSTM network for skeleton-based action recognition. In *The IEEE Conference on Computer Vision and Pattern Recognition (CVPR)* (pp. 1227–1236).

Sigal, L., Balan, A. O., & Black, M. J. (2009). Humaneva: Synchronized video and motion capture dataset and baseline algorithm for evaluation of articulated humanÂ motion. *International Journal of Computer Vision, 87*(1), 4.

Simonyan, K., & Zisserman, A. (2014). Two-stream convolutional networks for action recognition in videos. In *Proceedings of the 27th International Conference on Neural Information Processing Systems, NIPS'14* (Vol. 1, pp. 568–576), Cambridge, MA, USA: MIT Press.

Singh, S., Velastin, S. A., & Ragheb, H. (2010). Muhavi: A multicamera human action video dataset for the evaluation of action recognition methods. In *2010 Seventh IEEE International Conference on Advanced Video and Signal Based Surveillance (AVSS)* (pp. 48–55).

Song, S., Chandrasekhar, V., Cheung, N.-M., Narayan, S., Li, L., & Lim, J.-H. (2014). Activity recognition in egocentric life-logging videos. In *Asian Conference on Computer Vision*, (pp. 445–458). Springer.

Starner, T., & Pentland, A. (1995). Real-time American sign language recognition from video using hidden markov models. In *Proceedings of the International Symposium on Computer Vision, 1995* (pp. 265–270).

Sung, J., Ponce, C., Selman, B., & Saxena, A. (2012). Unstructured human activity detection from RGBD images. In *ICRA* (pp. 842–849).

Tabia, H., Gouiffès, M., & Lacassagne, L. (2012). Motion histogram quantification for human action recognition. In *2012 21st International Conference on Pattern Recognition (ICPR)* (pp. 2404–2407).

Tenório, R. (2019). Hand-talk. https://handtalk.me/.

Todo, T. (2018). *Seer: Simulative emotional expression robot* (pp. 1–2).

Tombari, F., Salti, S., & Di Stefano, L. (2010). Unique shape context for 3D data description. In *Proceedings of the ACM Workshop on 3D Object Retrieval, 3DOR'10*, New York, NY, USA (pp. 57–62). ACM.

Trimech, I., Maalej, A., & Essoukri Ben Amara, N. (2017). *3D facial expression recognition using nonrigid CPD registration method* (pp. 478–481).

Tsalakanidou, F., Tzovaras, D., & Strintzis, M. (2003). Use of depth and colour eigenfaces for face recognition. *Pattern Recognition Letters, 24*(9–10), 1427–1435.

Turk, M., & Pentland, A. (1991). Eigenfaces for recognition. *Journal Cognitive Neuroscience, 3*(1), 71–86.

Valentin, J., Kowdle, A., Barron, J. T., Wadhwa, N., Dzitsiuk, M., Schoenberg, M. et al. (2018). Depth from motion for smartphone ar. *ACM Transactions on Graphics, 37*(6), 193:1–193:19.

Varol, G., Romero, J., Martin, X., Mahmood, N., Black, M. J., Laptev, I. et al. (2017). Learning from synthetic humans. In *CVPR*.

Vemulapalli, R., Arrate, F., & Chellappa, R. (2016). R3dg features: Relative 3D geometry-based skeletal representations for human action recognition. *Computer Vision and Image Understanding, 152*, 155–166.

Vieira, A. W., Nascimento, E. R., Oliveira, G. L., Liu, Z., & Campos, M. F. M. (2012). Stop: Space-time occupancy patterns for 3d action recognition from depth map sequences. In L. Alvarez, M. Mejail, L. Gomez & J. Jacobo (Eds.), *Progress in pattern recognition, image analysis, computer vision, and applications*, Berlin, Heidelberg (pp. 252–259). Berlin, Heidelberg: Springer.

Viola, P., & Jones, M. (2001). Rapid object detection using a boosted cascade of simple features. In *Proceedings of the 2001 IEEE Computer Society Conference on Computer Vision and Pattern Recognition, 2001. CVPR 2001* (Vol. 1, pp. I-511–I-518).

von Agris, U., Zieren, J., Canzler, U., Bauer, B., & Kraiss, K.-F. (2008). Recent developments in visual sign language recognition. *Universal Access in the Information Society, 6*(4), 323–362.

Vrigkas, M., Nikou, C., & Kakadiaris, I. (2015). A review of human activity recognition methods. *Frontiers in Robotics and AI, 2*(28).

Wan, J., Li, S. Z., Zhao, Y., Zhou, S., Guyon, I., & Escalera, S. (2016). Chalearn looking at people RGB-D isolated and continuous datasets for gesture recognition. In *2016 IEEE Conference on Computer Vision and Pattern Recognition Workshops (CVPRW)* (pp. 761–769).

Wan, J., Ruan, Q., Li, W., & Deng, S. (2013). One-shot learning gesture recognition from RGG-D data using bag of features. *Journal of Machine Learning Research, 14*, 2549–2582.

Wang, H., Chai, X., & Chen, X. (2016). Sparse observation (so) alignment for sign language recognition. *Neurocomputing, 175*, Part A:674–685.

Wang, H., Kläser, A., Schmid, C., & Liu, C.-L. (2013). Dense trajectories and motion boundary descriptors for action recognition. *International Journal of Computer Vision, 103*(1), 60–79.

Wang, H., Oneata, D., Verbeek, J., & Schmid, C. (2015). A robust and efficient video representation for action recognition. *International Journal of Computer Vision*, pp. 1–20.

Wang, H., & Schmid, C. (2013). Action recognition with improved trajectories. In *ICCV* (pp. 3551–3558).

Wang, J., Liu, Z., Wu, Y., & Yuan, J. (2012). Mining actionlet ensemble for action recognition with depth cameras. In: *IEEE Conference on Computer Vision and Pattern Recognition (CVPR)* (pp. 1290–1297).

Wang, L., Qiao, Y., & Tang, X. (2015). Action recognition with trajectory-pooled deep-convolutional descriptors. In *2015 IEEE Conference on Computer Vision and Pattern Recognition* (pp. 4305–4314).

Wang, P., Li, W., Ogunbona, P., Wan, J., & Escalera, S. (2018). RGB-D-Based human motion recognition with deep learning: A survey. *Computer Vision and Image Understanding, 171*, 118–139.

Wang, S., Liu, Z., Lv, S., Lv, Y., Wu, G., Peng, P., Chen, F., & Wang, X. (2010). A natural visible and infrared facial expression database for expression recognition and emotion inference. *IEEE Transactions on Multimedia, 12*(7), 682–691.

Wang, S. J., Yan, W. J., Li, X., Zhao, G., & Fu, X. (2014). Micro-expression recognition using dynamic textures on tensor independent color space. In *22nd International Conference on Pattern Recognition* (pp. 4678–4683).

Weise, T., Bouaziz, S., Li, H., & Pauly, M. (2011). Realtime performance-based facial animation. *ACM Transactions on Graphics, 30*(4), 77:1–77:10.

Yan, W.-J., Wu, Q., Liu, Y.-J., Wang, S.-J., & Fu, X. (2013). Casme database: A dataset of spontaneous micro-expressions collected from neutralized faces. In *2013 10th IEEE International Conference and Workshops on Automatic Face and Gesture Recognition (FG)* (pp. 1–7).

World Health Organization (2019). Deafness and hearing loss. https://www.who.int/news-room/fact-sheets/detail/deafness-and-hearing-loss.

Wu, D., Pigou, L., Kindermans, P. J., Le, N. D. H., Shao, L., Dambre, J., & Odobez, J. M. (2016). Deep dynamic neural networks for multimodal gesture segmentation and recognition. *IEEE Transactions on Pattern Analysis and Machine Intelligence, 38*(8), 1583–1597.

Xia, L., Chen, C., & Aggarwal, J. (2012). View invariant human action recognition using histograms of 3D joints. In *Computer Vision and Pattern Recognition Workshops (CVPRW), IEEE Conference on Computer Society* (pp. 20–27).

Yang, L., Onal Ertugrul, I., Cohn, J. F., Hammal, Z., Jiang, D., & Sahli, H. (2019). Facs3d-net: 3D convolution based spatiotemporal representation for action unit detection. In *International Conference on Affective Computing and Interactive Intelligence (ACII)*.

Yang, X., & Tian, Y. (2014). Effective 3D action recognition using eigenjoints. *Journal of Visual Communication and Image Representation, 25*(1), 2–11.

Yang, X., Zhang, C., & Tian, Y. (2012). Recognizing actions using depth motion maps-based histograms of oriented gradients. In *20th ACM International Conference on Multimedia, MM '12*, New York, NY, USA (pp. 1057–1060). ACM.

Yao, B., & Fei-Fei, L. (2010). Modeling mutual context of object and human pose in human-object interaction activities. In *IEEE Conference on Computer Vision and Pattern Recognition (CVPR)* (pp. 17–24).

Yao, B., Jiang, X., Khosla, A., Lin, A. L., Guibas, L., & Fei-Fei, L. (2011). Human action recognition by learning bases of action attributes and parts. In *International Conference on Computer Vision* (pp. 1331–1338).

Ye, M., Zhang, Q., Wang, L., Zhu, J., Yang, R., & Gall, J. (2013). *A Survey on Human Motion Analysis from Depth Data* (pp. 149–187). Berlin, Heidelberg: Springer.

Ye, Q., Yuan, S., & Kim, T.-K. (2016). Spatial attention deep net with partial PSO for hierarchical hybrid hand pose estimation. In B. Leibe, J. Matas, N. Sebe, & M. Welling (Eds.), *Computer vision—ECCV 2016* (pp. 346–361). Cham: Springer Int. Publishing.

Yin, L., Chen, X., Sun, Y., Worm, T., & Reale, M. (2008). A high-resolution 3D dynamic facial expression database. In: *2008 8th IEEE International Conference on Automatic Face Gesture Recognition* (pp. 1–6).

Yin, L., Wei, X., Sun, Y., Wang, J., & Rosato, M. J. (2006). A 3D facial expression database for facial behavior research. In: *7th International Conference on Automatic Face and Gesture Recognition, FGR'06*, Washington, DC, USA (pp. 211–216). IEEE Computer Society.

Yu, M., Liu, L., & Shao, L. (2016). Structure-preserving binary representations for RGB-D action recognition. *IEEE Transactions on Pattern Analysis and Machine Intelligence, 38*(8), 1651–1664.

Yuan, S., Ye, Q., Stenger, B., Jain, S., & Kim, T.-K. (2017). Bighand2.2m benchmark: Hand pose dataset and state of the art analysis. In *IEEE Conference on Computer Vision and Pattern Recognition (CVPR)* (pp. 2605–2613).

Zanfir, M., Leordeanu, M., & Sminchisescu, C. (2013). The moving pose: An efficient 3d kinematics descriptor for low-latency action recognition and detection. *ICCV*, (pp. 2752–2759).

Zbakh, M., Haddad, Z., & Krahe, J. L. (2015). An online reversed French sign language dictionary based on a learning approach for signs classification. *Pattern Recognition Letters, 67*, Part 1:28–38. Cognitive Systems for Knowledge Discovery.

Zhalehpour, S., Onder, O., Akhtar, Z., & Erdem, C. E. (2017). Baum-1: A spontaneous audio-visual face database of affective and mental states. *IEEE Transactions on Affective Computing, 8*(3), 300–313.

Zhang, D., & Shah, M. (2015). Human pose estimation in videos. In: *2015 IEEE International Conference on Computer Vision (ICCV)* (pp. 2012–2020).

Zhang, H.-B., Zhang, Y.-X., Zhong, B., Lei, Q., Yang, L., Du, J.-X., & Chen, D.-S. (2019). A comprehensive survey of vision-based human action recognition methods. *Sensors, 19*(5).

Zhang, X., Yin, L., Cohn, J., Canavan, S., Reale, M., Horowitz, A., Liu, P., & Girard, J. (2014). Bp4d-spontaneous: A high-resolution spontaneous 3D dynamic facial expression database. *Image and Vision Computing, 32*, 692–706.

Zhang, Z., Girard, J. M., Wu, Y., Zhang, X., Liu, P., Ciftci, U. et al. (2016). Multimodal spontaneous emotion corpus for human behavior analysis. In *IEEE Conference on Computer Vision and Pattern Recognition (CVPR)* (pp. 3438–3446).
Zhang, Z., Luo, P., Loy, C. C., & Tang, X. (2018). From facial expression recognition to interpersonal relation prediction. *International Journal of Computer Vision, 126*(5), 550–569.
Zhong, C., Sun, Z., & Tan, T. (2007). Robust 3D face recognition using learned visual codebook. In: *IEEE Conference on Computer Vision and Pattern Recognition* (pp 1–6).

Chapter 10
A Software Architecture for Developing Disease Registries

Hatem Bellaaj, Afef Mdhaffar, Sondes Mseddi, Moez Elloumi, Hajer Aloulou, Mongia Hachicha, Mohamed Jmaiel, and Bernd Freisleben

Abstract This paper proposes a novel software architecture for developing disease registries. Our proposal consists of five horizontal layers and two transversal layers. It takes into account several aspects of disease registries, such as various sources of data acquisition, security and privacy issues, data analytics, and user management tasks. Several use cases are presented: the Tunisian Fanconi Anemia Registry (TFAR), the Tunisian Gaucher Disease Registry (TGDR), the Tunisian Non-Hodgkin Lymphoma Registry (TNHLR), and the Tunisian Registry of Diabetes Type 1 (TRD-T1).

Keywords Disease registry · Data quality · Classification · Artificial intelligence · Interoperability · Analysis · Security · Privacy

---

H. Bellaaj (✉) · A. Mdhaffar · S. Mseddi · M. Elloumi · H. Aloulou · M. Hachicha
University of Sfax, Sfax, Tunisia
e-mail: hatem.bellaaj@redcad.org

A. Mdhaffar
e-mail: afef.mdhaffar@enis.tn

S. Mseddi
e-mail: sondes@medecinesfax.org

M. Elloumi
e-mail: moez.elloumi@rns.tn

H. Aloulou
e-mail: aloulouhajer2000@gmail.com

M. Hachicha
e-mail: mongia.hachicha@rns.tn

M. Jmaiel
Digital Research Center of Sfax, Sfax, Tunisia
e-mail: mohamed.jmaiel@redcad.org

B. Freisleben
Department of Mathematics and Computer Science, University of Marburg, Marburg, Germany
e-mail: freisleb@informatik.uni-marburg.de

N. Derbel and O. Kanoun (eds.), *Advanced Methods for Human Biometrics*,
Smart Sensors, Measurement and Instrumentation 40,
https://doi.org/10.1007/978-3-030-81982-8_10

10.1 Introduction

A disease registry is a software application managed by an organization that seeks to record all cases of a disease occurring in a given territory. Therefore, it is used to identify the true incidence of a disease in that territory.

The development of a disease registry is a quite expensive effort. Thus, research and health organizations are particularly interested in the management of serious and very costly diseases.

There are several global experiences in developing disease registries. For instance, the US Department of Health and Human Services publishes a list of official American diseases registries (National Institute of Mental Health, 2016). The list has 52 entries, including cancer pathologies such as breast cancer, colon cancer, and genetic cancer. Furthermore, the Orphanet report series (Orphanet, 2016) summarizes European experiences in disease registry development (until January 2016). The highest number of established registries is in France (134 registries). In Germany, there are 124 registries. 81% of the listed institutions are public affiliations. A summarized distribution of disease registries by coverage is given: only 74 global registries exist.

The main goal of disease registries is the collection and analysis of epidemiological information on frequency and distribution like incidence and prevalence of a specific disease (Gliklich et al., 2014). Disease registries allow us to identify the relationships between risk factors and a disease and consequently gain valuable knowledge about the origin of the disease. This is important for assessing the situation of care and for predicting framework conditions, preventive measures, and management plans. Thus, they play a crucial role in health policy decisions.

Despite the growing number of disease registries, several of them are limited with respect to different aspects. For example, data security and privacy is often not guaranteed. Some disease registries have issues in terms of completeness and accuracy. The problems become even more complex when we talk about interoperability for experience sharing and data aggregation.

In this contribution, we present a novel software architecture for developing disease registries to tackle these problems. The proposed software architecture consists of five horizontal layers and two transversal layers. It takes into account several aspects of diesease registries, such as various sources of data acquisition, data analytics, security and privacy issues, and user management tasks. Several use cases are presented: the Tunisian Fanconi Anemia Registry (TFAR), the Tunisian Gaucher Disease Registry (TGDR), the Tunisian Non-Hodgkin Lymphoma Registry (TNHLR), and the Tunisian Registry of Diabetes type 1 (TRD-T1).

This chapter is organized as follows. In Sect. 10.2, we discuss related work on disease registries. Section 10.3 presents our novel software architecture. In Sect. 10.4, we discuss our experiences using the software architecture in several use cases. Section 10.5 concludes the chapter and outlines areas for future research.

10.2 Related Work

This section presents related work on diseases registries. The majority of approaches focus on particular problems, such as security or interoperability. For example, Kumar and Lee (Onukwugha et al., 2012) are particularly interested in security issues when using a wireless network of medical sensors. Gliklich et al. (Gliklich et al., 2014) discuss methods of analysis, interpretation, and reporting of registry data to evaluate outcomes.

Bellgard et al. (Bellgard et al., 2013) presents several issues for dispelling myths about developing rare disease registries, such as interoperability and semantic issues. Furthermore, four levels of interoperability are defined by NATO (NATO, 2005).

Chaterjee et al. (Chatterjee, 2009) present a roadmap showing the evolution of technologies for disease registries. They include four main components: (1) systems, (2) interaction capabilities, (3) technological foundations, and (4) human elements.

Choquet et al. (Landgrebe and Smith, 2011) discuss different approaches to support the recording and exchange of health data in health information systems. Among them are: information models (Health Level Seven (HL7) (Kohl et al., 2010), OpenEHR (EN 13606) (Openehr, 2017), value sets and terminologies (LOINC) (Wang et al., 2002).

In the following, we discuss several issues treated in the literature. These issues are: (1) technology, (2) data, (3) knowledge, (4) analytics, (5) services, (6) security, and (7) sharing.

10.2.1 Technology

There are different data capturing technologies that can be used (Stone et al., 2007; El Emam et al., 2009; Pavluck et al., 2014) to interact with various sensors, ranging from simple clinical scanners to sensors on connected objects. Data may vary from simple pulse readings to the inclusion of all parameters in the surrounding environment.

There are two types of tools: invasive (Chatterjee, 2009) and noninvasive (Lymberis, 2003; Bandodkar, 2014) tools. Invasive tools require a larger skin break-in than a simple venipuncture. Non-invasive tools use observations such as temperature or imaging.

We can distinguish between tools that support the following measurements: breath analysis (Fleischera et al., 2002), biomechanics (Daynarda et al., 2001), human movement capture (Gavrila, 1999), blood flow (Gill, 1985), heart rate (Zuckerwar et al., 1993), eye movement (Bulling et al., 2011), blood alcohol level, body temperature, and blood pressure.

10.2.2 Data

There are three main axes of data processing in disease registries: (1) data quality, (2) data privacy, and (3) data classification. They are discussed below.

10.2.2.1 Data Quality

Data quality is related to the following criteria: (1) completeness (Schouten et al., 1994), (2) accuracy (Arts et al., 2002), (3) agreement (Roos et al., 1993), (4) plausibility (Löwel et al., 1991), (5) update. There are several methods for assessing the quality of data. These include comparison with standards, evaluation of data sources, evaluation of distribution, and verification of validity.

The Top 10 Health Technology Risk Report of 2015 (ECRI, 2014) describes a variety of technological risks that put patients at risk. The 2015 list includes: (1) alarm hazards, (2) inaccurate or missing data in electronic health records and other health IT systems, (3) poor administration of drugs and solutions, (4) inadequate reprocessing of endoscopes and surgical instruments. In general, artificial intelligence methods and statistical tools are used to ensure data quality.

10.2.2.2 Data Privacy

The issue of data privacy is complex. It is even more complex when the context of private data is health. In disease registries, privacy is viewed from two distinct angles: the first one is ethical, the second one is technical. The ethical angle could be solved by a written consent in terms of privacy for the processing of sensitive data of patients involved in research procedures (e.g., questionnaires, experiments, evaluations). This consent allows us to process data within the limits of a general authorization issued by the data protection authority. A case-by-case examination of the developed prototypes will be undertaken to determine whether such regulatory restrictions could limit the use of these technologies or whether a solution could be adopted to ensure compliance with privacy without limiting the operation of the applications. The technical angle is typically solved by IT security experts. They develop solutions that address the problems of protecting private data in e-health. These problems vary according to the state of the personal data in question: collection, storage, processing, and communication.

Gliklich et al. (2014) summarize the ethical and legal considerations that should be adopted by developers that deal with health information registries. Ingelfinger and Drazen (2004) explain that combining the institutional review board (IRB) with two levels of separation could solve problems related to privacy.

10.2.2.3 Data Classification

Data classification is the process of organizing data into categories or classes for better utilization efficiency, mainly in the process of knowledge extraction. For example, medical data can be classified according to pathology in order to extract common characteristics between patients of the same pathology, which will allow us to find correlations between patient characteristics and pathologies (Gliklich et al., 2014; Santoro et al., 2015; Klöppel et al., 2008).

There are two approaches to data classification: supervised (Klöppel et al., 2008) and unsupervised (Lee et al., 2004). The objective of supervised classification mainly consists of defining rules for classifying objects into classes based on qualitative or quantitative variables characterizing these objects. Unsupervised classification (also known as clustering) refers to a corpus of methods whose aim is to draw up or retrieve an existing typology characterizing a set of observations, based on characteristics measured in each of the observations.

The concept of classification is closely related to learning (Nguyen et al., 2014). The latter is used to replace the acquisition of classification rules with an expert in the field. For instance, an object may correspond to a patient described by a set of characteristics such as age, sex, weight, and the class could be a particular disease. Learning in this example would be to "learn" classification rules from a set of patient characteristics.

10.2.3 Knowledge

Transforming data into knowledge (i.e., relevant data) is a general goal of disease registries. Medical knowledge is the corpus of mental representation relating to diseases and medicine. Knowledge management is a complex process that requires taking into account several parameters, such as having a precise objective, being sustainable over time with a definition of a precise process of updating, conviction, involvement and dedication of participants, and taking into account the available tools.

Medical expertise typically refers to the ability of assessing all physical injuries suffered by a victim. The location of expertise expresses the identification of medical skills for a well-defined pathology.

A transfer of knowledge is valuable if the knowledge is reused. The publication of data will be used to reproduce the data in a different context. Publishing is achieved by (1) basic tools such as books, brochures, (2) computer tools like CD ROM, DVD, flash disk, and (3) collaborative interfaces such as forums, specialized web sites. Wyatt (MySQL, 2017) confirmed these hypotheses based on a survey for 402 US health professionals.

10.2.4 Analytics

Data evaluation using analytical and logical reasoning to examine each component of the provided data is an important task in disease registries. This form of analysis is only one of many steps to be taken when conducting a research experiment. Data from various sources is collected, reviewed, and analyzed to reach a conclusion. There are a several methods for analyzing specific data of diseases registries. Some of them include data mining (Breault et al., 2002) and statistical analysis (Onukwugha et al., 2014). Data mining allows the analysis of large datasets in order to discover models and use these models to predict the probability of future events. Data mining offers great potential for the healthcare industry to enable health systems and systematically use data and analysis to identify inefficiencies and best practices that improve care and reduce costs.

In the medical field, data analysis consists of collecting, regrouping, and processing information concerning patients in order to establish statistics relating to a pathology. The heterogeneity of data is a crucial problem in processing data, including research, selection, interpretation, analysis, and exploitation. As a result, information must be represented in a conceptual format that supports evolution and change.

10.2.5 Services

Services for actors such as patients or physicians and services consumed from other actors or service providers are essential components of disease registries. We distinguish between (1) diagnosis assistance, (2) patient monitoring, (3) predicting anomalies, (4) data analysis and social networks. They are discussed below.

10.2.5.1 Diagnosis Assistance

The objective of diagnosis assistance services is to set up tools to assist physicians in decision-making, whether for diagnosis or treatment. Different approaches are discussed in the literature.

Mougiakakou et al. (2006) present an integrated system for medical image archiving, management, diagnosis support, and telematic cooperation. The proposed tool ensures collaboration and provides support by computing tomography images for focal liver lesions diagnosis.

Jin et al. (2005) propose a Wireless Health Advanced Mobile Bio-diagnostic System (abbreviated as WHAM-BioS) to improve medical services and diagnosis accuracy. The authors design a new networked sensor array architecture for long-term periodic tele-assistance applications.

Ketata et al. (2007) propose a new approach for adjustment of membership functions, generation, and reduction of fuzzy rule base from data in the same time. The authors apply this approach for liver trauma diagnostics.

10.2.5.2 Patient Monitoring

This service is used to provide a remote follow-up of patients who suffer from chronic diseases. The goal is to provide patients with portable sensors to monitor changes in health parameters, such as blood pressure, heart rate, weight, and glucose levels.

Tai et al. (2014) propose a wide adoption of electronic health record systems to assist the establishment of computer-based substance use disorder patient registries (SUD). They prove that such linkage can facilitate the advancement of SUD treatment research efforts and improve patient care.

Bogousslavsky et al. (1988) present the Lausanne Stroke Registry with an analysis of 1000 consecutive patients having a first stroke. Several analyses were conducted. For example, 24 h ECG (Holter) monitoring was performed on selected patients.

10.2.5.3 Anomaly Prediction

As previously mentioned, this service is based on the use of the results of the monitoring service. It involves analyzing the data collected by monitoring to detect and predict crisis situations and act on time. For example, it would be useful to detect/predict nocturnal epileptic seizures for children and to alert parents to take necessary precautions. Indeed, a crisis of night epilepsy can lead to the death of the child (Duun-Henriksen, 2012).

Schwenk et al. (2014) explore the validity of sensor-derived physical activity (PA) parameters for predicting future falls in people with dementia. Diagnosis is based on medical history, clinical examination, cerebral imaging, and established neuropsychological tests to establish a registry for Alzheimer's disease.

10.2.5.4 Data Analysis and Social Networks

An important part of the data can be extracted from social media (e.g., Facebook, Twitter) and blogs. Indeed, Internet users are expressing themselves more and more on social networks and tend to reveal the secrets of their personal lives on social media. These data can be used to provide tele-assistance and more accurate diagnosis for patients suffering from mental illnesses, such as bipolar patients.

Schmidt et al. (2012) use social media to predict and track disease outbreaks. Kuehn (Ameen et al., 2012) discusses the use of social media to track foodborne illness.

10.2.6 Security

We distinguish between three levels of intrusions in e-health applications: (1) sensors, (2) web, and (3) mobile.

10.2.6.1 Sensor Security

Sensor networks offer economically viable solutions for a variety of medical applications. Confidentiality and security issues for sensor networks represent a rather complex research challenge. In sensor networks, each node represents a potential point of attack. The network can be scattered over a wide area, exposing sensors to hackers that capture and reprogram specific nodes.

Ameen et al. (2012) classify the problem of security in two categories: system and information security. For the first category, the authors argue that security safeguard measures should be applied in three levels: administrative, physical, and technical. For the second category, they point out that the system is responsible for providing the following security services on specified biomedical data: (1) data encryption, (2) data integrity, (3) authentication, (4) freshness protection.

10.2.6.2 Web Security

In the literature, there are several security flaws, including: (1) injections, (2) authentication and session management violations, (3) cross-site scripting, (4) unsecured direct referencing to an object, (5) cross-site request forgery (CSRF), (6) bad security configuration, (7) and invalid redirects.

Bellgard et al. (2013) explain that disease registeries have two levels of access control. The first one defines functional authorizations. The second one deals with the limitations of data access. The authors present several architectural layers: SSL-based encryption of the whole traffic, credentials for accessing accounts, and system behavior after login success and failure.

10.2.6.3 Mobile Security

The main security issues of mobile applications are: (1) reverse engineering ()especially for Android applications), (2) backup of sensitive data, (3) malicious applications running in parallel to detect behavior of users.

Arora et al. (2014) explain some rules for improving security of mobile health applications, such as (1) considering carefully which data needs to be transmitted and where it will be stored, (2) not storing data on a mobile device and storing it in a secure server for aggregation, (3) data encryption, (4) using a virtual private network (VPN) for data transfer, (5) authentication.

10.2.7 Sharing

In addition to data consumption in its basic state, there are several benefits of data sharing for disease registries. These include: (1) sharing the results of clinical trials, (2) generating therapeutic standards, (3) fostering collaboration among health professionals, (4) promoting innovation and future reuse of data (5) maximizing transparency and accountability, (6) improving and validating research, (7) reducing the cost of duplicating data collection, (8) increasing the impact and the visibility of the research.

Different methods exist to share medical data: (1) depositing them in a specialized data center, a data archive or a data bank, (2) depositing them in an institutional repository, (3) making them available online through a project or institutional site, (4) making them available informally on a peer-to-peer basis. In addition to the data, sharing also concerns the knowledge generated and the results of analyses.

Workman (Schmidt, 2012) discusses the concept of Patient Powered Patients registries (PPRs), where patients or families share a set of experiences with a disease or condition. He explains also how a network provides a shared infrastructure and standardized data collection across registries.

Gliklich et al. (Landgrebe and Smith, 2011) explain that we can go further: sharing data with personal health records, as well as clinical research or national public health efforts engaged a variety of providers of commercial applications.

10.3 Proposed Software Architecture

Figure 10.1 shows our proposed software architecture for developing disease registries. There are five horizontals layers: Service, Analytics, Knowledge, Data and Technology. These layers are crossed by two vertical layers: Security and Privacy, and Sharing. They should be satisfied by each horizontal layer. In the following, each layer will be discussed in more detail.

10.3.1 Technology Layer

Figure 10.2 shows the technology layer. It consists of six components:

1. Image sensors
2. Facilitator components
3. Rehabilitation components
4. Motion sensors
5. Sense sensors
6. Communication components.

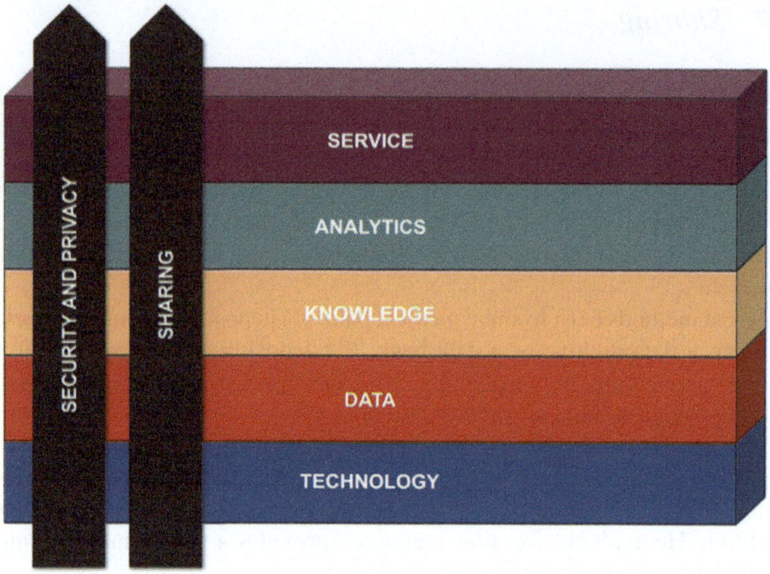

Fig. 10.1 Software architecture for disease registries

Communication components orchestrate the exchange between the different elements and with the upper layers.

10.3.1.1 Incoming Flow

The incoming flow consists of the signals sent by the sensors, the incoming streams of the cameras etc. The incoming flow must respect specific time, quality and frequency requirements. For each equipment, a list of minimum requirements should be satisfied, to ensure capturing the flow with specific needs.

Flow separation, sampling, and classification algorithms must be implemented taking into account the technical properties of the devices and the surrounding environment.

10.3.1.2 Outgoing Flow

The outgoing flow has two types of parameters:

- Quantitative parameters can be measured, like the blood pressure, the weight, or the heart rate.
- Qualitative parameters are similar to instructions. For instance, "please repeat the measurement in 10 min" and recommendations: "we must stop walking" are qualitative parameters.

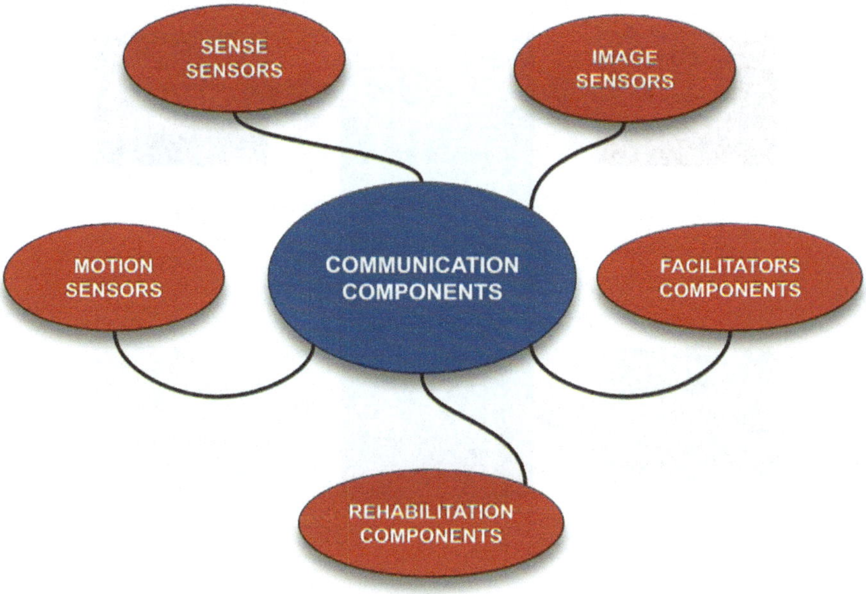

Fig. 10.2 Technology layer

Performance indicators are periodically monitored, such as speed, accuracy, and energy cost. The outgoing flow must respect international standards.

10.3.1.3 Interlayer Communication

The different tools of the technology layer can exchange pertinent information between them in order to collect filtered data or results of collaborative processing. This treatment is often associated with domains: distributed systems, artificial intelligence (multi-agent systems), and interoperability.

The communication between the different components must meet the criteria of privacy, security, processing time, etc. Different typologies can be adopted and evaluated in WiFi or Ethernet technolgies.

10.3.2 Data Layer

This layer provides the essential functionality for the management of data coming from the Technology layer, such as storag and archiving anonymized data. In particular, the Data layer is in charge of formatting data that could be supported by the upper layers of the proposed architecture. This will ensure a high level of abstrac-

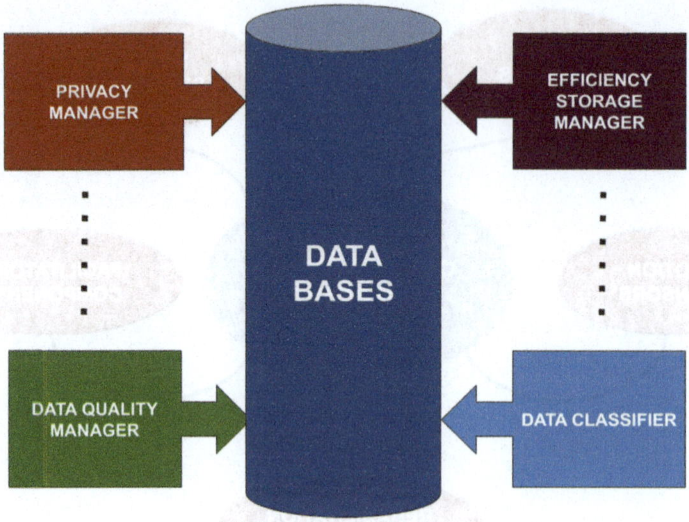

Fig. 10.3 Data layer

tion. This abstraction will facilitate the deployment of different services by means of an integration effort at the Technologies layer. In the following, some modules are presented.

Figure 10.3 illustrates the proposed architecture for the Data layer. It is composed of four modules: (1) privacy alert manager, (2) efficient storage manager, (3) data quality manager, and (4) data classifier. Other modules are be defined on the basis the physicians' requirements.

We propose to treat the privacy problem in three layers: (1) definition of a taxonomy model of sensitive personal data by physicians (and experts), (2) setting up an automatic data detection engine for information about the privacy of patients or the community. To do this, we can use cognitive search algorithms, ontologies and inference engines based on fuzzy logic, (3) implementation of techniques for anonymizing personal information identified to allow the processing of this data, while preserving the privacy of those involved.

10.3.3 Knowledge Layer

The most widespread modules that can be integrated into the Knowledge layer come from the follwing knowledge management steps: (1) identification, (2) collection, (3) organization, dissemination, (5) appropriation, (6) sharing, (7) creation.

Figure 10.4 illustrates the architecture of the Knowledge layer. These include: (1) data analyzer, (2) data advisor, (3) collaborative work tools, (4) expertise locator.

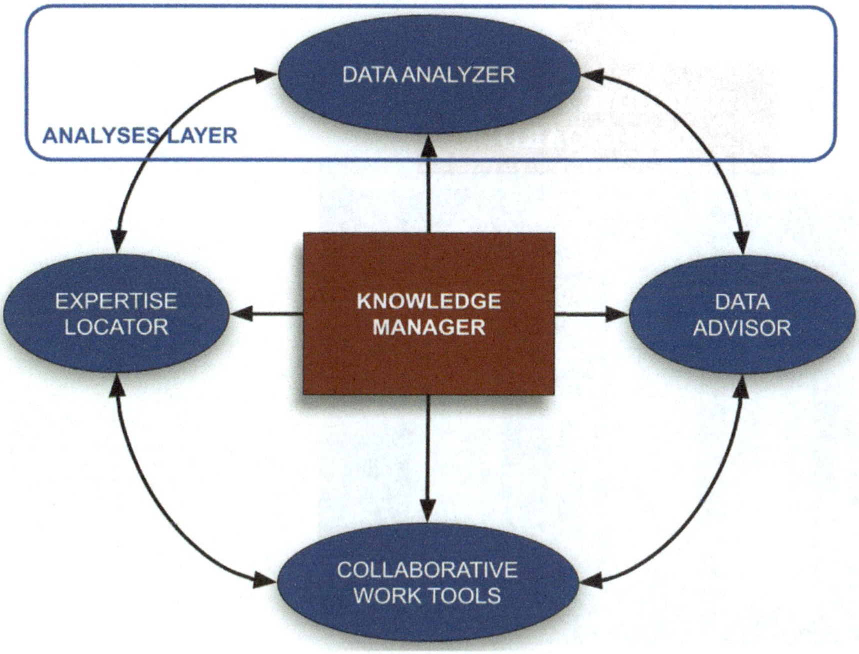

Fig. 10.4 Knowledge layer

Given the importance of the data analyzer, a specific layer is used. Its description is discussed in the next section.

10.3.3.1 Expertise Locator Module

This module can include search engines based on expert profiles (medical specialty, pharmaceutical, care, etc.). These profiles contain a description of the skills of the people to whom they refer. They will be created either explicitly or implicitly through the analysis of patient records.

10.3.3.2 Data Advisor Module

Knowledge modeling must allow for smooth navigation between concepts, indexing of documents, graphic illustration of phenomena or processes. This module consists of a collection of several platforms and collaborative tools, focusing on advanced concepts and targeting a fairly sophisticated target audience.

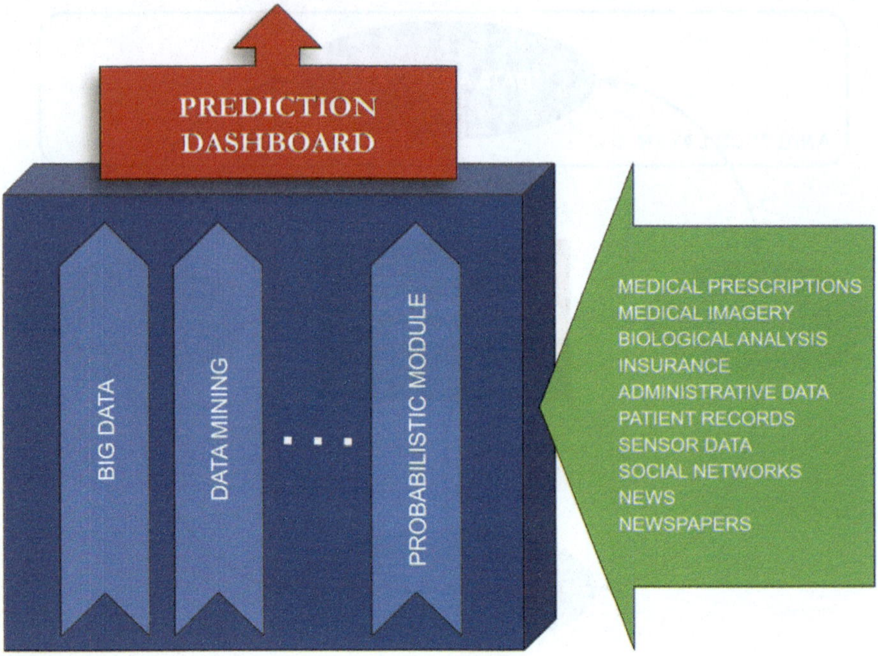

Fig. 10.5 Analytics layer

10.3.4 Analytics Layer

Figure 10.5 illustrates the functional architecture of the Analytics layer. The different entries from the lower layers are: (1) medical prescriptions, (2) MRI, scanner, radiology or cameras, (3) biological analysis results, (4) insurance data, (5) patient records, (6) administrative data, (7) sensor data, (8) data from publications in social networks, (9) news, and (10) scientific journals. In addition to the modules of Big Data and Data Mining, we can use fundamental probabilistic theory and mathematical logic. Expected outputs are essentially a dashboard allowing predictions and a fast decisions.

10.3.5 Service Layer

Figure 10.6 summarizes the internal architecture of the Service layer. There are two types of services: consumed and returned. The first one includes applications and infrastructure providers. The second one is in charge of prediction anomalies, tele-assistance, data analysis, social networks, creation of reference, sharing of expertise, and diagnostic assistance.

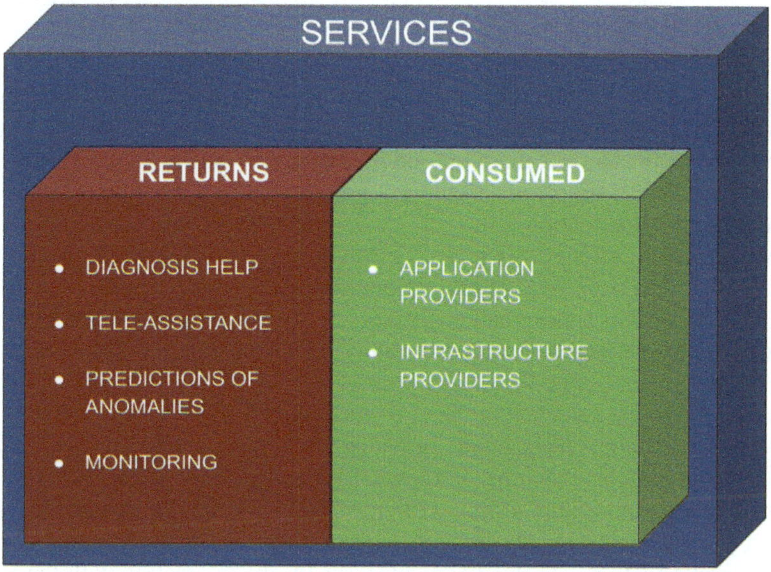

Fig. 10.6 Service layer

10.3.6 Security and Privacy

A transversal layer is provided in the proposed architecture. This layer will enforce the security and privacy requirements throughout the various layers.

We distinguish three levels of intrusions: (1) sensors, (2) web, and (3) mobile. For each level, a charter of good practice is set up for development and mechanisms for warning and secure operation.

For the web, we propose the design and implementation of intrusion detectors around the data themselves. Alerts are triggered by intelligent data analysis systems.

For the mobile part, we propose techniques of obfuscation, class encryption associated with a mechanism of blocking, expiration of sessions and detection of intrusion. The research is particularly interesting in terms of response time and intelligent analysis of the user's actions.

10.3.7 Sharing

A second transversal layer, called Sharing layer, is provided to ensure exchange between the layers. We consider sharing and notification algorithms ensuring efficient storage, a real-time exchange and, above all, a dynamic rights management mechanism. The dynamicity constitutes the optimization of the algorithms of usual engineering.

10.3.8 Interactions

The proposed architecture offers the possibility of interacting with several categories of actors, such as patients, physicians, and administrative agents. Patients and physicians are the main actors of the proposed platform.

Interaction with patients is ensured through the layers Technology and Service. The first type of interaction allows the acquisition of the different types of data, whether clinical or through connected objects. The second type of interaction concerns the use of services provided by the platform, in particular access to its medical data, monitoring of its state of health, receipt of instructions and from its doctors, etc. For example, the patients have the opportunity to open channels for direct interaction with their physicians, either online for immediate consultation or offline by sending new relevant documents related to their medical records.

Interaction with doctors is mainly provided by the Service and Knowledge layers. Indeed, the creation and sharing of knowledge is one of the major objectives to be achieved by the proposed platform. Although the knowledge layer is mainly fed by the data and analyzes carried out on these data, the role of the physician is crucial in the validation and use of this knowledge. As for the interaction with the Service layer, it is a matter of making the doctor benefit, both in his or her diagnostic work and in his or her treatment of the services offered, such as help with diagnosis.

The proposed architecture opens the door for interaction to different actors, other than doctors and patients, to set up services or exchange data and knowledge with other actors. In particular, service providers who will come to interface with the Technology layer in order to ensure the import of data, especially in applications requiring the use of connected objects providing data in an unusual format, could be envisaged.

10.4 Use Cases

The proposed software architecture is defined on the basis of several experiences and observations collected during the development of several disease registries: Tunisian Fanconi Anemia registry (TFAR), Tunisian Gaucher Disease Registry (TGDR), and Tunisian Non-Hodgkin Lymphoma Registry (TNHLR). Currently, it is applied to developing the Tunisian Diabetes Type 1 Registry (TDR-T1). The latter is in the testing and development phases of the Analytics and Service layers. Other tests are performed by applying artificial intelligence concepts to (1) the diagnosis process of liver trauma severity, (2) a comparison between parameters that can predict mortality by staphylococcus aureus, and (3) diagnosis and treatment assistance for acute dehydration.

Given the accessibility and interoperability requirements, the platform engine is implemented as a Java Web application. Figure 10.7 illustrates our particular usage of the proposed software architecture.

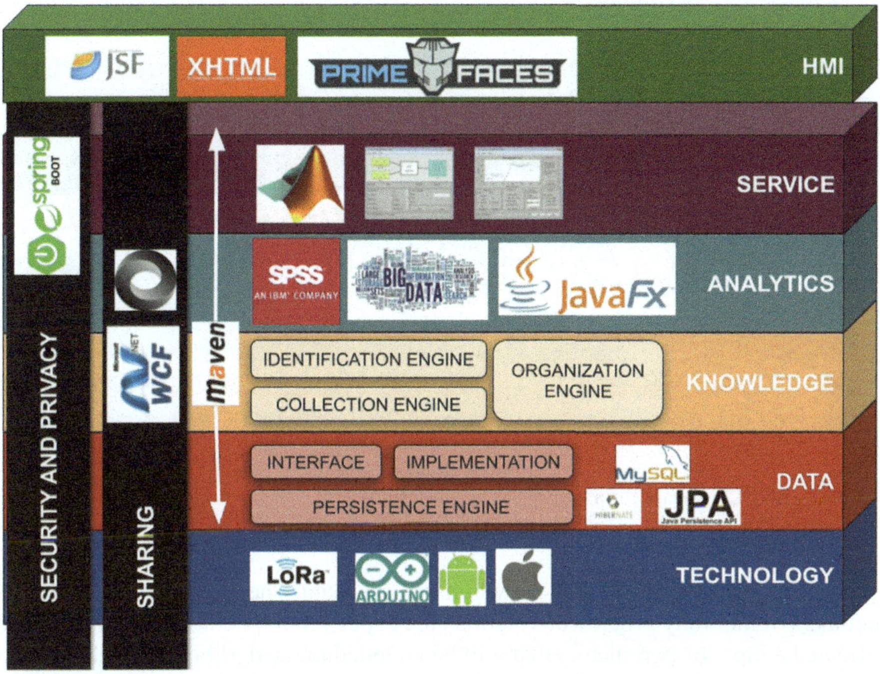

Fig. 10.7 Using the proposed software architecture

Data structure persistence is ensured by Hibernate (Hibernate, 2017) as an object-relational mapper and MySQL (MySQL, 2017) as a relational database. Thus, we reduce database coupling with the application and protect the development of future changes in the domain model's storage system.

Additional components are also used, such as Spring Security for improved security features, Apache Tomcat to deploy our Java Servlets, JSPs, JSFs files and Apache Maven for project dependency management, building and deployment.

Additional layers have been implemented to interface with different technologies, like the use of web services SOAP or JSON.

The Tunisian Fanconi Anemia Disease Registry, the Tunisian Diabetic Type 1 Disease Registry, and the Tunisian Gaucher Disease Registry have more than 200 fields in their electronic disease forms, and the Tunisian Non-Hodgkin Lymphoma Disease Registry (TNHLR) has up to 1000 fields. The use of the AccordionPanel of the Primefaces framework is highly recommended for monitoring transactions in the database. Figure 10.8 illustrates a part of the disease form of TNHLR.

Secure access to the web workspace is implemented through a HTTPS channel. The use of valid certificates enforces the execution of workflows within the service execution engine. Private data are encrypted using the RSA encryption algorithm (Rivest et al., 1978). We distinguish between four groups of users: (1) administrator, (2) referent, (3) participant, and (4) guest (see Fig. 10.9). Administrators have the

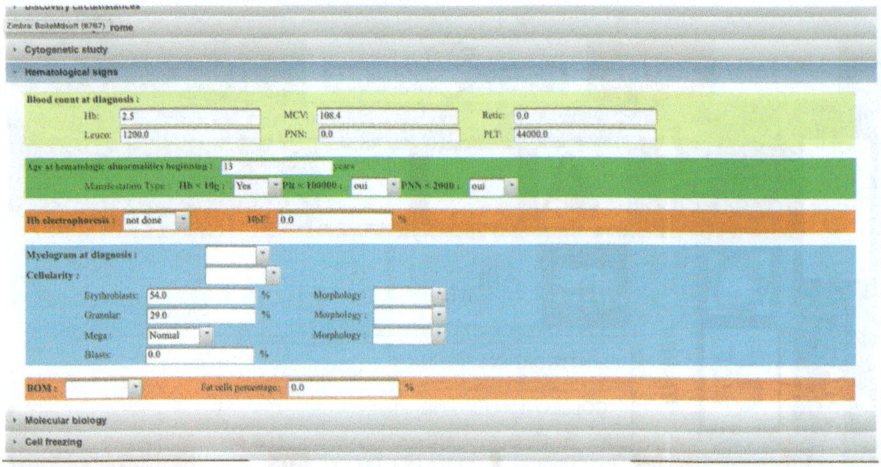

Fig. 10.8 TNHLR disease form

rights to create users and access the whole registry data. The referent doctors are the members of the study group who establish the paper version of the electronic disease form and define the parameters that will be included in the dashboard. Each referent member has the modification rights for electronic disease forms (EDFs) included by him/her. He/She also has reading rights for the other EDFs and access to the whole statistic values. Each participant member can access only his/her EDF and some statistical values authorized by the administrators. These rules can be changed from one registry to another. Given the sensitive nature of the majority of the stored data, all actions are saved, which helps to monitor everything that happens in the system.

The filling of an EDF may require the intervention of one or more specialists. For example, in the case of Fanconi anemia, several specialties are considered such as hematology, cytogenetic, and marrow transplantation. In this case, the rights of modification are managed by the administrator according to the pre-established rules of the study group.

Figure 10.10 illustrates the life cycle of the EDF. After creating the EDF by one specialist and validated by an administrator, it is in operating mode. Included data are used to draw the dashboard. Once modified, it will be in standby mode until validated by the administrator. Corresponding data are not used in this case. Once validated, the form returns to standard operating mode. The EDF can be deleted by its creator or assigned owner or the administrator. After validation, it remains in the database for tracability reasons but will not be considered in dashboard. A backup database is always set up to avoid problems of overflow or bad manipulations.

In order to avoid the risks associated with reverse engineering in mobile applications, we opted for manual obfuscation. In addition, no data of a private nature is saved on the device.

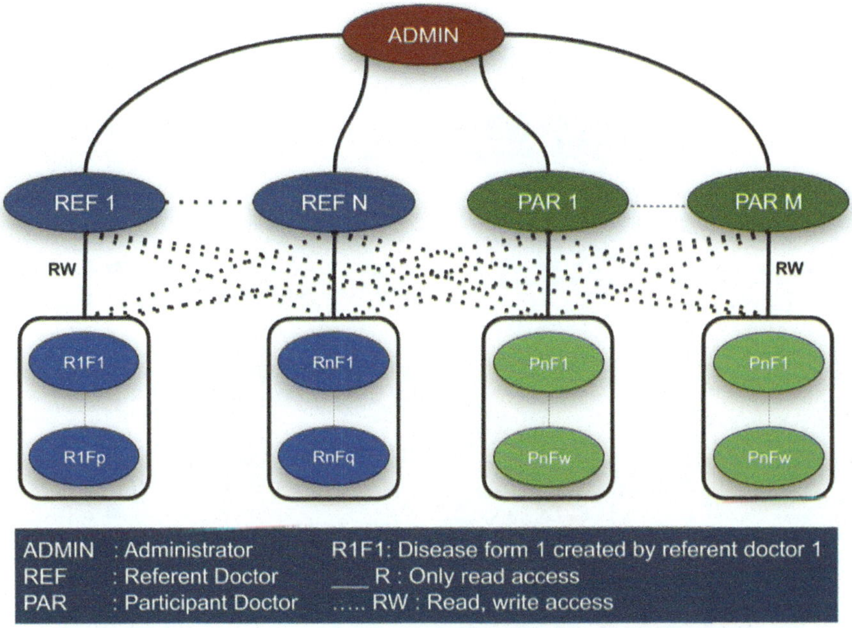

Fig. 10.9 Rights

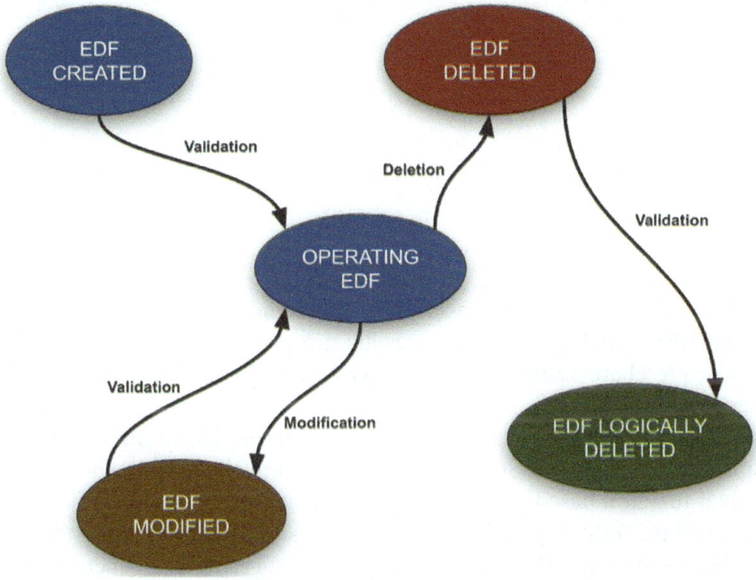

Fig. 10.10 Electronic Disease Form (EDF) life cycle

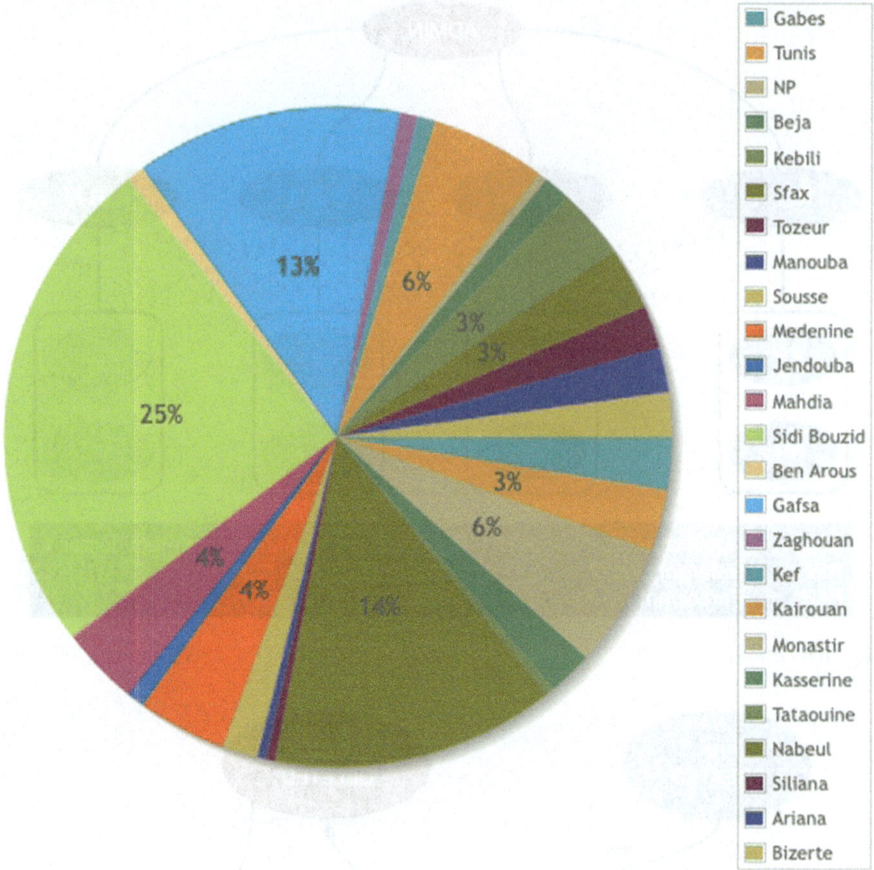

Fig. 10.11 Pie chart in TFAR

The data is analyzed using basic SQL commands on the database, Microsoft Excel, and SPSS (IBM, 2017). The results are presented in the form of curves, histograms, pie charts, percentages and crisp values. JavaFX and JFreeChart are used (see Fig. 10.11).

After the establishment of each registry, it is important to discuss the value of the Kaplan-Meier estimator S(t). The Kaplan-Meier estimator allows us to estimate the survival function based on lifetime data. S(t) is a decreasing staircase function, continuous on the right. Figure 10.12 shows the result example given by SPSS (IBM, 2017). The curve shows the evolution of overall survival as a function of age for TFAR registry. The median age of survival is 17 years 5 months.

A comparative interoperability study was conducted with tests of interaction between existing systems. The work carried out included the most relevant work and research trends to solve the problem of heterogeneity (Hammami et al., , 2014(@).

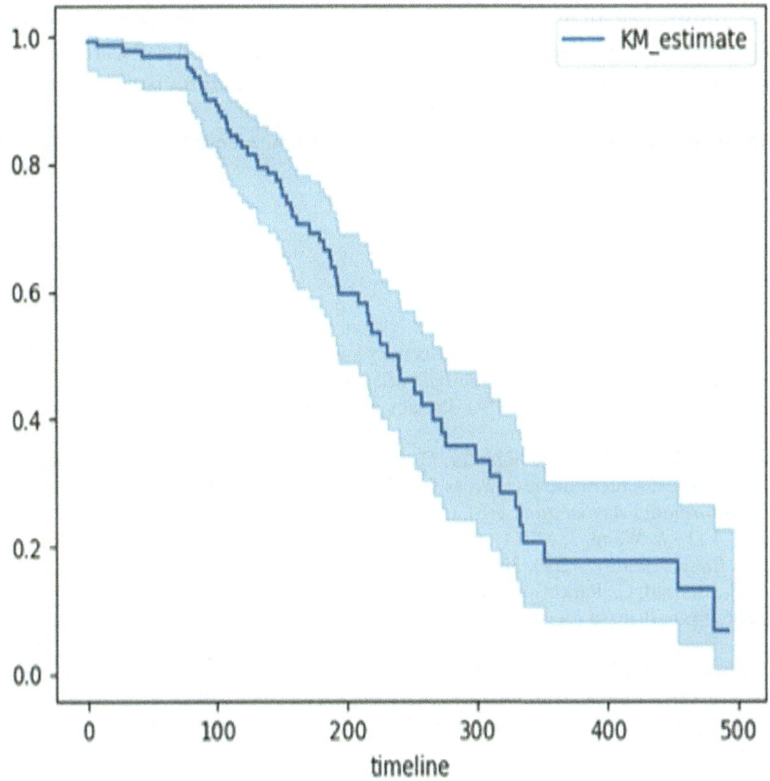

Fig. 10.12 Kaplan-Meier for TFAR registry

The study was extended to the semantic web, and a new approach has been proposed (Hammami et al., 2016).

10.5 Conclusion

In this contribution, we discussed several issues related to disease registries and proposed a novel software architecture for developing disease registries. The architecture consists of five horizontal layers: (1) Technology layer, (2) Data layer, (3) Knowledge layer, (4) Analytics layer, (5) Services layer, and two transversal layers: (1) Security and Privacy layer, and (2) Sharing layer. Several use cases were presented: (1) Tunisian Fanconi Anemia Registry (TFAR), (2) the Tunisian Gaucher Disease Registry (TGDR), (3) Tunisian Non-Hodgkin Lymphoma Registry (TNHLR), and (4) Tunisian Registry of Diabetes type 1 (TRD-T1). The proposed architecture has been validated by health professionals.

In the future, we plan to develop a disease registry generator allowing us to automatically generate disease registries based on the proposed software architecture. Moreover, we aim to design a general architecture for such registry generators.

Acknowledgements This work is supported by the German Academic Exchange Service (DAAD) (Transformation Partnership: Theralytics Project).

References

Ameen, M. A., Liu, J., & Kwak, K. (2012). Security and privacy issues in wireless sensor networks for healthcare applications. *Journal of Medical Systems*, *36*(1), 93–101.

Arora, S., Yttri, J., & Nilsen, W. (2014). Privacy and security in mobile health research. *Alcohol Research*, *36*(1), 143–151.

Arts, D. G., De Keizer, N. F., & Scheffer, G. J. (2002). Defining and improving data quality in medical registries: A literature review, case study, and generic framework. *Journal of the American Medical Informatics Association*, *9*(6), 600–611.

Bandodkar, A. J., & Wang, J. (2014). Non-invasive wearable electrochemical sensors: A review. *Trends in Biotechnology*, *32*(7), 363–371.

Bellgard, M., Beroud, C., Parkinson, K., Harris, T., Ayme, S., Baynam, G., et al. (2013). Dispelling myths about rare disease registry system development. *Source Code for Biology and Medicine*, *8*, 21.

Bogousslavsky, J., Van Melle, G., & Regli, F. (1988). The lausanne stroke registry: Analysis of 1,000 consecutive patients with first stroke. *Stroke*, *19*(9), 1083–1092.

Breault, J. L., Goodall, C. R., & Fos, P. J. (2002). Data mining a diabetic data warehouse. *Artificial Intelligence in Medicine*, *26*(1–2), 37–54.

Bulling, A., Ward, J. A., Gellersen, H., & Troester, G. (2011). Eye movement analysis for activity recognition using electrooculography. *IEEE Transactions on Pattern Analysis and Machine Intelligence*, *33*(4), 741–753.

Chatterjee, S., & Price, A. (2009). Healthy living with persuasive technologies: Framework issues, and challenges. *Journal of the American Medical Informatics Association*, *16*(2), 171–178.

Daynard, D., Yassi, A., Cooper, J. E., Tate, R., Norman, R., & Wells, R. (2001). Biomechanical analysis of peak and cumulative spinal loads during simulated patient-handling activities: A substudy of a randomized controlled trial to prevent lift and transfer injury of health care workers. *Applied Ergonomics*, *32*(3), 199–214.

Duun-Henriksen, J. (2012). Detection and prediction of epileptic seizures, Ph.D. thesis, Technical University of Denmark.

ECRI. (2014). Top 10 health technology hazards for 2015 a report from health devices. https://www.ecri.org/Documents/White_papers/Top_10_2015.pdf. Last visit: 01/07/2020.

El Emam, K., Jonker, E., Sampson, M., Krleza-Jerić, K., & Neisa, A. (2009). The use of electronic data capture tools in clinical trials: Web-survey of 259 Canadian trials. *Journal of Medical Internet Research*.

Fleischer, M., Simon, E., Rumpel, E., Ulmer, H., Harbeck, M., Wandel, M., et al. (2002). Detection of volatile compounds correlated to human diseases through breath analysis with chemical sensors. *Sensors and Actuators B: Chemical*, *83*(1–3), 245–249.

Gavrila, D. M. (1999). The visual analysis of human movement: A survey. *Computer Vision and Image Understanding*, *73*(1), 82–98.

Gill, R. W. (1985). Measurement of blood flow by ultrasound: Accuracy and sources of error. *Ultrasound in Medicine & Biology*, *11*(4), 625–641.

Gliklich, R. E., Dreyer, N. A., & Leavy, M. B. (2014). Registries for evaluating patient outcomes: A user's guide. Rockville (MD): Agency for Healthcare Research and Quality (US).

Hammami, R., Bellaaj, H., & Kacem, A. H. (2014). Interoperability for medical information systems: An overview. *Health and Technology*, *4*(3), 261–272.

Hammami, R., Bellaaj, H., & Kacem, A. H. (2016). A novel approach for semantic web service discovery. *WETICE, 2016*, 250–252.

Hibernate. (2017). Hibernate. http://www.hibernate.org, Last visit: 01/07/2020.

IBM. (2017). SPSS. https://www.ibm.com/analytics/us/en/technology/spss/, Last visit: 01/07/2020.

Ingelfinger, J. R., & Drazen, J. M. (2004). Registry research and medical privacy. *The New England Journal of Medicine, 1;350*(14), 1452–1453.

Jin, M. H., Lee, R. G., Kao, C. Y., Wu, Y. R., Hsu, D. F., Dong, T. P., & Huang, K. T. (2005). Sensor network design and implementation for health telecare and diagnosis assistance applications. In *Proceedings of the 11th International Conference on Parallel and Distributed Systems*.

Ketata, R., Bellaaj, H., Chtourou, M., & Ben Amer, M. (2007). Adjustment of membership functions, generation and reduction of fuzzy rule base from numerical data. *Malaysian Journal of Computer Science, 20*(2).

Klöppel, S., Stonnington, C. M., Chu, C., Draganski, B., Scahill, R. I., Rohrer, J. D., et al. (2008). Automatic classification of MR scans in Alzheimer's disease. *Brain, 131*(Pt 3), 681–9.

Kohl, C. D., Garde, S., & Knaup, P. (2010). Facilitating secondary use of medical data by using OpenEHR archetypes. *Studies in Health Technology and Informatics*, *160*, 1117–21.

Landgrebe, J., & Smith, B. (2011). The HL7 approach to semantic interoperability. In *International Conference on Biomedical Ontologies*, pp. 140–146.

Lee, J. S., Chu, I. S., Heo, J., Calvisi, D. F., Sun, Z., Roskams, T., et al. (2004). Classification and prediction of survival in hepatocellular carcinoma by gene expression profiling. *Hepatology*, *40*(3), 667–76.

Löwel, H., Lewis, M., Hörmann, A., & Keil, U. (1991). Case finding, data quality aspects and comparability of myocardial infarction registers: Results of a south German register study. *Journal of Clinical Epidemiology*, *44*(3), 249–260.

Lymberis, A. (2003). Smart wearable systems for personalised health management: Current R&D and future challenges, engineering in medicine and biology society. In *Proceedings of the 25th Annual International Conference of the IEEE*.

Mougiakakou, S. G., Valavanis, I., Mouravliansky, N. A., Nikita, A., & Nikita, K. S. (2006). Diagnosis: A telematics enabled system for medical image archiving, management and diagnosis assistance, imagining systems and techniques. In *Proceedings of the 2006 IEEE International Workshop on Imaging*.

MySQL. (2017). MySQL. https://www.mysql.com/. Last visit: 01/07/2020.

National Institute of Mental Health. (2016). *List of Registries*.

NATO. (2005). NATO Intelligence, Surveillance, and Reconnaissance (ISR) Interoperability Architecture (NIIA), NATO UNCLASSIFIED (Allied Engineering).

Nguyen, D. H., & Patrick, J. D. (2014). Unsupervised machine learning and active learning in classification of radiology reports. *Journal of the American Medical Informatics Association*, *21*(5), 893–901.

Onukwugha, E., Yong, C., Hussain, A., Seal, B., & Mullins, C. D. (2014). Security issues in healthcare applications using wireless medical sensor networks: A survey. *Sensors*, *12*(1), 55–91.

Onukwugha, E., Yong, C., Hussain, A., Seal, B., & Mullins, C. D. (2014). Concordance between administrative claims and registry data for identifying metastasis to the bone: An exploratory analysis in prostate cancer. *BMC Medical Research Methodology*, *14*, 1.

Openehr. (2017). openehr. http://www.openehr.org/. Last visit: 01/07/2020.

Orphanet. (2016). Rare disease registries in Europe. Orphanet report series. http://www.orpha.net/orphacom/cahiers/docs/GB/Registries.pdf. Last visit: 01/07/2020.

Pavluck, A., Chu, B., Flueckiger, R. M., & Ottesen, E. (2014). Electronic data capture tools for global health programs: Evolution of LINKS, an Android. *Web-Based System, PLoS Neglected Tropical Diseases, 8*(4)

Rivest, R. L., Shamir, A., & Adleman, L. M. (1978). A method for obtaining digital signatures and public-key cryptosystems. *Communications of the ACM, 21*(2), 120–126.

Roos, L. L., Mustard, C. A., Nicol, J. P., McLerran, D. F., Malenka, D. J., Young, T. K., et al. (1993). Registries and administrative data: Organization and accuracy. *Medical Care, 31*(3), 201–12.

Santoro, M., Coi, A., Lipucci, Di Paola, & M., Bianucci, A. M., Gainotti, S., Mollo, E., Taruscio, D., Vittozzi, L., & Bianchi, F. (2015). Rare disease registries classification and characterization: A data mining approach. *Public Health Genomics, 18*(2), 113–22.

Schmidt, C. W. (2012). Trending now: Using social media to predict and track disease outbreaks. *Environmental Health Perspectives, 120*(1), a30–a33.

Schouten, L. J., Straatman, H., Kiemeney, L. A., Gimbrere, C. H., & Verbeek, A. L. (1994). The capture-recapture method for estimation of cancer registry completeness: A useful tool?. *International Journal of Epidemiology, 23*(6), 1111–1116.

Schwenk, M., Hauer, K., Zieschang, T., Englert, S., Mohler, J., & Najafi, B. (2014). Sensor-derived physical activity parameters can predict future falls in people with dementia. *Gerontology, 60*(6), 483–92.

Stone, A., Shiffman, S., Atienza, A., & Nebeling, L. (2007). *The science of real-time data capture: Self-reports in health research*. Oxford University Press.

Tai, B., Hu, L., Ghitza, U. E., Sparenborg, S., Veldhuisen, P. V., & Lindblad, R. (2014). Patient registries for substance use disorders. *Substance Abuse and Rehabilitation, 5*, 81–86.

Wang, A. Y., Sable, J. H., & Spackman, K. A. (2002). The SNOMED clinical terms development process: Refinement and analysis of content. In *Proceedings of the AMIA Symposium* 845–849.

Workman, T. A. (2013). Engaging patients in information sharing and data collection: The role of patient-powered registries and research networks, agency for healthcare research and quality (US).

Wyatt, J. (1991). Use and sources of medical knowledge. *The Lancet, 338*(8779), 1368–1373.

Zuckerwar, A. J., Pretlow, R. A., & Baker, D. A. (1993). Development of a piezopolymer pressure sensor for a portable fetal heart rate monitor. *IEEE Transactions on Biomedical Engineering, 40*(9), 963.

Part IV
Biomedical Characteritics

Chapter 11
3D Textures Analysis Based on Features Extraction

Samah Yahia, Yassine Ben Salem, and Mohamed Naceur Abdelkrim

Abstract The Decimal Descriptor Patterns (DDP) operator is a recent method of analyzing textures. Based on structural analysis of pattern distribution, the DDP presents an excellent description of the spatial structure of image texture. In this chapter, two successful applications of this operator are highlighted. The first one consists in Multiple Sclerosis (MS) Lesions detection from 3D Magnetic Resonance (MR) brain images tissue. The second application is 3D face analysis. Different experimentations conducted on a two well-know three-dimensional databases. For comparison, two best known operators of texture measures that are considered as references in image analysis are chosen: the Grey Level Co-occurrence Matrix (GLCM) and the Local Binary Patterns (LBP). Several tests of texture classification are performed in the same conditions using the classifier multiclass Support Vector Machines (SVM). The performance of the three operators is tested in front of illumination, inhomogeneity, contrast, and noise level variations with a various and a large number of textures. Experimental results show clearly the robustness of the DDP operator.

Keywords DDP · Textures · 3D MR brain images · 3D face analysis · GLCM · LBP · SVM

11.1 Introduction

Texture has been widely used in human life since it provides useful information that appeared on the surface of every object.

S. Yahia (✉) · Y. B. Salem · M. N. Abdelkrim
Research Laboratory Modeling, Analysis and Control of Systems (MACS),
University of Gabes - Tunisia National Engineering School of Gabes (ENIG), Gabes, Tunisia
e-mail: ysamah4@gmail.com

Y. B. Salem
e-mail: bensalemy73@yahoo.fr

M. N. Abdelkrim
e-mail: Naceur.abdelkrim@enig.rnu.tn

N. Derbel and O. Kanoun (eds.), *Advanced Methods for Human Biometrics*,
Smart Sensors, Measurement and Instrumentation 40,
https://doi.org/10.1007/978-3-030-81982-8_11

The most common use of texture is to help everyone to identify different objects in daily life. Texture is also often involved in many important real life applications such as medical image processing, remote sensing, wood species recognition, in textile industry etc.

Such situation has encouraged extensive researches to be conducted on texture, such as texture analysis and texture classification under the computer vision field.

In this contribution, we conducted a research study on texture classification. Features extraction is the most important phase in the procedure of texture classification. In others words, to success the classification, it is necessary to apply an efficient analysis texture method.

Since some decades, numerous approaches were developed to describe the content of textures. The strategy of texture characterization can be categorized in four main classes.

The first class groups approaches that consider the texture as the realization of a random field. The aim is to describe the image using probabilistic pattern like Markoviens patterns (Chellappa et al., 1985). Furthermore, several approaches rely on a statistic analysis of spatial distributions of grey levels such as the Grey Level Co-occurrence Matrix (GLCM) (Haralick & Shanmugam, 1972; Haralick et al., 1973).

Another kind class of textures analysis consists in the frequency filtering of image content using filtrate algorithms such as wavelet decompositions (Arivazhagan & Ganesan, 2003). The last class of textures analysis groups the geometric approaches. Methods of this class are based on a structural analysis of patterns organization. We can cite as example the Local Binary Patterns (LBP) (Ojala et al., 1996, 2002).

This chapter covers three efficient methods for the analysis of textures in the 3-dimensional space. Recently, The Decimal Descriptor Patterns (DDP), an efficient operator was introduced on (Yahia et al., 2016). The Local Binary Pattern (LBP) (Ojala et al., 1996, 2002) and the Grey Level Co-occurrence Matrix (GLCM) methods (Haralick et al., 1972, 1973) which are successfully used in various applications (Yahia et al., 2016, 2018a, b) and the Decimal Descriptor Pattern (DDP) which is a new promising method are compared.

The performance of these three approaches is tested with a large number of 3D textures and evaluated in front of illuminations variation, inhomogeneity, contrast, and noise level variations, surfaces and textures rotation.

The chapter is organized as follows: Sect. 11.2 introduces the three used methods DDP, LBP and GLCM. Section 11.3 presents the different experiments performed and the obtained results. Some concluding remarks are given in Sect. 11.4.

11.2 Methods of Texture Measures

11.2.1 Decimal Descriptor Patterns (DDP)

11.2.1.1 Presentation of the DDP Method

This technique was introduced on (Yahia et al., 2018a). The objective of the Decimal Descriptor Pattern (DDP) approach is to analyze the texture repartition, extract a set of attributes and construct a feature vector that describes the texture image.

Features extraction using the DDP operator can be summarized on three main steps:

- The first step consists in extracting of patterns from the texture image and calculating features that describe each pattern to obtain a first features vector.
- The second step corresponds to calculating of the maximum value, the minimum value and the mean value of the first features vector.
- The third step is construction of the final features vector by assigning a DDP code to each value of the first features vector.

To obtain an efficient features vector, the DDP operator proposes to label the entire pixels image. The first step consists in computing from the 3×3 neighborhood of each pixel three textural measures: the mean, the maximal and the minimal values.

This way, the DDP operator provides a more robust way for describing patterns in a texture. As illustrate the fig., if we consider an image Img composed of patterns of 3*3 pixels, the DDP technique defines to each pattern a set of Els elements ($E_{ls} = \{E_{l1}, E_{l2}..., E_{l9}\}$). This set correspond respectively of pixel intensities El1, ..., El9, the neighborhoods pixels.

Then, three elements E_{lm1}, E_{lm2} and E_{lm3} are extracted from the set E_{ls} and defined as the following way:

$$E_{lm1} = \max E_{l1}, E_{l2}..., E_{l9} \tag{11.1}$$

$$E_{lm2} = \min E_{l1}, E_{l2}..., E_{l9} \tag{11.2}$$

$$E_{lm3} = \text{ mean } E_{l1}, E_{l2}..., E_{l9} = (E_{l1} + E_{l2}..., E_{l9})/9 \tag{11.3}$$

Consequently, the different attributes obtained from all the image texture are concatenated together in one vectorV_1 (Fig. 11.1).

The second step consists in computing the maximum ($MaxV$), the minimum value ($\min V$) and the mean value ($\text{mean } V$) of this vector V_1. Using ($\max V$), ($\min V$) and ($\text{mean } V$), a code is assigned to each value of the vector V_1.

The code is an integer in this set $S = \{0, 1, 2, 3, 4, 5, 6, 7, 8, 9, 10\}$.

To do this, the vector V1 is decomposed in 8 intervals and only 3 values are considered to calculate codes as it is detailed in Table 11.1. For instance, the minimal

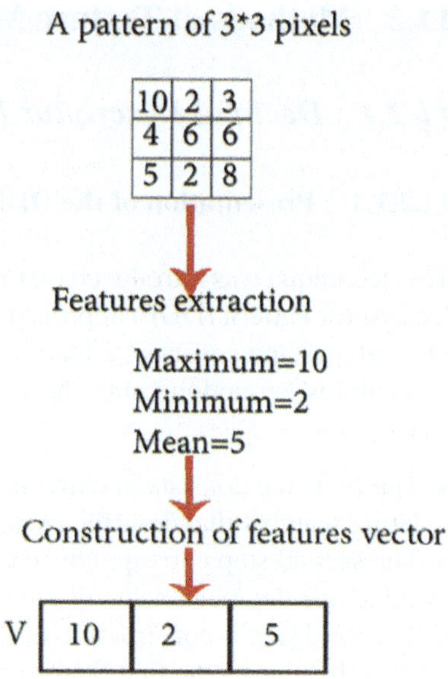

Fig. 11.1 Construction of the features vector using the DDP approach

Table 11.1 Calculation of the DDP code

Values and intervals of the features vector	Values
min	0
]min , min+$\frac{1}{4}$(mean-min)]	1
] min+$\frac{1}{4}$(mean-min), min+$\frac{1}{2}$(mean-min)]	2
] min+$\frac{1}{2}$(mean-min), min+$\frac{3}{4}$(mean-min)]	3
] min+$\frac{3}{4}$(mean-min), mean [	4
mean	5
] mean , mean+$\frac{1}{4}$(max-mean) [	6
] mean+$\frac{1}{4}$(max-mean), mean+$\frac{1}{2}$(max-mean)]	7
]mean+$\frac{1}{2}$(max-mean), mean++$\frac{3}{4}$(max-mean)]	8
]mean+$\frac{3}{4}$(max-mean), max]	9
max	10

value of the vector V is coded by 0, the maximal value by 10 and all values that are in $]\min V, (\text{mean } V + \min V)/4]$ by 1.

So the final feature vector is a succession of decimals values.

In order to highlight the DDP approach, an example has been presented in Yahia et al. (2018b) illustrated in Figs. 11.2, 11.3, 11.4 and 11.5. Given an image Img, the DDP operator starts by labeling its pixels. So the maximum, the minimum and the mean value from the 3×3 neighborhood of each pixel are computed as it illustrated in Fig. 11.2.

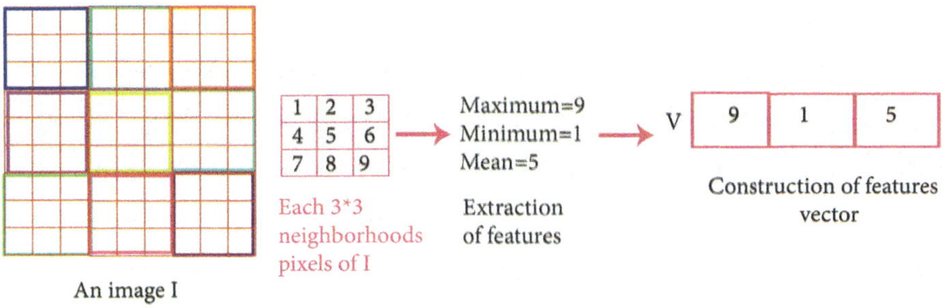

Fig. 11.2 Features extraction from a 3 × 3 pixels

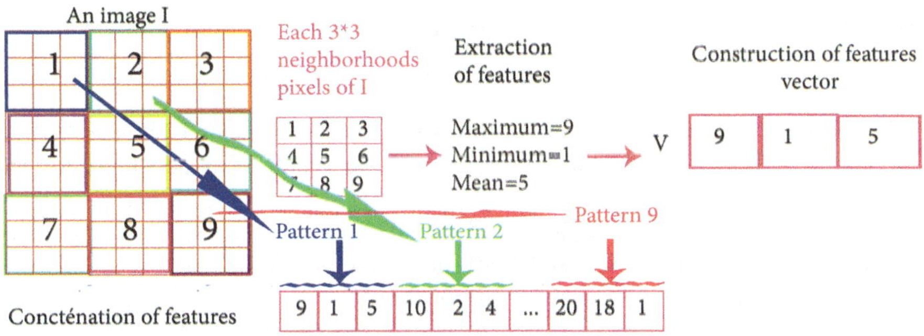

Fig. 11.3 Features extraction from all the image

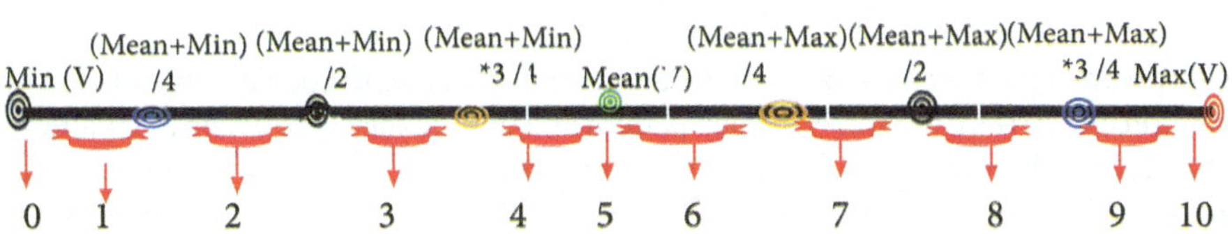

Fig. 11.4 DDP codes

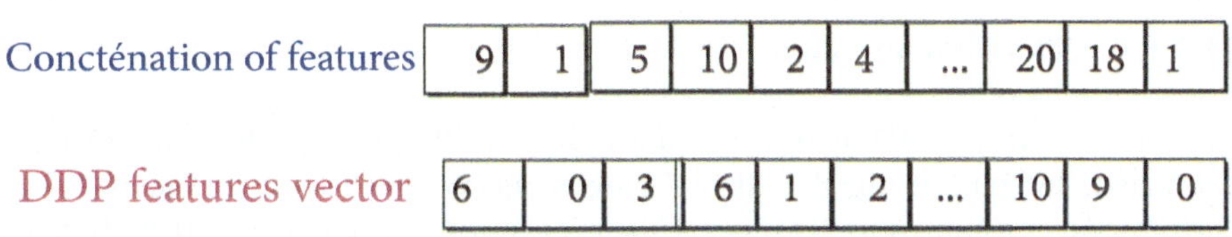

Fig. 11.5 Construction of the DDP features vector

Figure 11.3 illustrates the features extraction step from the image I_{mg} and the obtained vector V_1.

The second step consists in calculating the maximum value (max V), the minimum (min V) and the mean value (mean V) of this vector V_1.

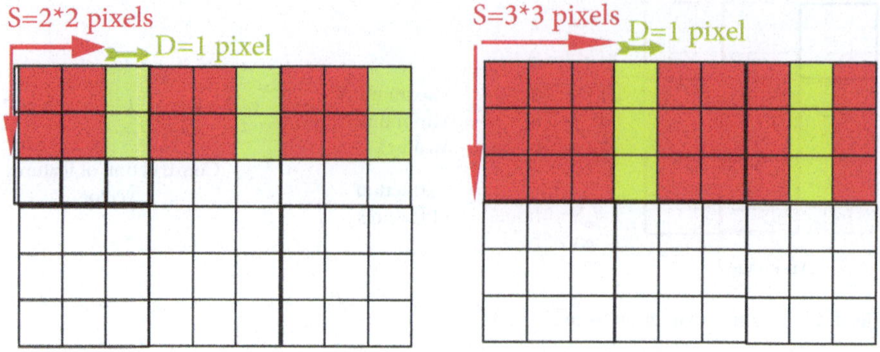

Fig. 11.6 Examples of DDP (S, D): **a** DDP(2, 1), **b** DDP(3, 1)

In this example, max $V(V_1) = 20$, min $V(V_1) = 1$ and mean $V(V1) = \frac{1}{27}(9 + 1 + 5 + \cdots + 1) = 8$. Using these three values, the vector V_1 is decomposed in 8 intervals. The DDP code is assigned as it shown Fig. 11.4 and Table 11.1.

Figure 11.5 presents the vector V_1 and the DDP features vector. In this example, the minimum of V_1 is 1 so it is coded by 0 and the mean of V_1 is 8 so it is coded by 5.

The above steps are summarized in Algorithm 1.

11.2.1.2 Parameters of the DDP Operator

Two parameters which can affect the calculation of the DDP code are introduced on Yahia et al. (2018a): the size S of each pattern and the distance D between two patterns (Fig. 11.6). The variation of these two parameters allows the use of neighbourhoods of different sizes. It determines the size of the DDP features vector and consequently influences in the classification results.

11.2.1.3 3D-DDP Operator

In literature, several 3D approaches are based on 2D plane analysis for features extraction. In some medical studies for example, the analysis of 3D textures are performed using numerous 2D slices (Cuadra et al., 2002; Zhao et al., 2012; Michoux et al., 2015; Ortiz et al., 2013).

The purpose consists in exploiting the maximum of information from the 3D images. Experiments results shown in Sect. 11.3 demonstrate the efficiency of this proposition. In Fig. 11.7 an example of 2D plane extracted from a brain volume is plotted.

Therefore, to analyze 3D textures, we suggest extending the proposed DDP operator by segmenting the volume in 2D planes (parallels slices), 3D image is considered as a succession of 2D parallels planes (Fig. 11.8). In each plan, the 2D- DDP is

Algorithm 1: Algorithm of the DDP approach.

Input: an image I
Output: Vector of features
// Features extraction

for $t \leftarrow 1$ **to** $sizeof(I)$ **do**
  Extract a pattern N (a 3x3 neighborhood points) ;
  Calculate the three features: Feature1=Maximum of (P);
  Feature2= Minimum of (P);
  Feature3= mean of (P);
  Insert Feature1, Feature2 and Feature3 in a vector V;
end
// Calculate the maximum, the minimum and the mean of V
for $i \leftarrow 1$ **to** $size\ of(V)$ **do**
  Calculate max= $\max(V)$;
  min= $\min(V)$;
  mean= mean (V);
end
//Assign to each value of the vector V a code
for $j \leftarrow 1$ **to** $size\ of\ V$ **do**
  if $j = \min(V)$ **then**
    $code = 0$
  end
  if $j \in [\min, \min + \frac{1}{4}(mean - \min)]$ **then**
    $Code = 1$
  end
  ;
  ....
  if $j = \max$ **then**
    $code = 10$
  end
  Insert the code in the features vector VF
end

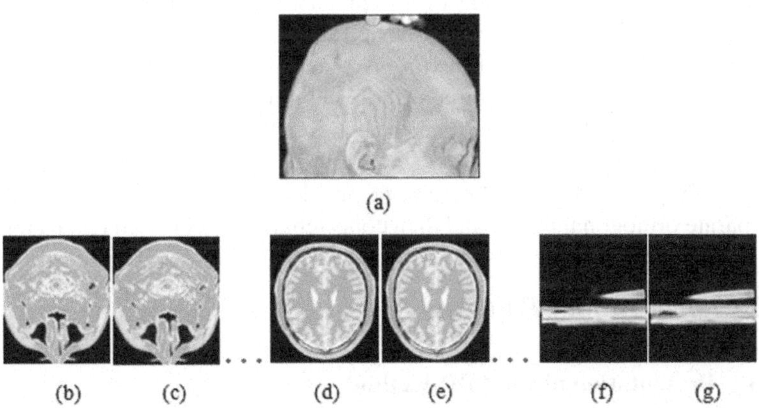

Fig. 11.7 Examples of 2D images 2D segmented from a 3D brain image in 180 slices: **a** 3D image, **b** slice 1 **c** slice 2, **d** slice 100, **e** slice 101, **f** slice 179, **g** slice 180

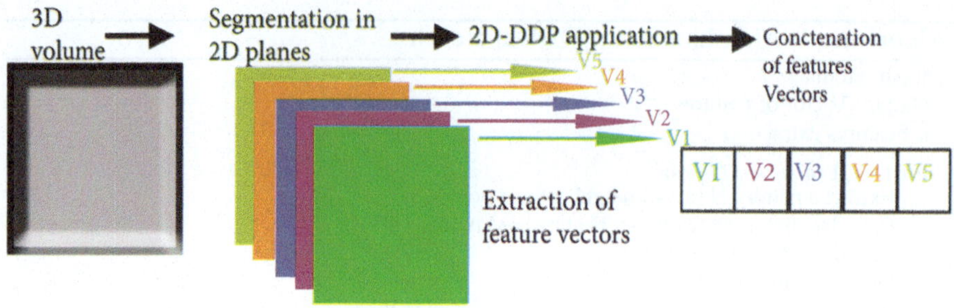

Fig. 11.8 3D-DDP operator

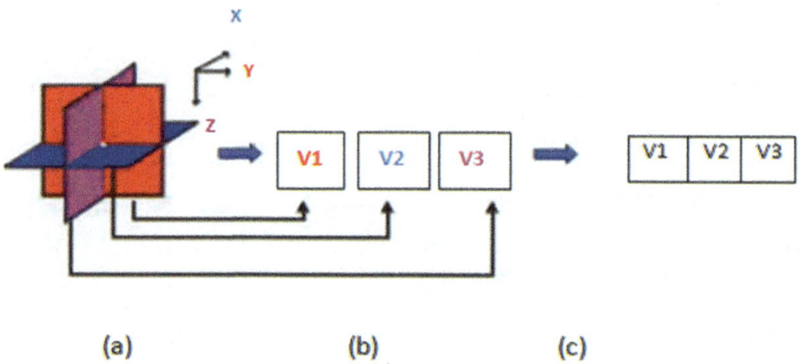

Fig. 11.9 DDP-TOP approach

calculated and then results obtained from all the planes are concatenated to provide a feature vector that describes the volume (Yahia et al., 2018b).

The Decimal Descriptor Patterns From Three Orthogonal Planes (DDP-TOP) The DDP-TOP operator is another way extension of the basic DDP. It is introduced in Yahia et al. (2018b), to analyze 3D texture. We suggested the concatenation the basic Decimal Descriptor Patterns on three orthogonal planes (DDP-TOP): XY, XZ and YZ, considering only the co-occurrence statistics in these three directions (shown in Fig. 11.9).

Consequently, the DDP-TOP operator considers the feature distributions from three separate orthogonal planes and then concatenates the obtained results together.

11.2.2 Local Binary Patterns

11.2.2.1 Presentation of the LBP Method

The LBP texture analysis operator is introduced by Ojala et al. (1996, 2002). It is a very popular and powerful means of texture description. Among its properties are its robustness against monotonic gray-scale changes and computational simplicity.

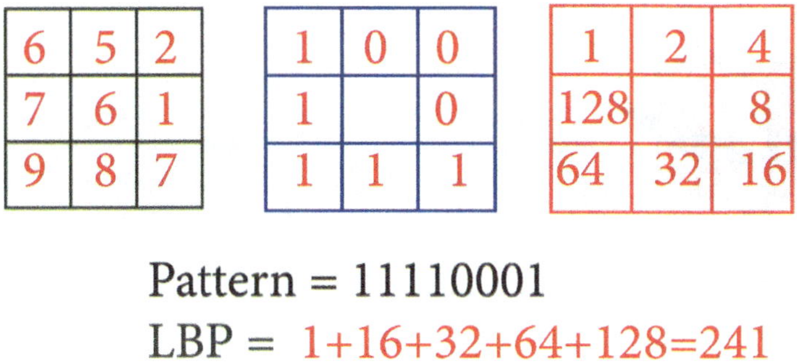

Fig. 11.10 Example of LBP calculation

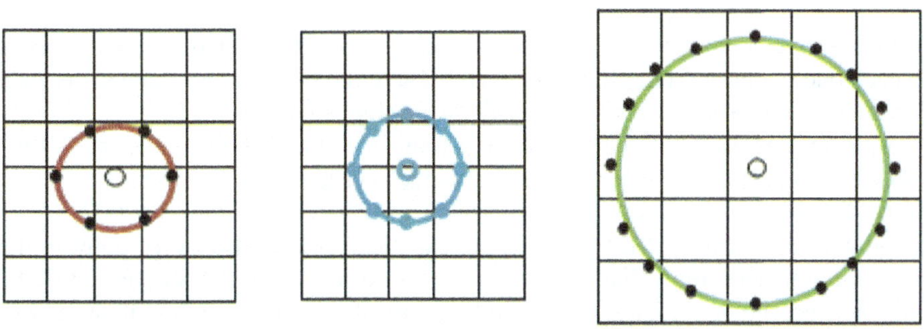

Fig. 11.11 Three examples of circular neighborhoods: ($R = 1$, $P = 6$), ($R = 1$, $P = 8$), ($R = 2$, $P = 16$)

The basic LBP operator thresholds the 3×3 neighborhood of each texture pixel by the value of the center pixel. Each pixel is compared by the center pixel and if the result of subtraction is positive the is coded by one else it is coded by zero. After thresholding, the obtained values are then multiplied by the weights given to the corresponding pixels (1, 2, 4, 16, 8, 16, 32, 64 and 128) and finally summed to calculate the binary number or the LBP code of this texture unit. The histogram constructed using these $256(2^8)$ labels may be used as a texture descriptor (Fig. 11.10).

Various extensions to this original operator are proposed. For example, by considering neigborhoods of different sizes this extension allows the use of any radius and number of pixels in the neighborhood. See Fig. 11.11 for an example of circular neighborhoods.

11.2.2.2 3D-LBP

The 3D-LBP operator

The LBP-TOP (Local Binary Patterns histograms from Three Orthogonal Planes) operator is based on modeling the 3D textures using three orthogonal planes extracted

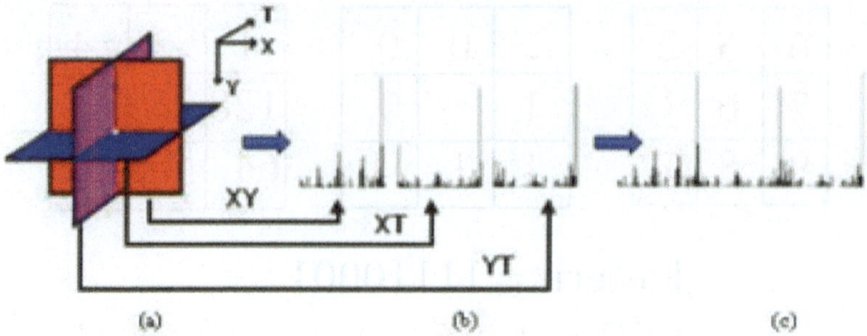

Fig. 11.12 LBP-TOP

from the image (Zhao et al., 2007). These three planes intersected in a center pixel. In each one the basic 2D-LBP is calculated. The three histograms obtained are then concatenated in order to describe the 3D image (Fig. 11.12).

Another 3D form of 3D-LBP operator is founded in the literature called the VLBP (Volume Local Binary Patterns). This technique is based on a 3D volume composed of the different neighbor's points of image slices (Zhao & Pietikäinen, 2007).

Neighboring points of the volume, are sampled, thresholded every point in the neighborhood with the value of the center pixel to calculate the VLBP code. The obtained thresholded values are multiplied with the binary weights of the corresponding pixel to sum up the result. A VLBP(T, NP, R) code describes the spatial structure of the local volume texture according to this formula:

$$V = v(v_0, \ldots v_q, \ldots v_{3p+1}) \tag{11.4}$$

VLBP code is est calculated using this equation:

$$VLBP_{L,P,R} = \sum_{q=0}^{3P+1} v_q 2^q \tag{11.5}$$

with:
V_q: substraction results.
q : value index in the volume V.
P: number of point in the circle.
R: circle rayon.
V: pixel of the volume.
L: Time interval.
An example of VLBP application is illustrated in Fig. 11.13 (Zhao & Pietikäinen, 2007).

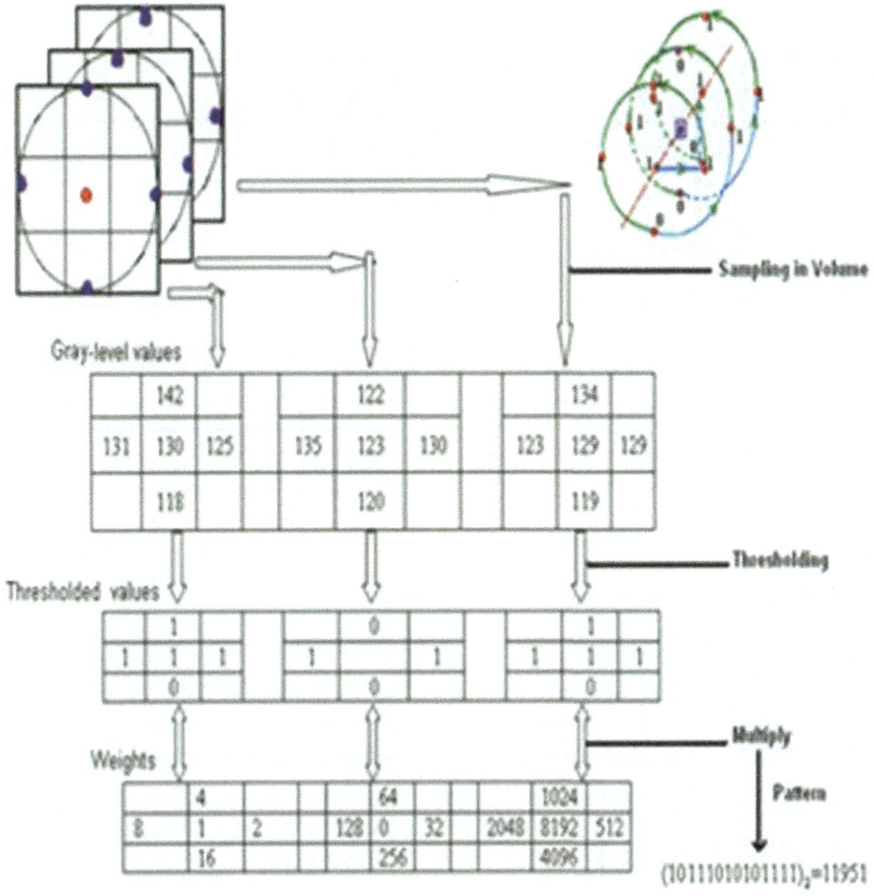

Fig. 11.13 Example of VLBP(1, 4, 1) application

11.2.3 *Grey Level Co-occurrence Matrix Method*

11.2.3.1 Presentation of the GLCM Method

The Grey Level Co-occurrence Matrix (GLCM) (Haralick & Shanmugam, 1973; Haralick et al., 1972) was introduced by Haralick. It represents one of the most known texture analysis methods. The idea is measuring texture features basing on the construction of a matrix that is called the Co-occurrence Matrix (Haralick & Shanmugam, 1973; Haralick et al., 1972). It is a square matrix, formed by the frequencies of appearance of a pixels pairs at a certain direction and a distance (Fig. 11.14).

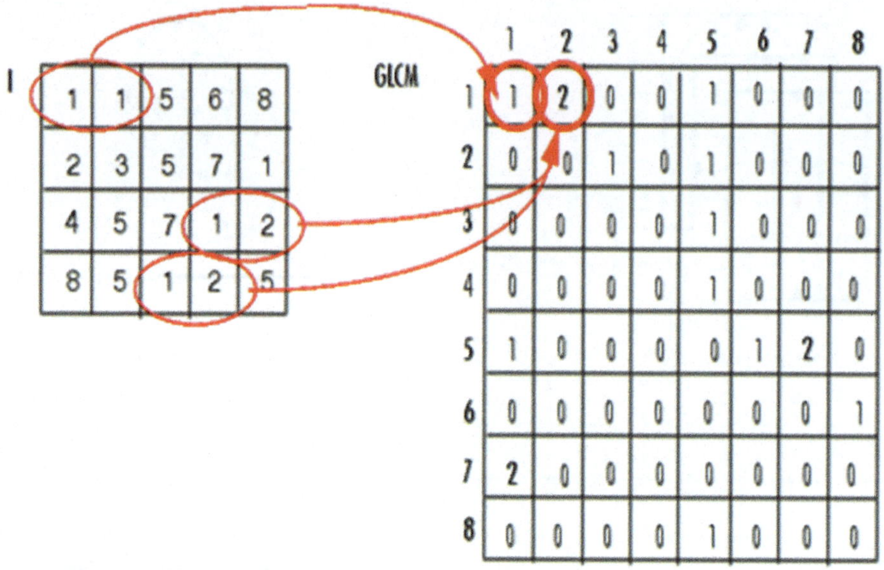

Fig. 11.14 Example of an image Img and its GLCM

Considering CM a co-occurrence matrix, N_b the number of the grey levels used, d_r a particular direction and d the distance that separates pixels, the size of G_M will be $N \times N$. The co-occurrence matrix CM is defined by this equation:

$$CM_{p,\theta}(i, j) = \sum_{x=1}^{N_b} \sum_{y=1}^{N_b} \begin{cases} 1 \; if \; f(x, y) = i \text{ and } f(x + dx, y + dy) = j \\ 0 \; otherwise \end{cases} \quad (11.6)$$

In literature four directions for the position angle were considered: the horizontal position ($d_r = 0°$), the vertical direction (90°), the right diagonal position direction (45°) and the left diagonal direction (135°) (Fig. 11.15).

The offset parameter that affects significantly the features vector is described by this set: ($[0, d], [-d, d], [-d, 0], [-d, -d]$). Once the GLCM matrix is constructed, there are 14 features that can be calculated from this matrix and can describe the texture (Haralick & Shanmugam, 1973). We note that the most used attributes in literature are the contrast, the correlation, the energy and the homogeneity.

$$Energy = \sum_{i=0}^{N-1} \sum_{j=0}^{N-1} G_{ij}^2 \quad (11.7)$$

$$\mu_i = \sum_{i=0}^{N-1} iG(i) \quad (11.8)$$

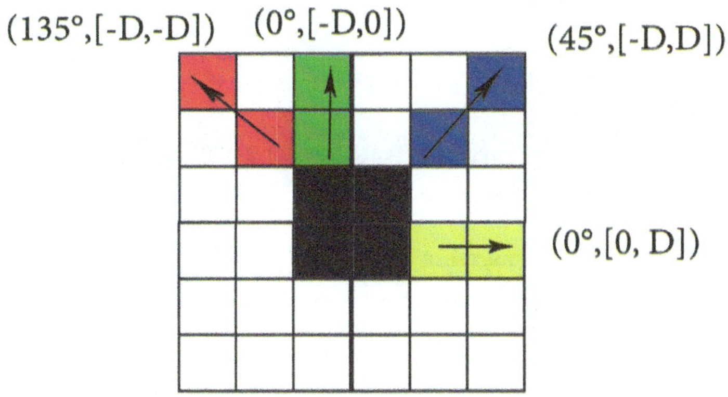

Fig. 11.15 Offset varying in distance and orientation in GLCM calculation

$$\mu_j = \sum_{j=0}^{N-1} jG(j) \tag{11.9}$$

$$\sigma_i = \sum_{i=0}^{N-1} (i - \mu_i)^2 G(i) \tag{11.10}$$

$$\sigma_j = \sum_{i=0}^{N-1} (j - \mu_j)^2 G(j) \tag{11.11}$$

$$Correlation = \sum_{i=0}^{N-1} \sum_{j=0}^{N-1} \frac{(1 - \mu_i)(1 - \mu_j)}{\sigma_i \sigma_j} G_{ij} \tag{11.12}$$

$$Contrast = \sum_{i=0}^{N-1} \sum_{j=0}^{N-1} (i - j)^2 .Gij \tag{11.13}$$

$$Homogeneity = \sum_{i=0}^{N-1} \sum_{j=0}^{N-1} \frac{G_{ij}}{1 + |i - j|} \tag{11.14}$$

11.2.3.2 3D-GLCM Operator

The 3D-GLCM proposition is detailed in Chen et al. (2009). It consists in calculating the triplet (i_1, i_2, i_3) grey level repetition instated of pairs (i_1, i_2) that are separated by a distance d_s and follow a direction d_i. The 3D-GLCM becomes:

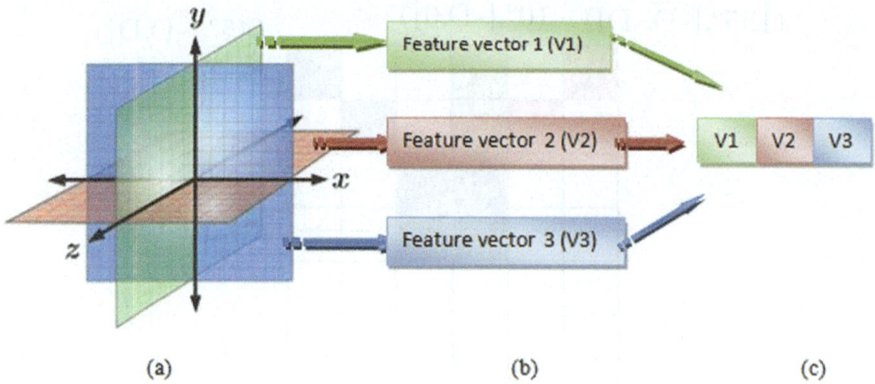

Fig. 11.16 GLCM-TOP method: **a** Extraction of 3 orthogonal planes, **b** Extraction of features using 2D GLCM, **c** Concatenation of features vectors

$$(i_1, i_2, i_3, d, d_r) = N\left\langle \begin{matrix} ((x_{n_1}, y_{n_1}), (x_{n_2}, y_{n_2}), (x_{n_3}, y_{n_3})) \\ \in M_1 \times M_2 \times M_3 \end{matrix} \middle| \begin{matrix} \max(|x_{n_1} - x_{n_2}|, |y_{n_1} - y_{n_2}|) = d \\ \max(|x_{n_2} - x_{n_3}|, |y_{n_2} - y_{n_3}|) = d \\ \mathrm{dr}((x_{n_1}, y_{n_1}), (x_{n_3}, y_{n_3})) = d_r \\ \mathrm{I}(x_{n_1}, y_{n_1}) = i_1 \\ \mathrm{I}(x_{n_2}, y_{n_2}) = i_2 \\ \mathrm{I}(x_{n_3}, y_{n_3}) = i_3 \end{matrix} \right\rangle \tag{11.15}$$

The Volumetric GLCM (VGLCM) operator is proposed by Tsai et al. (2007). The three dimensional matrices are composed by the appearance frequencies of grey values pairs within a moving box and separated by a displacement $s = (s_x, s_y, s_z)$.

So, in this 3D form, each pixel can move in 13 directions and has 27 neighbors (Tsai et al., 2007). Once the 3D GLCM are defined, Haralick's textural features can be calculated.

$$G(i, j) = \sum_{x=1}^{x} \sum_{y=1}^{y} \sum_{y=1}^{y} \begin{cases} 1 & \text{if } f(x, y, z) = i \\ & \text{and } f(x + dx, y + dy, z + dz) = j \\ 0 & \text{otherwise} \end{cases} \tag{11.16}$$

The Grey Level Co-occurrence Matrix from Three Orthogonal Planes (GLCM-TOP): Recently we suggested a new 3D operator basing on the 2D Grey Level Cooccurence Matrix (Yahia et al., 2018b). The aim was to develop a new 3D approach computationally simple. For this reason, only the co-occurrences on three orthogonal planes (GLCM-TOP) were considered.

The basic 2D-GLCM is applied in each plane and the obtained results from each plane are concatenated to form one feature vector which describes the 3D texture (Fig. 11.16).

11.3 Experiments and Results

11.3.1 Databases

11.3.1.1 Brainweb Database

The Simulated Brain Database (SBD) provides a realistic MRI data volumes that were produced by an MRI simulator (Cocosco et al., 1997). It contains simulated brain MRI data organized on two models: normal and Multiple Sclerosis Lesions.

These data can be used to evaluate the performance of different image analysis methods. The 3-dimensional data volumes have been simulated with different parameters and characteristic: three sequences (T_1, T_2 and PD- weighted), a variety of slice thicknesses (1, 3, 5, 7 and 9 mm), noise levels (0, 1, 3, 5, 7 and 9%), and levels of intensity non-uniformity (0, 20 and 40%). These data are may be interactively explored in three orthogonal views: transversal, sagittal and coronal.

The advantages of using this database are: the parameters may be independently controlled and the different answers are known a priori in the experiment.

11.3.1.2 University of Milano Bicocca 3D Face Database

The University of Milano Bicocca 3D face database (UMB-DB) (Colombo et al., 2011) contains 143 subjects (98 males and 45 females). Images taken include occlusions such as scarves, telephone, eyeglasses and some facial expressions Fig. 11.17). The occlusions in this database can be caused by eyeglasses, hands, hats, scarves, hair and other objects (Fig. 11.18).

In the UMB database the textured images can be rotated with simple angles. Faces are taken with different illumination conditions (Figs. 11.17 and 11.18).

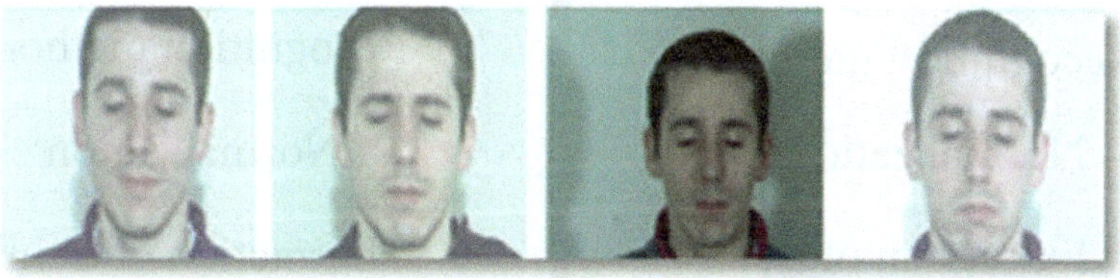

Fig. 11.17 Examples of facial expressions in the 3D-UMB database

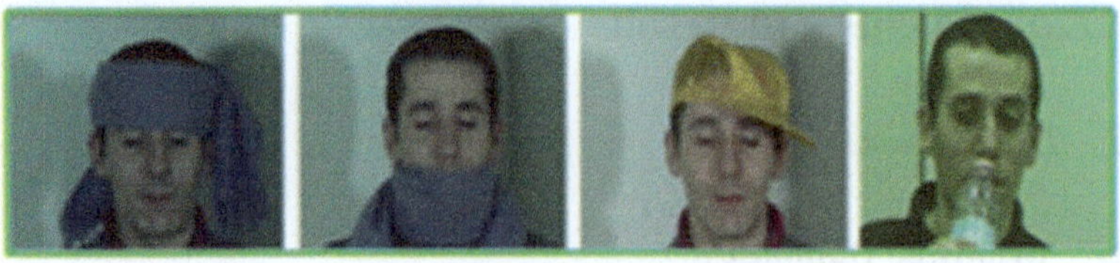

Fig. 11.18 Examples of occlusions in the 3D-UMB database

11.3.2 Phases of Simulation

In the following, we describe the phases of simulation and we present a set of experiments to validate the proposed approach. As it illustrated in Fig. 11.19, our work can be resumed in two main phases: train and test.

The objective of each classification process is to specify the class of the appurtenance of a given image. A learning phase is necessary for the classifier to learn the different characteristics of every class. Basing on this training, the classifier will be able to classify the image in the testing phase. To do this, a set of features are extracted from each image to obtain a description about this image. And basing on the features descriptor extracted, the image classification is conducted. So the texture classification phase is based on the comparison of these two features vectors: the train features vector and the testing features vector.

The train features vector is calculated by applying the analyze method to some simples that represent the different classes. The test features vectors are obtained using

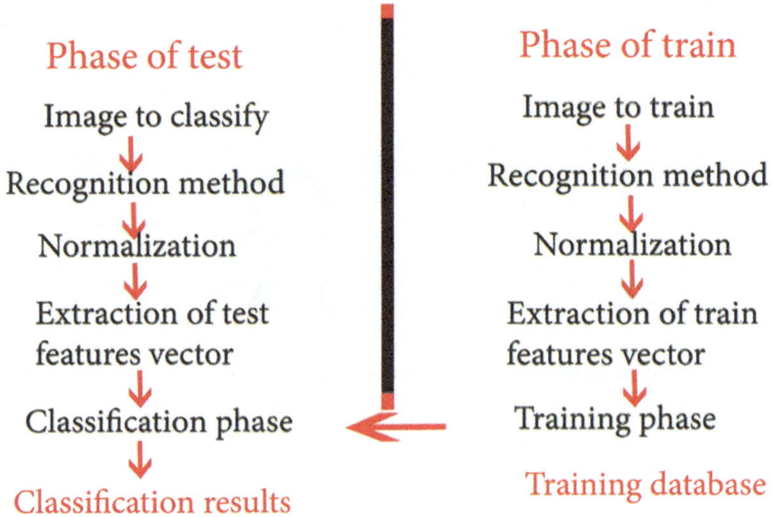

Fig. 11.19 Classification algorithm

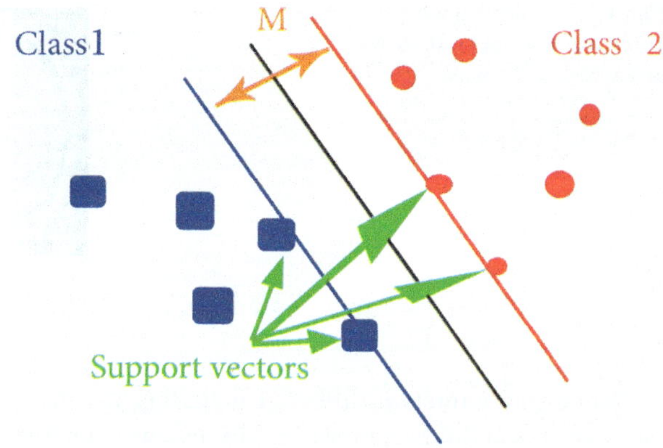

Fig. 11.20 SVM algorithm

the same method for each sample.The obtained values after the analyze method application require to be normalized. The normalization is necessary to obtain efficient results (Yahia et al., 2016, 2018a, b).

In all experiments, the texture classification process is conducted using the features descriptor extracted. The three used methods are applied using their basic parameters:

- The LBP: number of points = 8, rayon = 1 with a (0, 1) offset.
- The GLCM: the horizontal position $\hat{\mathrm{I}} = 0°$, offset = (0, 1) and the most used features: contrast, correlation, energy and homogeneity.
- The DDP: distance between patterns = 0 and size of patterns = 3.

For classification, the multi-class Support Vector Machine is chosen. In several studies, this classification approach is shown to be very effective and able to deliver higher precision (Zhao & Pietikäinen, 2012; Yahia et al., 2016, 2018a, b). This classifier is based on to searching a function f(x), that determines the decision boundary called also the hyperplane. This hyperplane as it plotted in Fig. 11.20, separates two classes.

11.3.3 3D MR Brain Images Analysis

The use of Magnetic resonance imaging (MRI) allows the diagnosis, the investigation and even the treatment of MS Lesions. MRI is often the only imaging modality used for imaging patients with MS.

Multiple sclerosis (MS) is an inflammatory and neurodegenerative disorder condition of the central nervous system. Multiple sclerosis is considered the most common immune-mediated disorder damaging the central nervous system (Berer & Krishnamoorthy, 2014). It can damage the white and grey matter of the brain, optic nerve and spinal cord.

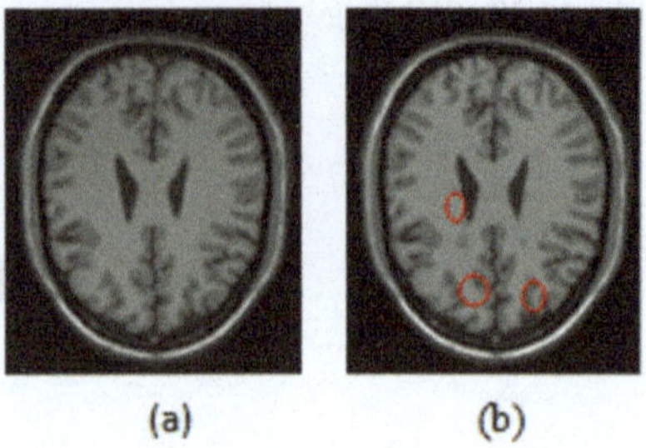

Fig. 11.21 Slices number 1000: **a** A normal brain, **b** A diseased MS Lesions

White matter tracts are affected, including also those of the spinal cord, the cerebral hemispheres and infratentorium. MS lesions is known as plaques, can form in white or grey matter of the brain in any location; thus, clinical presentations can be diverse (Fig. 11.21).

Multiple sclerosis (MS) Lesion brains attacks the central nervous system by disrupting the ability of parts of this system to communicate. A lesion is a part of a damaged tissue that can appear as white or dark spots in the white or in the grey matter. This disease may cause problems with balance, muscle control, vision and other basic body functions (Berer & Krishnamoorthy, 2014). These effects are usually different for everyone who has the disease. Some people can have trouble doing daily tasks and others can have mild symptoms and consequently don't need treatment. The main objective of simulations is to differentiate between MR brain images healthy and diseased. So, we achieve a classification of these images into two classes: normal brains and diseased brains.

11.3.3.1 Experiment 1: Test of Disease Progression

The objective of this experiment is to test the robustness of the classification method in disease progression detection. To do this, three "phantom" (anatomical model) for a human brains with "mild", "moderate" and "severe" MS Lesions and normal brains MR images were used. Basing on the disease progression, four classes are considered. A visual comparison of each class is illustrated in Fig. 11.22.

548 samples have been tested and 80 samples have been trained. For the features extraction, the DDP, GLCM and LBP methods have been used. for each method, the percentage of the successful classified images has been computed. The DDP operator was the best between the three methods (Fig. 11.23). It accounted 73.9%, the LBP operator leaded to a percentage of success of 59.6% and the GLCM operator reached to a percentage of success of 57.6%.

Disease diagnostics may be significantly influenced by images quality. Inhomogeneity and additive noise are among the major voxel intensity modifiers in MR images (Patel et al., 2016).

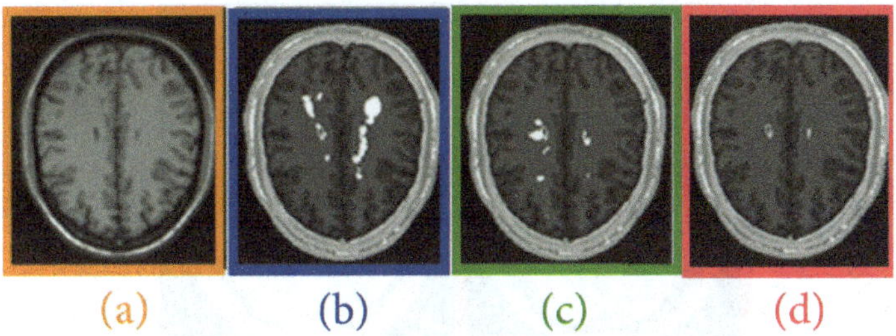

Fig. 11.22 Slices number 104: **a** A normal brain, **b** A severe MS Lesions, **c** A moderate MS Lesions, **d** A mild MS Lesions

Fig. 11.23 Experimental results of the classification of normal brains, moderate, mild and severe MS Lesions

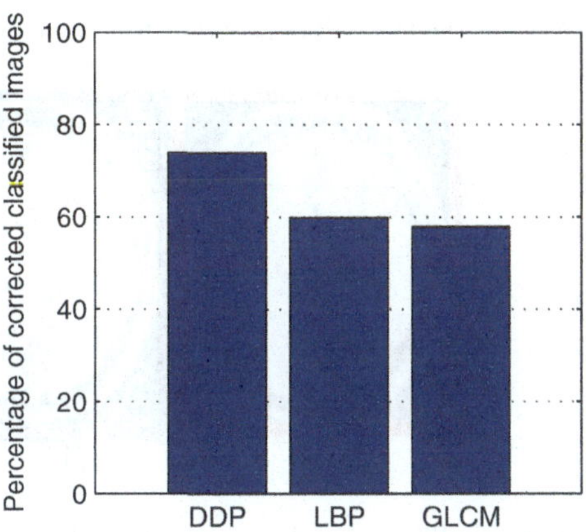

11.3.3.2 Experiment 2: Inhomogeneity Robustness

The robustness of the three methods in front of the in-homogeneity was analyzed in these simulations. We considered the three levels of inhomogeneities present in the MR images simulator: 0, 20 and 40%. Two experiments were performed for T_1 and T_2 weighted images of BrainWeb database having voxel dimension of $1.0 \times 1.0 \times 1.0$ mm.

These two sequences are among the principal processes that give the contrast in MR images (Patel et al., 2016) (Example in Fig. 11.24).

T_1 recovery indicates the result of spin lattice interactions. T1 time determines how long a tissue takes after a radiofrequency pulse to recover back to the longitudinal plane. T_2 decay indicates the result of the interactions of the small magnetic fields between protons. T_2 time determines the amount of time which takes a tissue for the transverse magnetization to decay. However, T_1 and T_2 are independent of one another, and T_2 times of tissue are shorter than T_1 times (Patel et al., 2016).

In these experiments, 6 samples that were segmented in 180 slices of normal and diseased images were used. So, in total, 120 slices were applied for training. 1080 slices were considered for testing the performance of the training process.

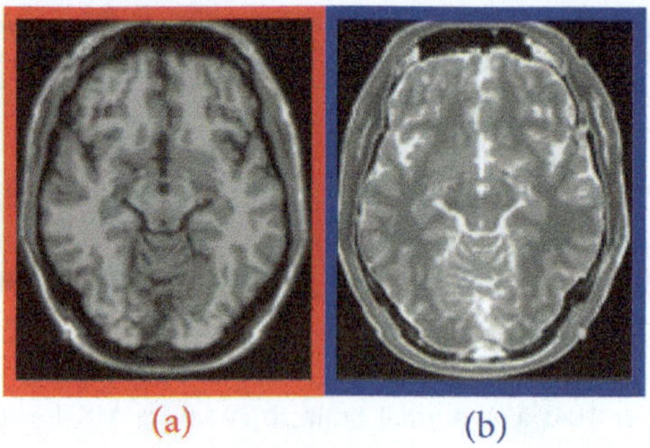

Fig. 11.24 Example of sequences of MS brain, slice = 60, Noise = 0% and in-homogeneity = 0%: **a** T1-weighted MR brain image, **b** T2-weighted MR brain image

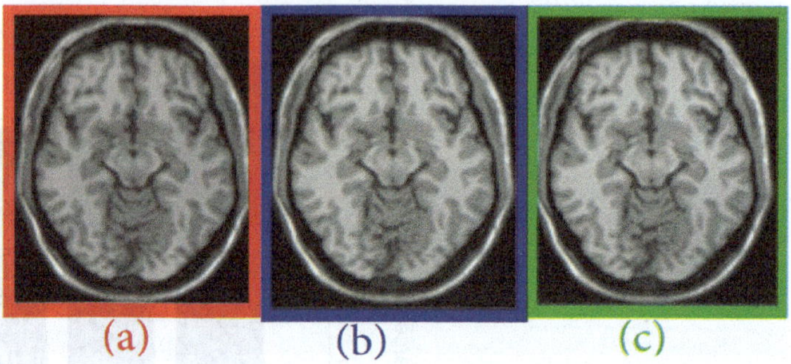

Fig. 11.25 Slice 60 of T1- weighted MS Lesions brain with a noise level = 0: **a** Intensity of in-homogeneity = 0%, **b** Intensity of in-homogeneity = 20%, **c** Intensity of in-homogeneity = 40%

An example of simulated T_1-weighted MR images MS Lesions brain under varying conditions of inhomogeneity is presented in Fig. 11.25.

Results obtained in Fig. 11.26 and its details presented in Table 11.2 revealed that the DDP was the best between the three methods. The DDP presents high robustness to inhomogeneity for any level of in-homogeneity. It gave 98.97% with the sequences of T1 and 93.24% with T_2. The LBP provided 87.77% with sequences of T1 and 89.72% T2. We obtained with GLCM operator 52.96% with the sequences of T_1 and 90.18% with T_2.

We can conclude that the GLCM is the most sensitive technique of in-homogeneities whereas the LBP and the DDP techniques showed stability to any level of inhomogeneities.

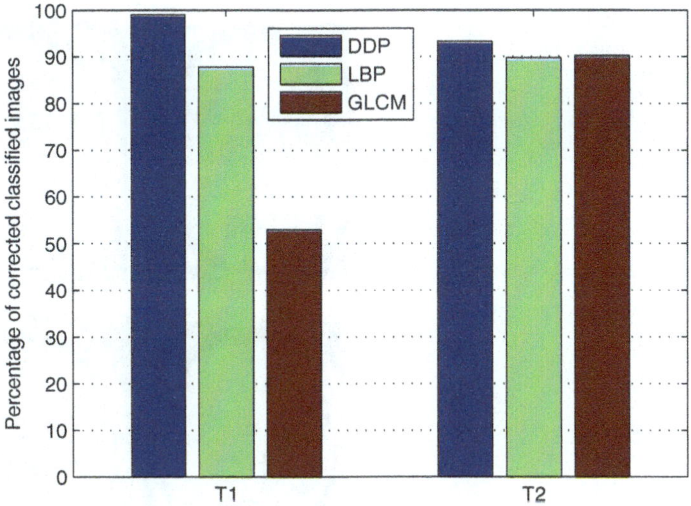

Fig. 11.26 Performance of the classification using the simulated T1- and T2-weighted MR images under varying conditions of in-homogeneity (noise level = 0% and slice thickness = 1 mm)

Table 11.2 Number of corrected classified image among 180 images under varying in-homogeneity conditions

	Slices	Diseased			Healthy		
	Percentage of in-homogeneity	0%	20%	40%	0%	20%	40%
T1	DDP	175	176	177	180	179	180
	LBP	162	164	168	149	153	152
	GLCM	106	103	126	106	78	53
T2	DDP	175	174	175	161	161	161
	LBP	170	162	162	150	164	161
	GLCM	135	172	173	171	171	152

11.3.3.3 Experiment 3: Noise Robustness

The robustness of the three methods in front of the noise is analyzed in this experiment. As it shown in Fig. 11.26 different degrees of noise are considered. 12 simulated image volumes having voxel dimension of $1.0 \times 1.0 \times 1.0$ mm have been used by the combination of different noise levels. Each images has been tested with all possible levels of noise in the Brainweb simulator. Consequently, 2160 images have been considered in the testing phase and 240 image in the training phase. The classification has been performed with two classes: diseased and healthy. So each class contained 1080 images.

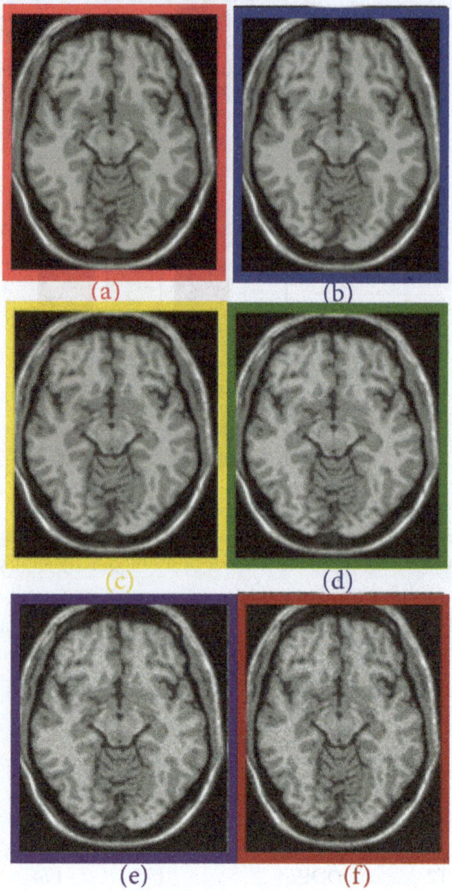

Fig. 11.27 Example of T_1 weighted MS brain slice 60: **a** Noise = 0%, **b** Noise = 1%, **c** Noise = 3%, **d** Noise = 5%, **e** Noise = 7%, **f** Noise = 9%

With increasing noise, the quality of the image classification decreases. As it shown in Table 11.3 these results hold with increased noise up to 9%, a level that presents a challenge for most of the existing studies (Cuadra et al., 2002).

The DDP operator leaded to the best results (Fig. 11.27): the highest percentage of images correctly classified is 97.12% with the sequences of T_1 and 94.21% with the sequences of T_2. The GLCM approach was the most sensitive to the noise between the three approaches. It provided 50.50% with the sequences of T_1 and 85.83% with sequences of T_2 . The LBP method was also not stable to the noise: we obtained 70.87% with sequences of T_1 and 79.07% with the sequences of T_2. The strength of the DDP was clearly evident in its robustness and stability to any level of noise (details in Table 11.3).

Table 11.3 Number of corrected classified image among 180 images under varying noise conditions

	Slices	Diseased						Healthy					
	Noise	0%	1%	3%	5%	7%	9%	0%	1%	3%	5%	7%	9%
T1	DDP	174	174	175	174	174	172	178	175	175	175	175	177
	LBP	130	150	135	105	125	93	154	145	123	136	114	121
	GLCM	154	154	144	42	46	158	26	26	36	151	136	18
T2	DDP	175	176	177	179	177	169	158	156	157	161	174	176
	LBP	159	165	128	140	115	163	145	136	140	143	162	112
	GLCM	175	172	158	156	171	137	156	160	163	135	107	164

11.3.4 3D Face Analysis

3D face analysis has numerous difficulties. For example, with variation of scene illumination face change and can decrease the percentage of recognition.

Besides, pose variation, face rotation, facial expressions and the use of some others components that may disappear basic facial features. This factors modify clearly faces and consequently affect person classification. In the 3D UMB database, the objective was to recognize persons faces. Each person had different expressions

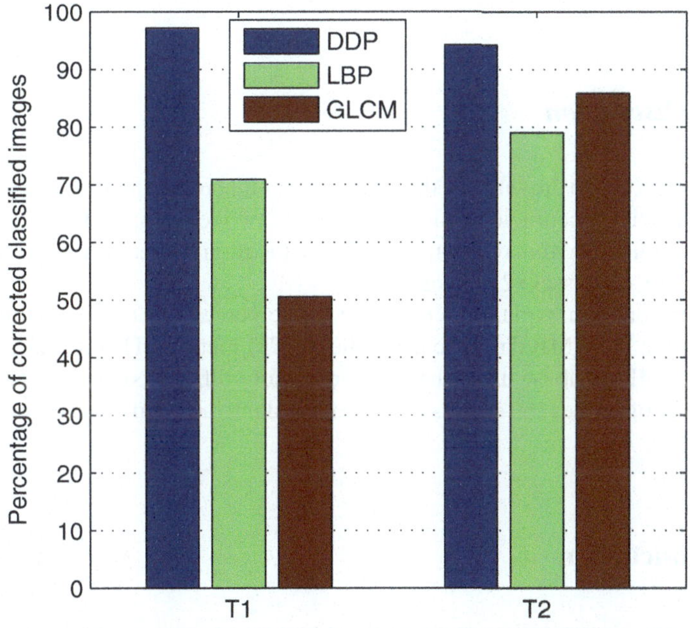

Fig. 11.28 Performance of classification using the simulated T_1 and T_2 weighted MR images under varying conditions of noise (Intensity in-homogeneity = 0% and slice thickness = 1 mm)

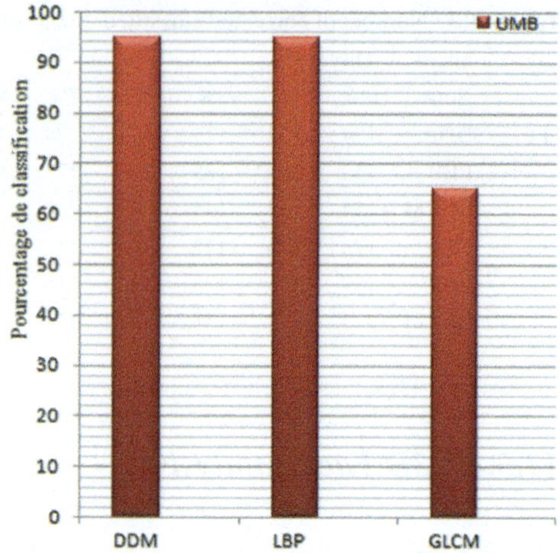

Fig. 11.29 Classification results using 3D-UMB database

(smile, angry, boring, neuter...) (Fig. 11.17) and represented by a class. We considered a large number of classes (60 classes). For each person, two images were used in the phase of train and five images in the phase of test.

The DDP operator accounted for almost 95% on the UMB database (Fig. 11.29). The GLCM approach reached to a percentage of success of 65% on the 3D-UMB database.

11.3.5 Discussion

The main observation drawn from these results was that using the Brainweb database; we obtained height accuracies. This is due to these two reasons: the textures of Brainweb were simulated within rotation and the classification process was performed with a few numbers of classes (2 classes).

These reasons can explain the decrease of the percentage of successful classification over the 3D-UMB database using the GLCM method. This is logical because on the 3D-UMB database, the classification is more difficult since we considered a large number of classes (60 classes) and we tacked into account different expressions for each face.

11.4 Conclusion

This work was a comparative study of three texture analysis methods, based on the extraction of features measures.

Numerous simulations were conducted using two well-know three-dimensional databases. These databases have variants characteristics. SVM was adopted to train and test the textures. The performance of the DDP approach was compared to that obtained with a well-known approach, the Grey Level Co-occurrence Matrix (GLCM).

A difficult classification problem of a large number of textures with different conditions of illuminations, surfaces and textures rotation was considered.

We performed a classification of MR images switch the degree of the disease appearance and we tested the robustness of the different methods in front of the noise, the in-homogeneity and contrast variations.

We tried also to analyze 3D faces taking into account different facial expressions and various occlusions. The performed simulations confirmed that the DDP is the most efficient descriptor between the three methods.

The most important property of the DDP operator is its computational simplicity, that makes it possible to analyse 3D textures. Equally important is its robustness against noise, in-homogeneity, contrast and illumination changes. We aim in our future work, to test the robustness of the DDP approach in others problems and applications.

References

Arivazhagan, S., & Ganesan, L. (2003). Texture classification using wavelet transform. *Pattern Recognition Letters, 24*, 1513–1521.

Berer, K., & Krishnamoorthy, G. (2014). Microbial view of central nervous system autoimmunity. *FEBS Letters, 588*(22)

Bouali, S., Buscarino, A., Fortuna, L., Frasca, M., & Gambuzza, L. V. (2012). Emulating complex business cycles by using an electronic analogue. *Nonlinear Analysis: Real World Applications, 13*, 2459–2465.

Chellappa, R., & Chatterjee, S. (1985). Classification of textures using Gaussian Markov random fields. *IEEE Transactions on Acoustics, Speech, and Signal Processing, 33*(4), 959–963.

Chen, W. S., Huang, R. H., & Hsieh, L. (2009). Iris recognition using 3D co-occurrence matrix. International Conference on Biometrics (pp. 1122–1131). Berlin, Heidelberg: Springer.

Cocosco, C. A., Kollokian, V., Kwan, R. K. S., Pike, G. B., & Evans, A. C. (1997). Brainweb: Online interface to a 3D MRI simulated brain database. In *Neuro Image*.

Colombo, C., Cusano, & Schettini. (2011). The University of Milano Bicocca 3D face database. University of Milano Bicocca.

Cuadra, M. B., Platel, B., Solanas, E., Butz, T., & Thiran, J. P. (2002). Validation of tissue modelization and classification techniques in T1-weighted MR brain images. In *International Conference on Medical Image Computing and Computer-Assisted Intervention*, (pp. 290–297).

Deepa, T, Muthalagu R., & Chitra. K. (2015). Mem base brain image segmentation and classification using SVM. *Journal of Engineering and Applied Sciences Asian Research Publishing Network (ARPN), 10*(8).

Haralick, R. M., & Shanmugam, K. (1973). Textural features for image classification. *IEEE Transactions on Systems, Man, and Cybernetics, 6*, 610–621.

Haralick, R. M., Shanmugam, K., & Dinstein, I. (1972). On some quickly computable features for texture. In *Proceedings of Symposium Computer Image Processing and Recognition* (Vol. 2, pp. 2–12). Columbia: University of Missouri.

Michoux, N., Guillet, A., Rommel, D., Mazzamuto, G., Sindic, C., & Duprez, T. (2015). Texture analysis of T2-weighted MR images to assess acute inflammation in brain MS lesions. *PloS One, 10*(12)

Ojala, T., Pietikäinen, M., & Harwood, D. (1996). A comparative study of texture measures with classification based on featured distributions. *Pattern Recognition, 29*(1), 51–59.

Ojala, T., Pietikäinen, M., & Maenpaa, T. (2002). Multiresolution gray-scale and rotation invariant texture classification with local binary patterns. *IEEE Transactions on Pattern Analysis and Machine Intelligence, 24*, 971–987.

Ortiz, A., Palacio, A. A., Górriz, J. M., Ramírez, J., & Salas-González, D. (2013). Segmentation of brain MRI using SOM-FCM-based method and 3D statistical descriptors. *Computational and Mathematical Methods in Medicine*.

Patel, S. M., Shah, S. M., & Sharma, P. (2016). Analytical study of brain MRI protocols, sequences and parameters for detection of abnormalities. *International Journal of Information, 6*(1/2)

Tsai, F., Chang, C. K., Rau, J. Y., Lin, T. H., & Liu, G. R. (2007). 3D computation of gray level co-occurrence in hyperspectral image cubes. In: *International Workshop on Energy Minimization Methods in Computer Vision and Pattern Recognition* (pp. 429–440). Berlin, Heidelberg: Springer.

Yahia, S., Ben Salem, Y., & Abdelkrim, M.N. (2016). 3D face recognition using local binary pattern and grey level co-occurrence matrix sciences and techniques of automatic control and computer engineering (STA). In *2016 17th International Conference on IEEE* (pp. 328–338).

Yahia, S., Ben Salem, Y., & Abdelkrim, M.N. (2018a). Classification of 3D magnetic resonance brain images using texture measures from orthogonal planes. *International Journal of Digital Signals and Smart Systems (IJDSSS)*.

Yahia, S., Ben Salem, Y., & Abdelkrim, M. N. (2018b). Texture analysis of magnetic resonance brain images to assess multiple sclerosis lesions. *Multimedia Tools and Applications (MTAP), 67*(3)

Zhao, G., Ahonen, T., Matas, J., & Pietikainen, M. (2012). Rotation-invariant image and video description with local binary pattern features. *IEEE Transactions on Image Processing, 21*(4), 1465–1477.

Zhao, G., Pietikäinen, M. (2007). Dynamic texture recognition using volume local binary patterns. *Dynamical vision* (pp. 165–177). Berlin, Heidelberg: Springer.

Chapter 12
Image Processing and Analysis for Decision Making Applied to Melanoma

Nadia Smaoui Zghal

Abstract Malignant melanoma is considered one of the terrible disorders causing death. The goal of modern dermatology is the early screening of skin cancer, aiming at reducing the mortality rate with less extensive treatment. In this context, this work focuses on the problem of an automatic melanoma diagnosis based on two approaches. The first system is based on the ABCD diagnostic rule widely accepted by clinicians. The second system is based on extracting features and then classifying based on Support Vector Machine.

Keywords Melanoma · ABCD rule · TDS score · SVM · PH2 database

12.1 Introduction

Because of the continuous increase in medical knowledge, doctors can no longer master the whole of medical knowledge even if they are specialists. This does not allow, in many cases, to recognize the diseases in order to determine the best therapeutic management. To get by, doctors often use other sources of information such as books, colleagues or the scientific literature. However, despite the large volume of easily accessible online information, finding a solution to the problem with a given patient remains a difficult task. Early on, in view of the rapidly increasing quality of the computer, namely memory, speed and computing power, Medical Decision Support Systems (MDSS) have emerged as potential solutions to the difficulty encountered by the doctors.

Generally, the MDSS are defined as a computer tools intended to provide clinicians with the necessary information describing the clinical situation of a given patient. These tools can also describe the appropriate knowledge to a given situation, correctly filtered and presented, with the aim of improving the quality of care and,

N. S. Zghal (✉)
Cybernics Team, Control and Energy Management Laboratory, ENIS, ISBS, University of Sfax, Sfax, Tunisia
e-mail: nadia.smaoui@isimg.tn

N. Derbel and O. Kanoun (eds.), *Advanced Methods for Human Biometrics*, Smart Sensors, Measurement and Instrumentation 40,
https://doi.org/10.1007/978-3-030-81982-8_12

consequently, the health of patients (Berner, 2009). MDSS continue to develop for all medical activities (screening, diagnosis, treatment and even prevention) and for the majority of medical specialties (acute and chronic diseases). These systems offer services for different classes of physicians (students, generalists, specialists) and for different modes of application (hospitals, medical practices, emergency services or resuscitation services). More recently, MDSS intended for the sick have been developed, so that these patients are better informed about their illness and about possible care with a view to shared decision-making. The MDSS are of different types, there are diagnostic decision support systems, therapeutic decision support systems aiming at providing the best treatment for a given patient and prescription assistance systems (PAS) which are tools for securing the prescription. The diagnostic decision support systems are presented under different categories: diagnostic systems that are based on a calculation of the score, diagnostic systems exploiting artificial intelligence and finally diagnostic systems that are based on 3D reconstruction allowing a good visualization of the anomaly.

In this chapter, the objective is to develop a medical diagnostic system aiming at helping dermatologists in their analyses concerning the cutaneous lesions. During our research, two categories were proposed namely the systems based on a calculation of the score and the systems based on artificial intelligence. For the two categories, the diagnostic support system consists essentially of 4 steps namely image acquisition, pre-processing, segmentation and finally the analysis and interpretation stage.

Segmentation is the first link in the chain of analysis leading to the study of the morphology of internal structures and the detection of lesions. Segmentation of normal or pathological anatomical regions of interest that is reliable and accurate resulting imaging systems is a primary goal in medical information processing in general.

In the field of dermatology, different segmentation approaches have been proposed in the literature for melanoma detection. Some researchers detect the lesion using thresholding and histogram analysis methods (Maglogiannis et al., 2009; Peruch et al., 2014). Some techniques only used regional information (Silveira et al., 2009; Alfed and Khelifi, 2017). Xie and Bovik (2013), Dalila et al. (2017) have the advantage of genetic algorithms and metaheuristic approaches. Hybrid methods based on the combination of thresholding algorithms with level methods or active contour models have also been exploited (Yu et al., 2017; Silveira et al., 2009; Masood et al., 2013). Recently, convolutional neural networks and wavelet neural networks demonstrate very good dermoscopic image segmentation (Yu et al., 2017; Sadri et al., 2017; Yuan et al., 2017). In Silveira et al. (2009), the authors evaluate six different algorithms: active contours, adaptive thresholding, sets of levels and regional information. They conclude that the best result is obtained with the adaptive snake algorithm (Silveira et al., 2009). In Maglogiannis et al. (2009), the authors proposed a histogram-based thresholding method to extract lesions from the background, that's to say the healthy skin regions. In Pennisia et al. (2016), the authors used the Delaunay triangulation to divide the image into several regions and extract the lesion from them.

Subsequently, interpreting systems must be able to provide appropriate support to physicians by providing adequate synthesis in their daily practice. Many achievements have appeared since about twenty years. These systems can be presented in two ways. The first way is based on adopting the same approach used by doctor to solve a given medical problem. The second most common way is based essentially on artificial intelligence algorithms.

For our research, the phase of interpretation of the features inspected by dermatoscopy and the detection of malignancy of pigmented spots is based on two systems of decision support. The first system is based on the ABCD diagnostic rule widely accepted by clinicians (Stolz et al., 1994). The second system is based on support vector machines or SVM (Vapnik, 2000).

The structure of this chapter is then as follows: the second section gives an idea about the disease which is skin cancer. Diagnostic support systems based on a calculation of the score and based on artificial intelligence will be presented successively in Sects. 12.3 and 12.4. The experimental study will be presented in Sect. 12.5. The chapter will be closed with a conclusion.

12.2 About Melanoma

Melanoma is the most dangerous form of skin cancer. It begins in the melanocytes (pigment-producing cells present in the outer layer of the skin). In the majority of cases, it is caused by the sun's ultraviolet rays or tanning beds, which produce mutations (genetic defects) forcing skin cells to develop rapidly and form malignant tumors. Experts estimate that about 90% of melanomas are caused by intense exposure to UV rays and sunburns over the course of a lifetime (Tadeusiewicz, 2010). Malignant melanoma is likely to become one of the most articular malignancies of the future, with an incidence rate ten times higher. In fact, since 1970, the frequency of malignant melanomas has increased considerably; In the United States, for example, this rate increases by approximately 4% each year. The main reason for the rapid increase in the number of melanoma cases is the excessive exposure to the ultraviolet (UV) rays of the sun and tanning beds. Due to the high frequency of skin cancer, dermatological oncology has become a very important branch of medical development. Today, progress is identical in both primary types of research on the pathogenesis of tumors (role of viruses or genes in tumor development) and in the development of computer-assisted diagnostic algorithms. These algorithms are intended to help dermatologists, especially inexperienced family physicians and young dermatologists with serious difficulties in the correct visual assessment of skin lesions (Tadeusiewicz, 2010; Ins, 2016).

12.3 Diagnostic Aid System Based on Score Computation

For a diagnostic aid system based on the score computation, 4 parts are presented namely the images acquisition, the pretreatment step, the segmentation step and finally the analysis step which depends in particular on the calculation of the score.

12.3.1 *Images Acquisition*

For this first application, we are based on PH2 database (Mendonca, 2013). PH2 is a dermoscopic images database acquired in the dermatology department of Pedro Hispano Hospital, Matosinhos, Portugal, under the same conditions of illumination. This images database contains a total of 200 dermoscopic images of melanocyte lesions, including 80 common nevi, 80 atypical nevi, and 40 melanomas. It also includes the medical annotation of all images, namely the medical segmentation of lesions performed by experts (Mendonca, 2013).

12.3.2 *Images Pretreatment*

In the case of dermatological images, pretreatment serves to correct illumination defects, eliminate noise and small spots and enhance contours and contrast as much as possible without degrading the lesion. A small idea about the influence of artifacts on the segmentation result is shown in Fig. 12.1.

The considered database consists of RGB images. Therefore, each component (R, G, B) is considered as a grayscale image and has been treated separately. The proposed pretreatment consists of 3 steps. A filtering step based on a median filter (Huang et al., 1979) which is a spatial filter has been applied to each component of the

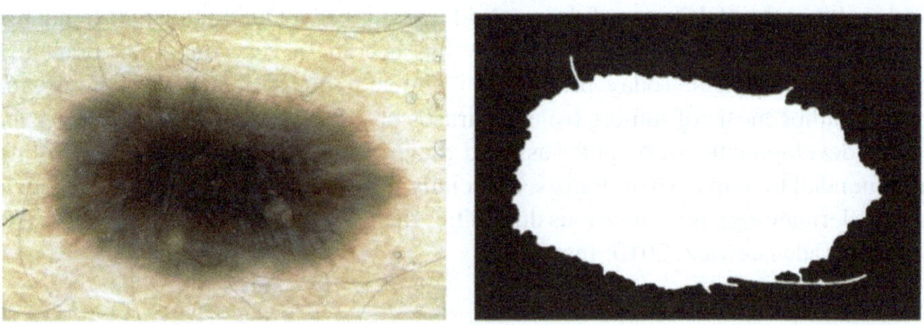

a) b)

Fig. 12.1 Influence of artifacts on segmentation result: **a** Original image, **b** segmented image

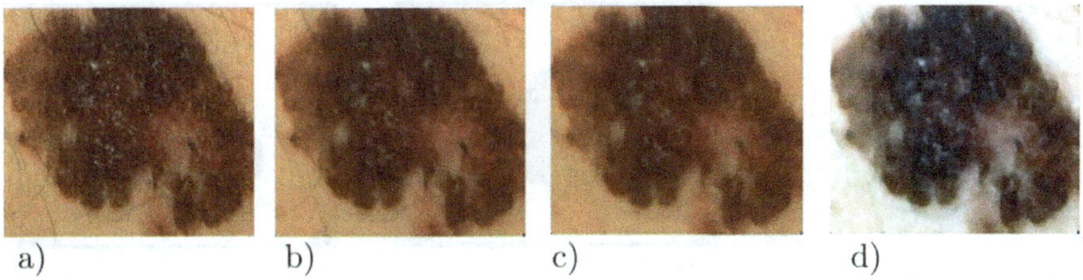

Fig. 12.2 Pretreatment steps: **a** Original Image, **b** Median filter application, **c** morphological closing application, **d** result of the histogram adjustment

image. This filter made it possible to calculate in each pixel of the image the median of the gray levels of the pixels of its window allowing, thus, to clean the image by eliminating certain defects (see Fig. 12.2b). Then, we carried out a morphological closing (Soille, 2004) which makes it possible to remove all the artifacts such as the hairs (see Fig. 12.2c). According to the images provided, it has been found empirically that the fixation of a 5×5 median filter and a disk-shaped structuring element with a diameter of 3 for the morphological closing permit to give better results. Subsequently, since the goal is to extract the lesion separately, the color relationship between the lesion and the healthy skin was improved by adjusting the intensity values for each component. The effect of adjusting the intensities on the histogram of the image is well presented in the Fig. 12.2d. At the end of each step, a simple concatenation recreates the colored images.

12.3.3 Lesion Detection

After the preprocessing step, we must remove the black frame introduced during the scanning process and detect the lesion. To achieve this goal, we need to define a binary mask showing the dark parts of the image (low gray values). In general, the binarization step based on the Otsu (Otsu, 1979) thresholding method, which classifies pixels into two classes, has some limitations because the lesion and edges have a fairly dark gray level. To overcome this gap, we use the multi-threshold Otsu (Liao et al., 2001) which is suitable to produce several classes and several thresholds from a single image. In fact, the dermatoscopic image converted to a grayscale image has essentially three parts: the edges (the darkest part), the lesion and the skin that is the lightest part. By defining the desired number of classes equal to three (the algorithm must look for two thresholds), we can obtain background pixels, intermediate pixels, and clear pixels.

In our work, the goal is to define a mask of the original image that presents only the lesion. To do that, we present two binary masks, the first is based on the first threshold found by the Multi-otsu thresholding algorithm (Fig. 12.3b), and the second threshold is based on the second one (Fig. 12.3c). So, in the first mask, we consider the dark

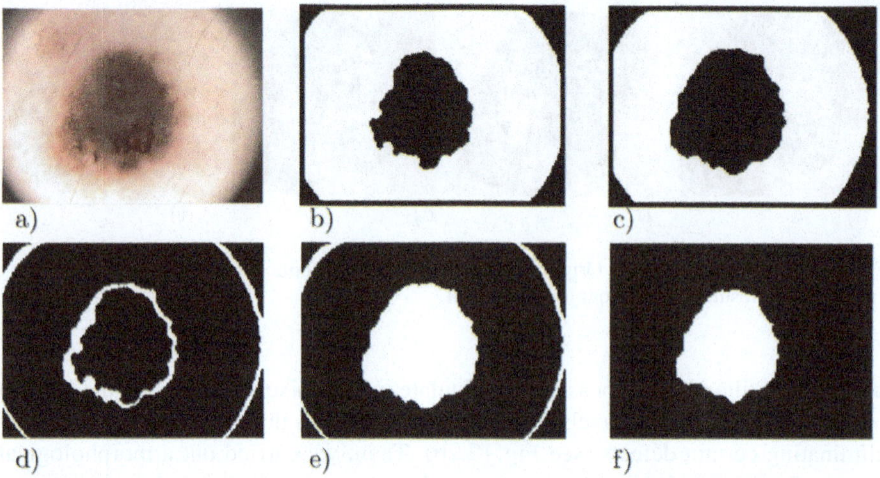

Fig. 12.3 Mask definition: **a** Original image, **b** first mask (with the first threshold), **c** second mask (with the second threshold), **d** Mask2-Mask1, **e** application of filling operator, **f** final mask

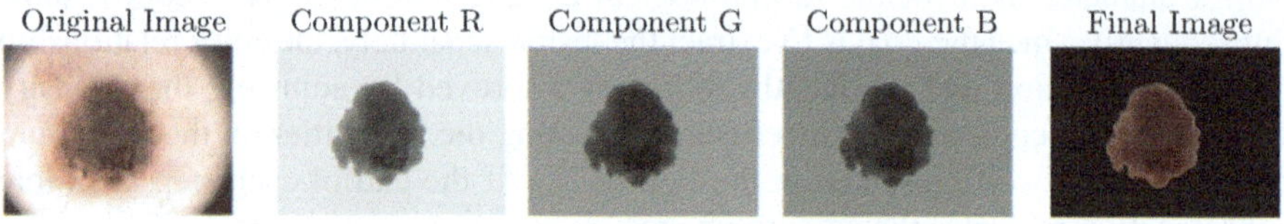

Fig. 12.4 Mask application on the 3 components

part of the image with the edges and in the second mask the lightest part. A simple subtraction between the two masks makes it possible to eliminate a good part of the edges and to keep the contour of the lesion as it appears in the Fig. 12.3d. Once the contour is detected, we perform the filling operation (Soille, 2004) from the central pixel. The obtained mask is presented in the Fig. 12.3e. The lesion is detected but the edges of the image persist. It is then enough to apply a simple erosion to eliminate the edges as indicates the Fig. 12.3e. Finally, the mask is superimposed on the three components to obtain the result shown in Fig. 12.4.

12.3.4 Interpretation of Medical Images

Early detection of melanoma is one of the biggest challenges in dermatological practice today. Based on the interpretation of the characteristics inspected by dermatoscopy and the detection of malignancy of pigmented spots, this first developed system is based on the ABCD diagnostic rule widely accepted by clinicians Stolz et al. (1994). The ABCD rule is introduced by Stolz et al. (1994) to assess the risk of

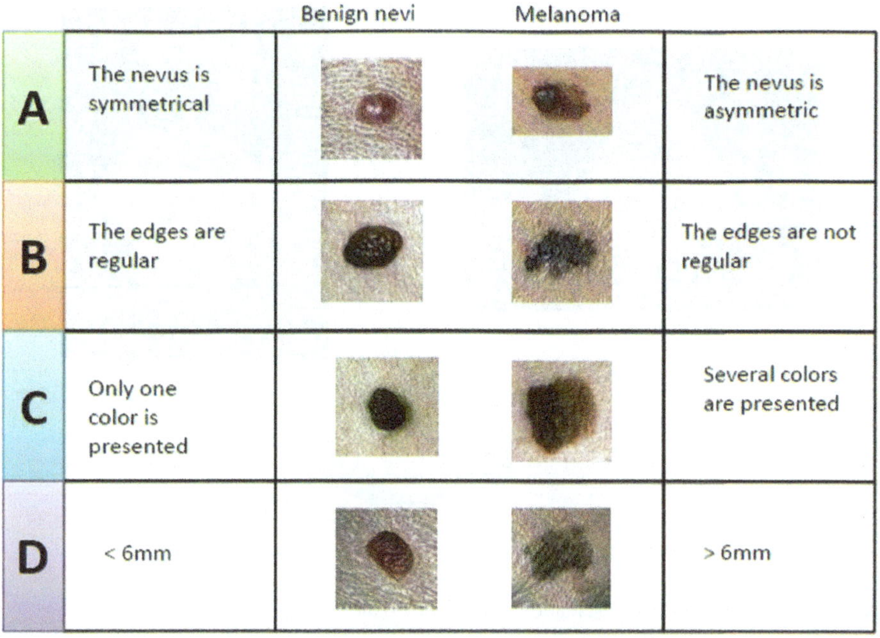

Fig. 12.5 Illustration of the ABCD rule

malignancy of a pigmented spot. This rule is a reliable method to provide an objective and reproducible diagnosis of skin cancer in addition to its speed of calculation.

Each letter of this rule corresponds to a visual characteristic taken into consideration to detect malignant lesions. Figure 12.6 and the Table 12.1 illustrate these concepts better (Fig. 12.5).

Once the four criteria are estimated, a score called the Total Dermatoscopic Score (TDS) is calculated according to the following equation. This score helps to differentiate between benign and malignant lesions.

$$TDS = 1.3 \times A + 0.1 \times B + 0.5 \times C + 0.5 \times D \tag{12.1}$$

12.3.4.1 Asymmetry

This criterion is considered as the main indicator indicating the malignancy (Turkar, 2012). This parameter is considered in terms of shape and color. To facilitate the task of calculating the asymmetry score, we propose to translate the lesion in the center of the image and rotate it to align the major axis horizontally as shown in Fig. 12.6. Now, it is easy to divide the image according to its major axis and minor axis as shown in Fig. 12.7.

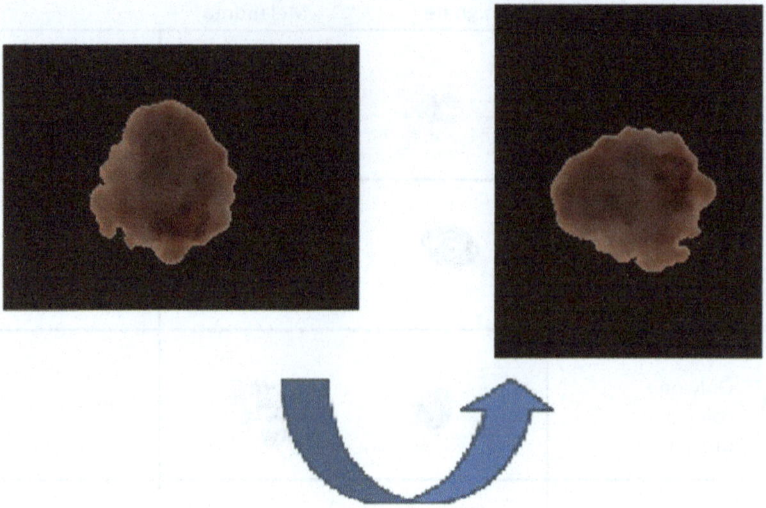

Fig. 12.6 Translation and rotation of the lesion

Table 12.1 Characteristics presentation of the ABCD rule

Criterion	Description	Score	Weight
A-Asymmetry	Evaluate asymmetry of contours and colors	0–2	1.3
B-Bordure	Evaluate the regularity of the border	0–8	0.1
C-Color	Evaluate the presence of one or more colors	1–6	0.5
D-Diameter	> 6 mm indicates the presence of melanoma	0.5–5	0.5

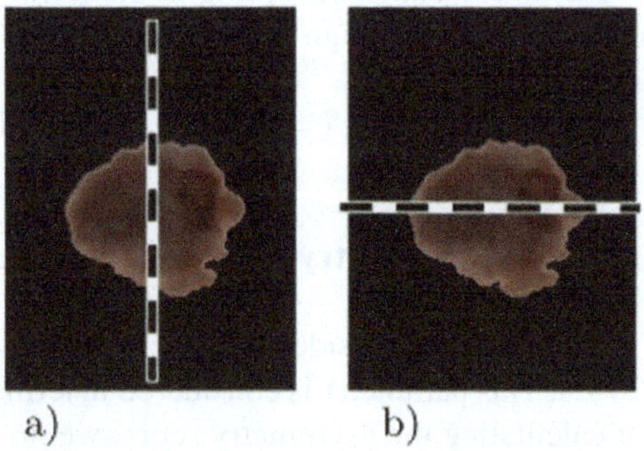

Fig. 12.7 Lesion division: **a** According to major axis, **b** according to minor axis

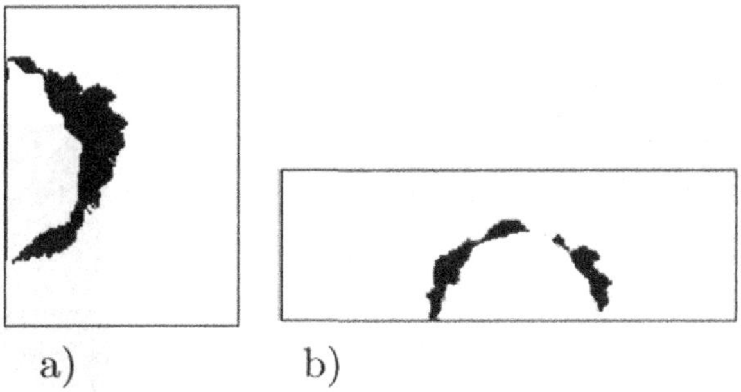

Fig. 12.8 Calculation of the non-overlapped region: **a** Folding left-right, **b** folding up and down

To measure asymmetry from the point of view of shape, we are based on the lesion folding respectively according to the minor axis and the major axis. Subsequently, we calculate the non-overlapped area between the two vertical parts (Fig. 12.8a) and horizontal (Fig. 12.8b). Therefore, the shape asymmetry score is calculated as follows:

$$\begin{aligned}
&\Delta A = 0 \text{ if } OVL(L, R) \leq T_0 \text{ and } OVL(U, D) \leq T_0 \\
&\Delta A = 1 \text{ if } OVL(L, R) \leq T_0 \text{ or } OVL(U, D) \leq T_0 \\
&\Delta A = 2 \text{ else}
\end{aligned} \tag{12.2}$$

with $OVL(L, R)$ and $OVL(U, D)$ respectively describe the non-overlapped area between the two right-left and upper-lower parts. T_0 has an empirically calculated threshold equal to 5% of the lesion area.

Now, the form asymmetry score A_{form} is calculated using the following equation:

$$A_{form} = \frac{\Delta A}{AL} \tag{12.3}$$

where AL is the total area of the lesion.

Now let's move on to calculate the asymmetry of color point of view. To do this, the histograms of the three components R, G, and B of each part of the lesion were calculated and normalized. Normalization is achieved by associating the 3 histograms of the components. Then the histogram of each part is divided by 3 as presented by the following equations:

$$hist1 = \frac{hist1}{3} \tag{12.4}$$

$$hist2 = \frac{hist2}{3} \tag{12.5}$$

Subsequently, the distance between the two normalized histograms is calculated based on distance Chisquare (Celebi et al., 2011) presented by the following equation:

$$A_{color} = \sum_{i=1}^{n} \frac{(hist1(i) - hist2(i))^2}{hist1(i) + hist2(i)} \tag{12.6}$$

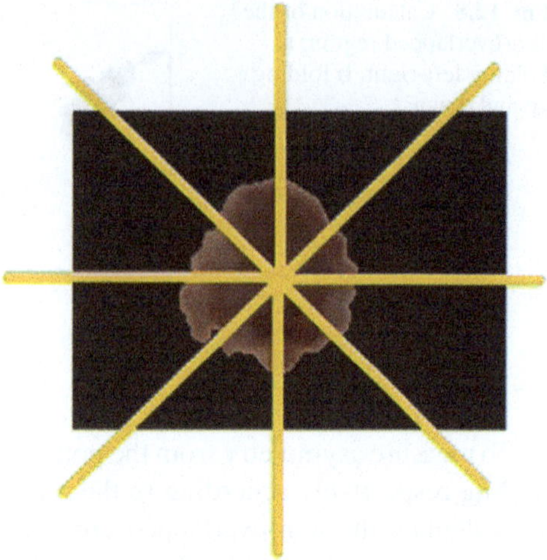

Fig. 12.9 Lesion divided into 8 sectors for calculating the parameter B

Finally, the asymmetry score A is calculated according to the following equation:

$$A = \frac{A_{form} + A_{color}}{2} \tag{12.7}$$

12.3.4.2 Irregularity of the Border

For the calculation of criterion B (Turkar, 2012), the lesions are divided into eight sectors as shown in Fig. 12.9. Within each sector, a strong and brutal cut of the pattern at the periphery receives a score of 1. On the other hand, a gradual, indistinct approach within each segment has a score of 0. Thus, the maximum irregular border score is 8, and the minimum score is 0.

To do this, the Euclidean distance (Westerink et al., 1991) is calculated for each pixel of the edge of the lesion based on the following equation:

$$Di = \sqrt{(xc - xi)^2 + (yc - yi)^2} \tag{12.8}$$

With (xi, yi) the coordinates of the considered pixel and (xc, yc) the coordinates of the center pixel.

The sum of all distances is calculated as follows:

$$Distance = \sum_{i=1}^{n} Di \tag{12.9}$$

where n is the number of pixels in the border.

The 8 sectors are then defined as follows:

$$\begin{aligned}
sector1 &= Distance(\theta \leq 45);\\
sector2 &= Distance(\theta > 45\, and\, \theta \leq 90);\\
sector3 &= Distance(\theta > 90\, and\, \theta \leq 135);\\
sector4 &= Distance(\theta > 135\, and\, \theta \leq 180);\\
sector5 &= Distance(\theta > 180\, and\, \theta \leq 225);\\
sector6 &= Distance(\theta > 225\, and\, \theta \leq 270);\\
sector7 &= Distance(\theta > 270\, and\, \theta \leq 315);\\
sector8 &= Distance(\theta > 315\, and\, \theta \leq 360);
\end{aligned} \tag{12.10}$$

Once the 8 sectors are defined, the standard deviation of each sector is calculated based on the following equation:

$$s = \left(\frac{1}{n-1}\sum_{1}^{n}(xi - \overline{x})^2\right)^{\frac{1}{2}} \qquad \overline{x} = \frac{1}{n}\sum_{i}^{n} xi \tag{12.11}$$

with n is the number of elements in the sector.

Finally, within each sector, if the standard deviation exceeds a certain threshold, the score B takes the value 1. This threshold, which is calculated empirically, is equal to the value 30.

In general, the border score is mainly between 3 and 8 in melanomas. Due to its low weight factor (0.1) the border score is not very relevant.

12.3.4.3 Color Indicator

To verify the presence of each of the colors: white, red, light brown, dark brown, gray-blue and black, we use the approach proposed by Grammatikopoulos et al. (2006). For a given image showing the lesion, it is necessary to examine the presence of the six candidate colors. To do this, the distance between the pixel value examined in the lesion and each color reference is calculated based on the Euclidean distance indicated by the following equation:

$$C_{k(i,j)} = \sqrt{(r_k - r_{ij})^2 + (g_k - g_{ij})^2 + (b_k - b_{ij})^2} \qquad k = 1..6 \tag{12.12}$$

$(i, j)$is the position of the pixel in the lesion. C_k is calculated as follows :

$$C_k = \frac{\sum C_k}{n} \qquad k = 1..6 \tag{12.13}$$

with n is the number of pixels in the lesion.

The existence of colors in the lesion depends on the comparison between C_k and a threshold called T_k. For each color, the threshold T_k is calculated as the distance between the maximum and the minimum reference values. The RGB codes chosen as reference points for each color are shown in the Table 12.2. The top rows are the

Table 12.2 RVB codes

White	Black	Red	Light Brown	Dark Brown	Gray-Blue
(1, 1, 1)	(0, 0, 0)	(0.784, 0.588, 0.392)	(0.784, 0.588, 0.392)	(0.588, 0.392, 0.392)	(0.588, 0.490, 0.588)
(0.960, 0.960, 0.960)	(0.039, 0.039, 0.039)	(1,0.196, 0.196)	(0.784, 0.392,0)	(0.490, 0.294, 0.294)	(0.490, 0.490, 0.588)
(0.921, 0.921, 0.921)	(0.078, 0.078, 0.078)	(0.784, 0, 0)	(0.784, 0.392, 0.196)	(0.392,0.196, 0.196)	(0.392, 0.392, 0.490)
(0.882, 0.882, 0.882)	(0.117, 0.117, 0.117)	(0.784, 0.196, 0.196)	(0.588, 0.392, 0.196)	(0.392, 0.196, 0)	(0.392, 0.490, 0.588)
(0.843, 0.843, 0.843)	(0.156, 0.156, 0.156)	(0.588, 0, 0)	(0.588, 0.392, 0)	(0.392, 0,0)	(0.196, 0.392, 0.588)
(0.803, 0.803, 0.803)	(0.196, 0.196, 0.196)	(0.588, 0.196, 0.196)	(0.588, 0.196, 0)	(0.196, 0,0)	(0, 0.392, 0.588)

codes for the base colors. The lines that follow are gradually less intensive color tones.

Take the example of the white color. The threshold T_0 is calculated as follows:

$$T0 = \sqrt{3 \times (1 - 0.803)^2} = 0.339 \tag{12.14}$$

The same procedure is performed for all six colors and the Color Score increments by 1 each time $C_k \leq T_k$.

12.3.4.4 Diameter Indicator

To calculate the diameter, the number of pixels along the length of the major axis of the lesion is transferred in millimeters as presented by the following equation:

$$M = \frac{25.4 \times length\,of\,major\,axis}{dpi} \qquad dpi = dots\ per\ inch\ = 96 \tag{12.15}$$

The diameter indicator is calculated as follows:

$$\begin{array}{ll} M < 2\,\text{mm} & \Longrightarrow D = 0.5 \\ 2\,\text{mm} \leq M < 3\,\text{mm} & \Longrightarrow D = 1 \\ \vdots & \\ 6\,\text{mm} \leq M < 7\,\text{mm} & \Longrightarrow D = 3 \\ \vdots & \\ 9\,\text{mm} \leq M < 10\,\text{mm} & \Longrightarrow D = 4.5 \\ M > 10\,\text{mm} & \Longrightarrow D = 5 \end{array} \tag{12.16}$$

Finally, the TDS score is calculated as presented in the Eq. 12.1.

12.4 Diagnostic Aid System Based on Machine Learning

For diagnostic aid systems based on machine learning, 4 parts are presented namely acquisition, pre-processing, segmentation and finally interpretation.

12.4.1 Images Acquisition

For this second application, we also relied on the PH2 database for research and analysis purposes, to facilitate comparative studies with segmentation and classification algorithms for dermoscopic images (Mendonca, 2013).

12.4.2 Pretreatment of Dermatoscopic Images

The second idea of pretreatment that seems simpler has been proposed. This step is divided into two parts. The first one is a simple morphological closing operation that is repeated three times with a structuring element in the shape of a line of size 3 and with different degrees of orientation (45, 90 and 135). Once reached, the second step aims at improving the contrast to obtain a clearer view of the lesion and, subsequently, a better interpretation (see Fig. 12.10).

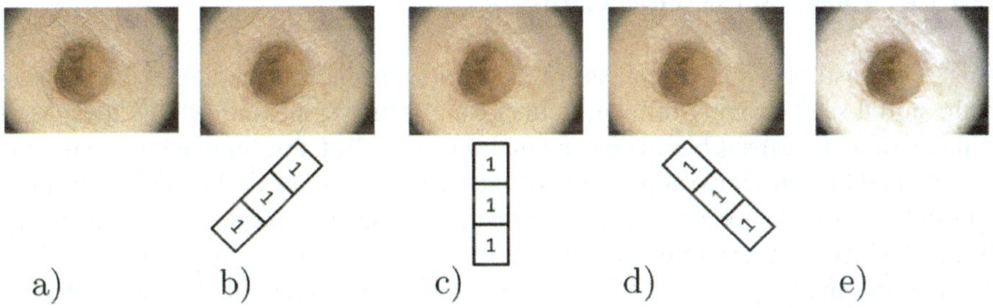

Fig. 12.10 Pretreatment steps: **a** Original Image, **b–d** application of the morphological closing with different structuring elements, **e** results of the histogram adjustment

12.4.3 Segmentation of Lesion Based on Region Growing Method

To make a good classification of the cutaneous lesions by distinguishing the cancerous nevi from the benign lesions, it is necessary to isolate the lesion of the healthy skin which surrounds it in each color image. For this, the region growing method was adopted because of its simplicity and robustness with regard to the acquisition conditions, especially in dermatoscopic images. The transition between the lesion and the surrounding skin is smooth and difficult to detect with precision. The region growing method is an iterative method of segmentation based on a region approach whose general principle has been proposed by Zucker (1976).

In the literature, several studies have been based on region growing algorithm for the detection of skin lesions. Fondón et al. (2007) have described a method based on the selection of the seed point by the mouse. Fatichah et al. (2010) chose the center of an homogeneous area as the seed and set the threshold in a random manner. This manual and random choice of the seed and the threshold can have a very bad influence on the segmentation result. Another approach based on a fuzzy region growing algorithm has been proposed by Amaliah et al. (2009).

In this work, the presented region growing method is a simple and automatic method and well adapted to our application. Indeed, it relies on an automatic choice of the threshold and the germ and on the gray level as being a criterion of similarity.

First, this technique consists of selecting one (or more) starting pixel(s). Regions gradually grow around their starting pixel according to a double criterion: homogeneity and adjacency. The process can be applied to one or more regions. Each region requires an initialization phase and an iterative phase (PreethaL et al., 2012).

Figure 12.11 gives a clear idea about this method.

12.4.3.1 Choice of the Initial Germ

This step constitutes the critical part of the algorithm since the choice of the position of the initial seed strongly influences the result of the segmentation. The most used initialization of germs is based on a manual selection. But, this human intervention is not solicited for automated applications. Since the lesion is always darker than healthy skin and only one region needs to be segmented, the seed is automatically selected as the pixel with the darkest gray level and the most similar to its neighbors. Figure 12.12 explains the method used and Fig. 12.13 presents the result of this method on a dermatoscopic image.

12.4.3.2 Region Growing

This step consists in growing the region by agglomerating the neighboring pixels. These pixels are chosen in such a way that the homogeneity of the region is main-

Fig. 12.11 Region growing principle: **a** Original image with the first seed presented in red, **b**, **c** Progress of the region growing algorithm until obtention of the final results in (**d**)

tained. For this, a homogeneity indicator is defined and the neighboring pixels are added to the region if this indicator remains true. Thus, the growth stops when the addition of the pixels breaks the homogeneity.

The application of this algorithm includes a specification of two variables that are the size of the window fixed in our case at 3×3 and the threshold which is calculated as follows:

⋆ Step 1: automatically calculate S which is the optimal threshold based on the method of triangle (Zack et al., 1977). This method makes it possible to construct a line between the peak of the histogram of the image and the farthest end. The threshold is then defined as the point of maximum distance between the line and the histogram as presents Fig. 12.14.

⋆ Step 2: Calculate the average moy of the region given by the following equation:

$$moy = \frac{\sum value\ of\ all\ pixels\ of\ the\ region}{N} \tag{12.17}$$

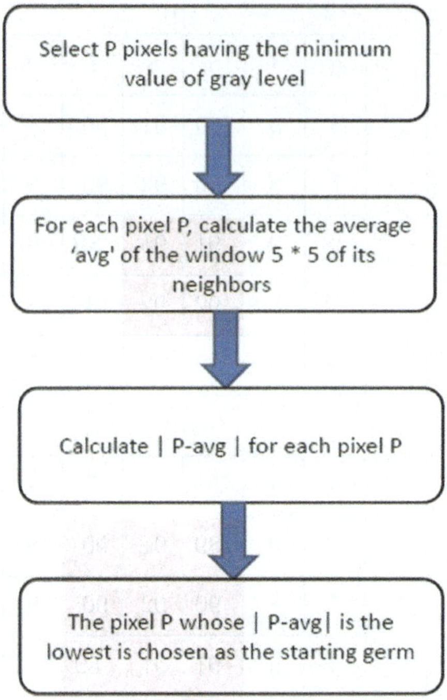

v1	v2	v3	v4	v5
v6	v7	v8	v9	v10
v11	v12	**P**	v13	v14
v15	v16	v17	v18	v19
v20	v21	v22	v23	v24

Fig. 12.12 Flowchart of the automatic method of selecting the first germ

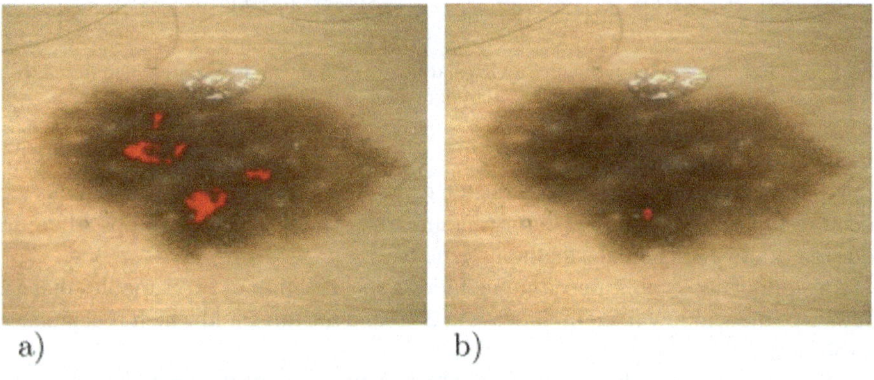

Fig. 12.13 Choice of the initial germ: **a** Selecting pixels with the darkest gray level, **b** germ detection

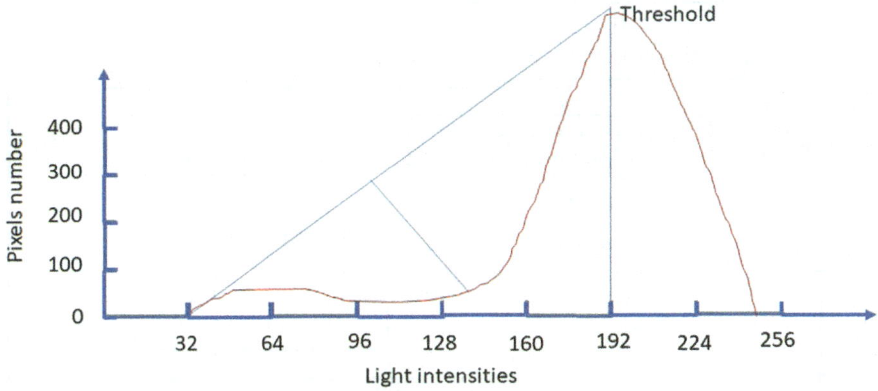

Fig. 12.14 Triangle method principle

⋆ Step 3: Calculate the threshold defined as the difference between the region mean and the optimal threshold as shown by the following equation:

$$Threshold = |avg - S| \tag{12.18}$$

The growth process follows now the next steps:

• Step 1: We add the neighboring pixels of the seed belonging to a window of size 3×3 pixels to a list of neighbors.

• Step 2: We calculate the difference between the gray level value of each neighboring pixel and the average of the region.

• Step 3: We add the neighboring pixel of the germ, having the smallest difference, to the region and mark it as a seed.

• Step 4: We update the average Avg, the threshold, and the difference of the gray level.

• Step 5: We return to Step 1 each time the difference is below the threshold. Hence the approach used is to grow the region around the initial pixel. This aggregation stops only when this difference exceeds a certain threshold which is described by the following equation:

$$|I(x, y) - avg| \geq Threshold \tag{12.19}$$

Figure 12.15 describes the process.

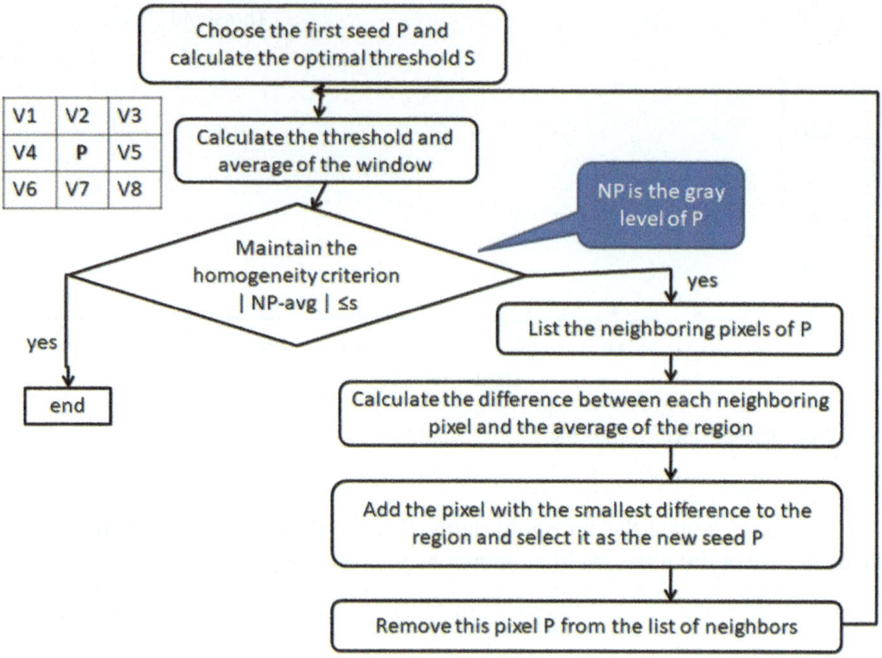

Fig. 12.15 Principle of region growing segmentation

12.4.3.3 Application of the Regions Growing Method

As mentioned before, the research was based on PH2 database (Mendonca, 2013). Recall that the presented processes are applied for each component R, V, and B.

For the PH2 database, the images are essentially presented in three parts, namely the lesion, the healthy skin and the black frame introduced during the digitization process.

As we notice the frame and the lesions are the darker parts of the image. So we have the idea of eliminating the framework. To achieve this goal, we need to define a binary mask showing the dark parts of the image (low gray values). These parts are the black frame and the lesion. A simple morphological operation (filling) eliminates the lesion and maintains the frame. It is now sufficient to perform a subtraction operation between the original image and our mask to find the result presented in Fig. 12.16d.

All that remains now is to apply the region growing algorithm from the found seed. Figure 12.17 shows the result found.

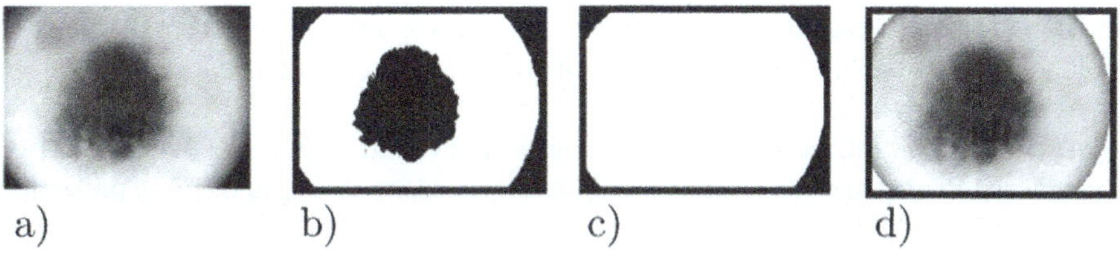

Fig. 12.16 Black frame elimination: **a** Original image, **b** mask with dark areas, **c** mask presenting the frame, **d** resulted image

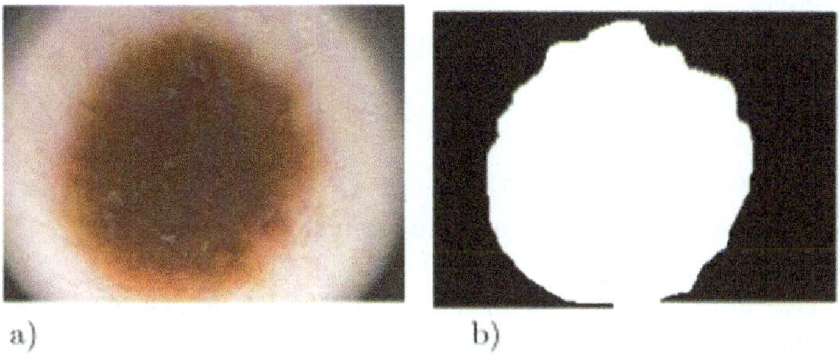

Fig. 12.17 Application of the region growing method on an image of the PH2 database: **a** original image, **b** resulted image

12.4.4 Skin Lesion Analysis

In this second part, the classification is based on the SVM classifier.

12.4.4.1 Features Extraction

Many different image features have been proposed for the analysis of medical images, including intensity, geometric, morphological characteristics ... The selected characteristics for the analysis of the detected lesions are based on knowledge and experience properties and appearance differences between benign and malignant skin lesions. The proposed characteristics must be identifiable, translatable, invariant to rotation and scale, noise-resistant and reliable.

For this purpose, we focus on the shape features that correspond to the diagnostic algorithm based on the ABCD rule. We extract information on asymmetry, border irregularity, and color. For the Diameter indicator, the ABCD rule specifies that most melanoma lesions have a diameter greater than 6 mm. This does not take into account small-diameter melanomas, especially as their incidence is increasing. Indeed, in Joanna (2016) and Abbasi et al. (2004), the authors specify that approximately 14 % of the melanomas detected in the world are of small diameter. We, therefore, have the idea of excluding the diameter as a distinctive element between benign lesions and melanomas. Thus, our system offers better opportunities to detect small diameter melanoma if it is able to achieve good sensitivity using the other features.

– **Asymmetry**

Asymmetry is the most relevant indicator as evoked in Sect. 12.3.4.1. The two characteristics that are taken into account in this case are A_{form} and A_{color} presented respectively by the Eqs. 12.3 and 12.6.

– **Border Irregularity**

As we have shown in Sect. 12.3.4.2, benign lesions are usually characterized by smooth edges. In contrast, malignant skin lesions are characterized by irregular edges with sharp edges.

The index of irregularity is based on the following equation (Bhardwaj, 2014):

$$B = \frac{4\pi A}{p^2} \tag{12.20}$$

with A and p respectively the area and the perimeter of the lesion.

– **Colors**

As we have introduced in Sect. 12.3.4.3, a skin lesion can have a diversity of colors that are essentially white, red, light brown, dark brown, blue gray, and black. Melanomas are generally characterized by three or more colors, and for about 40% of melanoma cases, even five or six colors can be presented (Joanna, 2016). To set the color score, the presence of each color counts for 1 point. Thus, the minimum color score is 1 and the maximum is 6. Based on the work presented in Ojala et al. (1996), we extracted 4 characteristics based on the color variation across 4 different channels including red, green, blue and intensity (presented by the Eq. 12.21) using respectively Eqs. (12.22), (12.23) and (12.24).

$$I = \sqrt{R^2(x, y) + V^2(x, y) + B^2(x, y)} \tag{12.21}$$

$$F_1 = \log \frac{\sigma(R)}{\mu(R)} \tag{12.22}$$

$$F_2 = \log \frac{\sigma(V)}{\mu(V)} \tag{12.23}$$

$$F_3 = \log \frac{\sigma(B)}{\mu(B)} \tag{12.24}$$

where mu represents the average value and $sigma$ indicates the standard deviation.

Table 12.3 presents the values extracted for two examples presenting respectively a benign lesion and a melanoma.

Table 12.3 Characteristic values showing malignant and benign lesions

Features	benign lesion	Melanoma
F_1	−0.9092	0.5109
F_2	−0.9099	0.5227
F_3	−0.9044	0.5393
F_4	−8.8883	0.6025
B	0.8757	0.6571
A_{color}	0.0187	0.0301
A_{form}	1.6772	8.9613
C	2	4

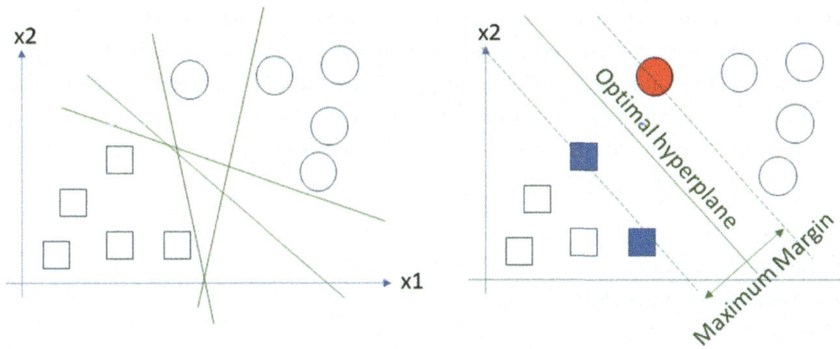

Fig. 12.18 Optimum hyperplane with maximum margin

12.4.4.2 Classification

Once, the set of discriminant features is selected, a classifier is designed for the evaluation of cutaneous lesions. In this work, the classification is based on the support vector machine (SVM).

Support vector machine is a supervised learning classification method, it was introduced by Vapnik in (2000). This algorithm is based on two phases: the first phase consists in storing the classes of the learning samples "Learning phase" . For the second phase, it's about knowing the class of the new individual "test phase".

The SVM principle is to find a linear separator between the data points of two different classes. This kind of separator is called hyperplane. So, the closest points that are used for the determination of the hyperplane called support vectors.

Nevertheless, if there is a separating hyperplane between the learning data, then there is an infinity. Among these hyperplanes, the SVM algorithm looks for one who is the best. To solve the separation problem, the goal of the SVM is to find the optimal hyperplane which maximizes Margin. The latter is defined by the smallest distance between the vectors and the hyperplane. Figure 12.18 presents an optimal hyperplane with a maximum margin:

For the case of two classes, the principle is summarized to an optimization problem.

$$\min_{\omega,b,\xi} \frac{\omega^T}{\omega} + C\sum_{i=1}^{h} \xi_i \tag{12.25}$$

which verifies

$$y_i(\omega^T \phi(xi) + b) \geq 1 - \xi_i; \quad \xi_i \geq 0; \quad i = 1..h \tag{12.26}$$

with $\{y_i, y_i \in \{-1, +1\}\}_{i=1..N}$ present the labels classes in the learning base $\left\{x_i, x_i \in R^N\right\}_{i=1..N}$

Φ presents the kernel function for representing the data in a hyperplane, ω presents a normal vector to the hyperplane that translates the decision to the edges. The constant $C > 0$ presents a penalty parameter giving the separation error by the hyperplane. Finally ξ presents the variable that penalizes the objective function (Lopes, 2009). The scalar product $\Phi(x_i)^T \Phi(x_i)$ is usually replaced by a kernel $K(x_i, x_j)$ allowing to minimize the algorithmic cost. The kernel function K has special properties. The 4 most used kernel families are (Lopes, 2009):

- Linear: $K(x, y) = x^T y$
- polynomial: $K(x, y) = (x^T y + m)^p$, $p \in \mathbb{N}$ and $m > 0$
- Radial Basis Functions (RBF): $K(x, y) = \exp(-\gamma \left\|x - y\right\|^2)$
- Sigmoid: $K(x, y) = \tanh(\gamma, x^T y + m)$

This classifier is chosen for two reasons; firstly, it is considered to be the tool that achieves the best results in the class of linear classifiers, and second, it is very effective when few learning data are available.

Thus, the extracted features are applied to the SVM classifier to be classified into two different classes (melanoma and non-melanoma). The process concerns two phases which are the learning phase and the test phase.

- Learning phase: the characteristics of melanoma and non-melanoma images are transmitted to the classifier for learning. For this phase, we chose 20 images representing a melanoma and 20 images with benign lesions.
- Test phase: A set of unknown tests of characteristics describing an image is inserted and the knowledge acquired during the training phase will classify the unknown image into melanoma and non-melanoma.

Original image	Detected lesion	Score TDS	Result	Expert opinion
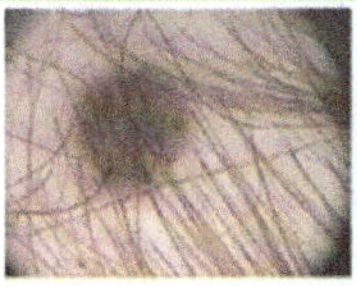		(A= 1, B=2 , C=2, D=1.5; TDS=3.25)	benign	benign
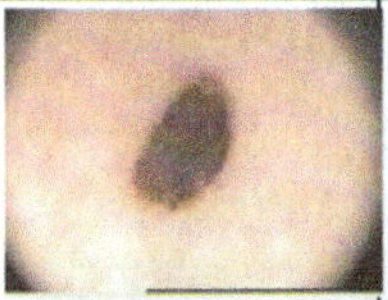	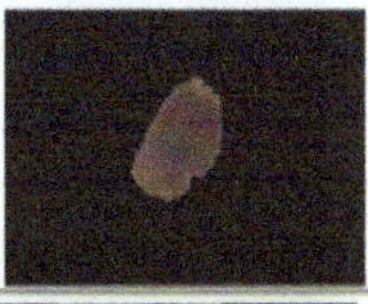	(A= 0.5, B=1 , C=2, D=1.5 ; TDS=2.5)	benign	benign
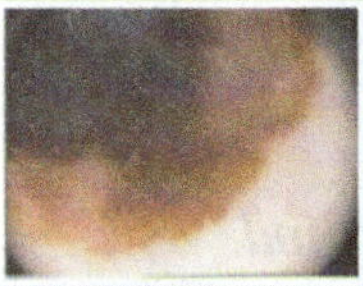		(A= 2, B=3 , C=3, D=3 ; TDS=5.9)	Melanoma	Melanoma
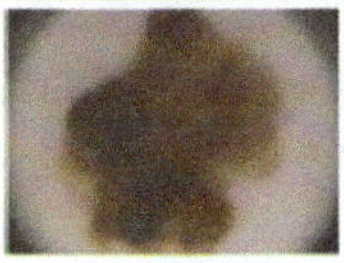	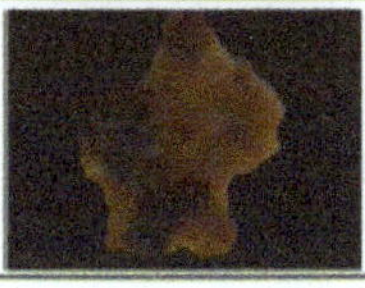	(A= 2, B=2 , C=3, D=3.5; TDS=6.05)	Melanoma	Melanoma

Fig. 12.19 Some results of cutaneous lesions based on the ABCD rule

12.5 Experimental Results and Discussion

12.5.1 Approach Based on the MultiOtsu Principle

The approach based on the principle of MultiOtsu is applied to the PH2 database. The mask of the cutaneous lesion is created, and then applied to the pretreated image to obtain the segmented image. The entry image of the cutaneous lesion is effectively segmented. Then the ABCD rule is applied to produce the TDS score that gives an idea of the malignancy of the lesion. Figure 12.19 presents a random selection of dermatoscopic images with the corresponding segmentation and classification results. Expertise is also presented in each case to evaluate the proposed approach.

Recall that the PH2 database contains 40 melanomas, 80 common nevi and 80 atypical nevi. The Table 12.4 summarizes the calculated values of the three classes namely benign (B), suspicious (S) and melanoma (M).

Table 12.4 Calculated values of the three classes of lesions

Class	Benign (B)	Suspect (S)	Melanoma (M)
TP	74	64	36
FN	6	16	4
FP	11	10	8
TN	109	110	152

Table 12.5 Classification results based on SVM

Class	Melanoma	Non-melanoma	Total
Melanoma	18	2	20
Non-melanoma	2	138	140

12.5.2 Approach Based on the Region Growing Method

The Table 12.5 presents the results of the classification obtained with the algorithm based on the method of region growing in a confusion matrix, knowing that we have 20 melanomas and 140 non-melanomas.

12.5.3 Evaluation and Discussion

For the evaluation of our work, we compared the proposed approachs to recent work in terms of sensitivity and specificity. These two metrics are presented by the following equations:

$$Sensitivity = 100 \times \frac{TP}{TP + FN} \tag{12.27}$$

$$Specificity = 100 \times \frac{TN}{TN + FP} \tag{12.28}$$

with: TP = number of true positives, TN = number of true negatives, FN = number of false negatives, FP = number of false positives.

The Table 12.6 presents the results found by the two proposed approaches and recent work.

According to the table, the region growing algorithm and the SVM classifier show good performance for the detection and classification of skin lesions.

Table 12.6 Evaluation and comparison with recent work

Method	Sensitivity	Specificity	References
RNA	90.86	96.11	Ozkan and Koklu (2017)
SVM	**94.10**	77.4	Jain et al. (2015)
SVM	84.00	94.00	Riaz et al. (2014)
TDS	87.5	92.46	MultiOtsu approach
SVM	90	**98**	Region growing approach

12.6 Conclusion

The purpose of the work presented in this chapter was to detect skin cancer based on two approaches. The first approach is based on MultiOtsu thresholding and the ABCD rule which is a classic method for the classification of skin lesions. The second approach is based on region growing segmentation and SVM classification. The characteristics extracted in this second case are based on the ABCD rule themselves. The classification results found indicate the performance of the two methods. But work based on the ABCD rule remains subjective work. Indeed, strategies to recognize suspicious lesions may have little impact on the diagnosis of small melanomas. The ABCD rule does not recognize the existence of melanoma of small diameter, that is to say, a melanoma less than 6 mm in diameter. For this reason, reliance on other form features provides a better interpretation of skin lesions.

Acknowledgements This research is funded by Vietnam National Foundation for Science and Technology Development (NAFOSTED) under grant number 102.99–2013.06.

References

Abbasi, N.R., Shaw, H.M., Rigel, D.S., Friedman, R.J., McCarthy, W.H., Osman, I. A, Kopf, A. W. A, & Polsky, D. (2004). Early diagnosis of cutaneous melanoma: revisiting the ABCD criteria. *The Journal of American Medical Association, 292*, 771–776.

Alfed, N., & Khelifi, F. (2017). Bagged textural and color features for melanoma skin cancer detection in dermoscopic and standard images. *Expert Systems with Applications, 90*, 101–110.

Amaliah, B.a, Fatichah, C., Widyanto, M.W. (2010). ABCD feature extraction of image dermoscopic based on morphology analysis for melanoma skin cancer diagnosis. *Journal Ilmu Komputer dan Informasi, 3*, 82–91.

Berner, E. (2009). *Clinical decision support systems: State of the art*. Agency for Healthcare Research and Quality U.S. Department of Health and Human Services, (pp. 1–26).

Bhardwaj, A., Bhatia, J.S. (2014). An image segmentation method for early detection and analysis of melanoma. Journal of Dental and Medical Sciences, 13: 18–22.

Celebi, M. E. A., Stoecker, W. V. A., & Moss, R. H. (2011). Advances in skin cancer image analysis. *Computerized Medical Imaging and Graphics, 35*, 83–84.

Dalila, F., Zohra, A., Reda, K., & Hocine, C. (2017). Segmentation and classification of melanoma and benign skin lesions. Optik - Int. *Journal for Light and Electron Optics, 140*, 749–761.

Fatichah, C. A, Amaliah, B. A, & Widyanto, M. R. (2009) . Skin lesion detection using fuzzy region growing and ABCD feature extraction for melanoma skin cancer diagnosis. In: *Proceedings of Industrial Informatics Seminar.*

Fondon, I., Serrano, C., & Acha, B. (2007) . Segmentation of skin cancer images based on multistep region growing. In *Proceedings of the IAPR Conference on Machine Vision Applications* (pp. 8–28).

Grammatikopoulous, G., Hatzigaidas, A., Papastergiou, A., Lazardis, P., Zaharis, Z., Kampitaki, D., & Tryfon, G. (2006). Automated malignant melanoma detection using matlab. In *5th WSEAS International Conference on Data Networks, Communications and Computers At Bucharest, Romania* (pp. 91–94).

Huang, T., Yang, G., & Tang, G. (1979). A fast two-dimensional median filtering algorithm. IEEE Trans. on Acoustics. *Speech, and Signal Processing, 27*, 13–18.

Ins Les traitements du mélanome de la peau. Cancer Info (2016)

Jain, S., Jagtap, V., Pise, N. (2015). Computer aided melanoma skin cancer detection using image processing. Procedia Computer Science, 48: 734–750.

Joanna, J. K. (2016) . Computer-aided diagnosis of micro-malignant melanoma lesions applying support vector machines.*BioMed Research International*, 1–8.

Liao, P.S., Chen, T.S., Chung, P.C. (2001) . A fast algorithm for multilevel thresholding. Journal of Information Science And Engineering, 17: 713–727.

Lopes, R. (2009). Analyses fractale et multifractale en imagerie mãcdicale: Outils, validations et applications. UniversitÃcde Lille 1.

Masood, A., Aljumaily, A. A., Hoshayr, A., & N., Masood, O. (2013). Fuzzy c mean thresholding based level set for automated segmentation of skin lesions. *Journal of Signal and Information Processing, 4*, 66–71.

Maglogiannis, I., & Doukas, C. N. (2009). Overview of advanced computer vision systems for skin lesions characterization. *IEEE Transactions on Information Technology in Biomedicine, 13*, 721–733.

Mendonca, T. (2013). Ph2, a dermoscopic image database for research and benchmarking. In *2013 35th Annual International Conference of the IEEE Engineering in Medicine and Biology Society (EMBC).*

Ojala, T., & Pietikãinen, M. A, & Harwood, D. A. (1996). A comparative study of texture measures with classification based on featured distributions. *Pattern Recognition, 29*, 51–59.

Otsu, N. (1979). A threshold selection method from gray-level histograms. IEEE Trans. on Systems. *Man, and Cybernetics, 9*, 62–66.

Ozkan, I. A., & Koklu, M. (2017). Skin lesion classification using machine learning algorithms. *Int. Journal of Intelligent Systems and Applications in Engineering, 4*, 285–289.

Pennisi, A. A., Bloisi, D. D. A., Nardi, D. A, Giampetruzzi, A. R. A, Mondino, C. A., & Facchiano, A. (2016). Skin lesion image segmentation using delaunay triangulation for melanoma detection. *Computerized Madical Imaging and Graphics, 52*, 89–103.

Peruch, F., Bogo, F., Bonazza, M. A, Cappelleri, V. M. A, & Peserico, E. (2014). Simpler, faster, more accurate melanocytic lesion segmentation through meds. *IEEE Transactions on Biomedical Engineering, 61*, 557–565.

PreethaL, M. M. S. J., Suresh, L. P., & Bosco, M. J. (2012) . Image segmentation using seeded region growing. In *International Conference on Computing, Electronics and Electrical Technologies.*

Riaz, F., Hassan, A., Javed, M. Y., & Coimbra, M. T. (2014). Detecting melanoma in dermoscopy images using scale adaptive local binary patterns. In *Engineering in Medicine and Biology Society (EMBC), 2014 36th Annual International Conference of the IEEE* (pp 6758–6761).

Sadri, A. R. A, Azarianpour, S., Zekri, M., Celebi, M. E. A, & Sadri, S. (2017). Wn-based approach to melanoma diagnosis from dermoscopy images. *IET Image Processing, 11*, 475–482.

Silveira, M., Nascimento, J. C., Marques, J. S., Marcal, A. R. S. A., Mendonca, T., Yamauchi, S., Maeda, J., & Rozeira, J. (2009). Comparison of segmentation methods for melanoma diagnosis in dermoscopy images. *IEEE Journal of Selected Topics in Signal Processing, 3*, 35–45.

Soille, P. (2004). Morphological image analysis. *GmbH* (vol. 2, pp. 1–15). Berlin Heidelberg: Springer.

Stolz, W., Reimann, A. L. G., & Cognetta, A. B. (1994). Abcd rule of dermatoscopy: A new practical method for early recognition of malignant melanoma. *European Journal of Dermatology, 4*, 521–527.

Tadeusiewicz, R. (2010). Place and role of intelligent systems in computer science. Computer Methods In Materials Science, 10: 193–206.

Turkar, V. (2012) . Melanoma decision support system for dermatologist. In *International Conference on Recent Trends in Information Technology and Computer Science* (pp. 28–30).

Vapnik, V. (2000). *The nature of statistical learning theory*. Springer (p. 314).

Westerink, P. H. A., Biemond, J., & Boekee, D. E. (1991). Subband coding of color images. *The Springer International Series in Engineering and Computer Science, 15*, 193–227.

Xie, F., Bovik, A.C. (2013). Automatic segmentation of dermoscopy images using self-generating neural networks seeded by genetic algorithm. Pattern Recognition, 46: 1012–1019.

Yu, L., Chen, H., Dou, Q., Qin, J., & Heng, P. A. (2017). Automated melanoma recognition in dermoscopy images via very deep residual networks. *IEEE Transactions on Medical Imaging, 36*, 994–1004.

Yuan, Y. A., Chao, M. A., & Lo, Y. C. (2017). Automatic skin lesion segmentation using deep fully convolutional networks with jaccard distance. *IEEE Transactions on Medical Imaging, 36*, 1876–1886.

Zack, G. W. A., Rogers, W. E. A., & Latt, S. A. (1977). Automatic measurement of sister chromatid exchange frequency. *The Journal of Histochemistry and Cytochemistry, 25*, 741–753.

Zucker, S. W. (1976). Region growing: Childhood and adolescence. *Computer Graphics and Image Processing, 5*, 382–399.

Chapter 13
Biomedical Computer Aided Design Systems: Application to Alzheimer Disease

Syrine Neffati, Okba Taouali, and Kais Bouzrara

Abstract Alzheimer's disease (AD) is a frequent disease among elderly people. There has been extensive research toward the early detection and diagnosis of this health problem in the past decade. The accuracy and precision of the diagnosis methods have been the mark of an accurate diagnosis. This work proposes a new Computer-Aided Design (CAD) system for AD diagnosis by analyzing Magnetic Resonance Images (MRIs) of the brain. First, we utilize a proposed Downsized Kernel Partial Least Square (DKPLS) technique as a feature extraction and dimensionality reduction method. Then, we perform the classification using artificial neural network (ANN) and we validate our work with k-fold cross validation (CV) approach. The suggested algorithm is entitled DKPLS-ANN. The algorithm is tested on a ADNI and OASIS datasets. The DKPLS-based classifier was found better performing than the existing methods.

Keywords Alzheimer · AD · Pattern recognition · KPLS · Downsized KPLS · Feature reduction · ANN

13.1 Introduction

Alzheimer's disease (AD) is the most dangerous form of dementia known across the globe. It is expected for the number of people affected with this disease to be increased and to meet the double in just the coming two decades (Motai et al., 2017).

S. Neffati (✉) · K. Bouzrara
Laboratory of Automatic Signal and Image Processing, Electrical Department, National Engineering School of Monastir, University of Monastir, Monastir, Tunisia
e-mail: syrine.neffati@gmail.com

K. Bouzrara
e-mail: kais_bouzrara@yahoo.fr

O. Taouali
Department of Computer Engineering, Faculty of Computers and Information Technology, University of Tabuk, Tabuk, Saudi Arabia
e-mail: otawali@ut.edu.sa

N. Derbel and O. Kanoun (eds.), *Advanced Methods for Human Biometrics*, Smart Sensors, Measurement and Instrumentation 40,
https://doi.org/10.1007/978-3-030-81982-8_13

Despite the scientific revolutions and discoveries, the AD is maintained unclear to humans. There is neither a clear understanding for the disease nor a plausible physio-pathological explanation about its progression (Kostic et al., 2013). Therefore, it become a necessity to diagnose this disease at an early stage which give the patient the possibility to stop the disease or to slow it in the worst cases (McCarville et al., 2006).

The key to the development of a treatment for Alzheimer's disease is the early diagnosis of the illness. In this respect, various MRI-based techniques were applied successfully to detect average AD. It is known that magnetic resonance images (MRIs) are very reliable for AD early diagnosis (Neffati et al., 2019a). MRI measures have also been used to predict or classify the disease class for patients. The MRI-based classification process could be essential in making a decisive diagnosis of AD in the future.

Computer Aided Design (CAD) systems are the appropriate automated process to assist in medical images analysis and to give a second objective diagnosis (Neffati et al., 2019a; Cheng et al., 2010; Erickson and Bartholmai, 2002). In fact, CAD systems were proven efficient in the diagnostic process; these systems reduced significantly the unneeded biopsies and thoracotomy operations (McCarville et al., 2006). CAD systems are well set up to manage the high-dimensional data issued from MRIs (Cao and Lin, 2015; Neffati et al., 2019c).

AD symptoms can be different among patients, but in general the most frequent manifestation of the illness is the worsening capacity of remembering recent information because the brain cells responsible of storing recent memories are the ones that gets first destroyed. Along, other symptoms manifest on the subject like: Change in attitude, sleeping problem, etc. (Khachaturian, 1985). The diagnosis of the disease is based mainly on the Cognitive Impairment (CI) test results and the medical history of the person (Khachaturian, 1985). MRI imaging along with other image modalities grant providing additional information about the brain to facilitate the detection of the shrinking cells of the gray matter. Hence, the use of CAD technologies become a necessity to trace any cerebral modifications in the brain and to help establishing a diagnosis at an early stage of the disease (El-Dahshan et al., 2014).

CAD systems are widely used in the medical field to help doctors establish a diagnosis not only for AD but also in a range of medical areas; such as heart disease, brain tumor, cancer research, etc. (Doi, 2007). Recently, researchers and scientists showed that classification of brain MRI is possible using machine learning (ML) combined with supervised techniques such as: Artificial Neural Network (ANN) and Support Vector Machine (SVM), and unsupervised techniques; such as fuzzy c-means and self-organization maps (SOM) (El-Dahshan et al., 2010).

In this work, we propose a new kernel-based technique. We use KPLS for feature reduction of the MRIs and ANN for image classification. We suggest a new Downsized KPLS (DKPLS) for biomedical CAD systems entitled DKPLS-ANN. The DKPLS-ANN method was successfully tested on a set of real AD MRIs. The parameters of DKPLS were optimized to better improve the detection phase using the multi-objective optimization (MOO). The chapter is organized as follows: Sect. 13.2 presents the methodology used in this work. Section 13.3, gives an idea about pre-

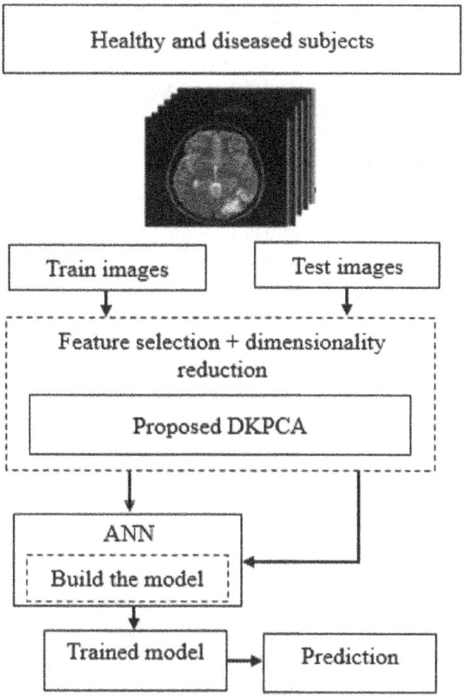

Fig. 13.1 Proposed methodology

vious works based on PLS and KPLS. Section 13.4, details the proposed DKPLS technique. Section 13.5, explains the used MOO method for the optimisation process. Section 13.6 presents the utilized classifier in this work. Section 13.7 puts forward the experimental results as well as the used MRI datasets. Finally, Sect. 13.8 concludes the chapter.

13.2 Proposed Methodology

The proposed work involves three stages. First, the data acquisition and preprocessing stage is performed on the Magnetic Resonance (MR) brain images. Second, feature selection technique is applied using the proposed DKPLS. Finally, we perform classification using the Artificial Neural Network (ANN). The sequential flow of these steps is described in Fig. 13.1 and the following sections describe these steps in detail.

13.3 Previous Works

13.3.1 Partial Least Square (PLS)

Principle

PLS approach extracts a set of vectors named Latent Components (LCs) from the original input/output data space and builds linear multi-variable regression model. Hence, the input matrix $X \in R^{N \times m}$ comprising N samples with m process variables and the output matrix $Y \in R^{N \times J}$ containing N observations with J quality variables.

$$X = \begin{pmatrix} x_1 \\ x_2 \\ \vdots \\ x_N \end{pmatrix} \in R^{N \times m}, Y = \begin{pmatrix} y_1 \\ y_2 \\ \vdots \\ y_N \end{pmatrix} \in R^{N \times J}$$

The PLS model aims to determine a set of vectors (the LCs); $T = [t_1; t_2; \ldots; t_l]$ and $U = [u_1; u_2; \ldots; u_l]$. The LCs give the closest representation of the variations of the input and output observations (Taouali et al., 2015). PLS technique projects the input and the output matrices to a low-dimensional space with a certain number l of latent variables. PLS technique represents the X and Y matrices as follow:

$$\begin{cases} X = TP^T + E \\ Y = UQ^T + F \end{cases} \tag{13.1}$$

where $P = [p_1; p_2; \ldots; p_l]$ and $Q = [q_1; q_2; \ldots; q_l]$ are respectively the loadings for X and Y. Therefore, the matrices E and F are the PLS residuals corresponding to the input matrix X and to the output matrix Y. We determine the number of LCs using the cross validation technique. This latter gives the maximum prediction based on data banned from the training (Qin, 2012).

Algorithm

The PLS model consists mainly of extracting each pair of the corresponding LCs as a linear combination of the input and output variables (Baffi et al., 1999). Rosipal (Rosipal and Trejo, 2001) presented the basic form of the PLS approach; this latter is built upon a nonlinear iterative partial least squares (NIPALS) algorithm. During each iteration, the weight vectors w_i and c_i are calculated. They introduce the weights of the input and output projections.

$$w_i = \frac{u_i^T X}{u_i^T u_i}, i = [1; 2; \ldots; l] \tag{13.2}$$

$$c_i = \frac{t_i^T Y}{t_i^T t_i}, i = [1; 2; \ldots; l] \quad (13.3)$$

During each iteration, the variables u_i, t_i, p_i and q_i are stored each in a column matrix. For each iteration, the results of the previous iteration are used as input, to assure that the scores are extracted from the input/output matrices. The standard phases of PLS can be summarized in four steps shown in algorithm 1.

Algorithm 1:

---

1. Mean-center and scale X and Y.
2. Use the NIPALS algorithm and compute the variables u_i, t_i, p_i and q_i.
3. Reduce X and Y by subtracting the computed latent vectors from them.
4. Return to step 2 and compute the next latent vector.

The chart presented in Fig. 13.2 illustrates the fundamental steps of PLS.

13.3.2 Kernel Partial Least Square (KPLS)

Principle

PLS is mainly restricted by its linearity. Kernel methods have come with the solution and solved this problem by mapping the data to nonlinear feature space with high dimensions (Taouali et al., 2015; Rosipal and Trejo, 2001; Neffati et al., 2019b). These methods determine the nonlinear latent variables with an approximately linear computational cost (Zhang and Ma, 2011). Accordingly, PLS theory can be applied to linear systems in combination with the kernel trick (Zhang and Ma, 2011). This new model is the Kernel PLS (KPLS) (Talukdar et al., 2018). KPLS is defined in the feature space, it extends linear PLS to its nonlinear kernelized form (Talukdar et al., 2018). Moreover, the aim of KPLS method is to map the input variables x_i, with $i = [1; 2; \ldots; N]$, into a feature space F using a nonlinear transformation as shown in eqnarray 13.4:

$$\Phi : x_i \in R^N \rightarrow \Phi(x_i) \in F \quad (13.4)$$

Under the previous conditions, and taking into account the curse of dimensionality (Neffati et al., 2019b), there is no possible way of computing the nonlinear mapping of the unfolded samples of the data. To surmount this problem, we set a Mercer kernel *k(.,.)* as the inner product of two mapped samples (Wang, 2012).

$$k(x_i, x_j) = \langle \Phi(x_i), \Phi(x_j) \rangle = \Phi(x_i)\Phi(x_j)^T \quad (13.5)$$

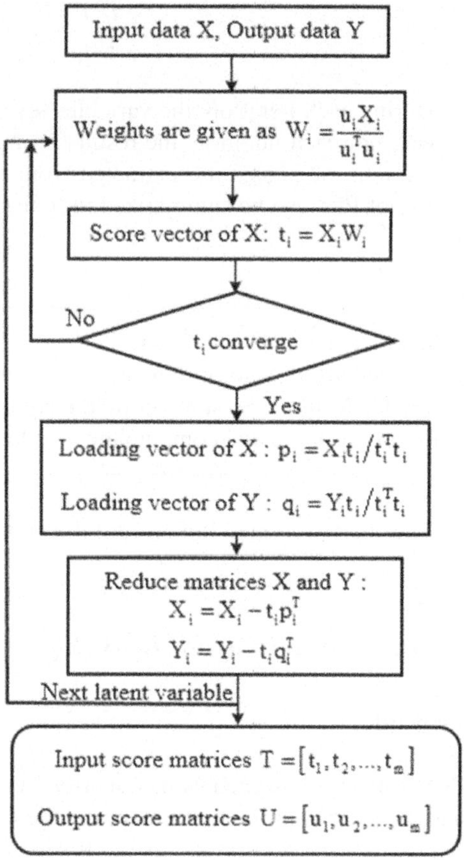

Fig. 13.2 Flowchart of PLS

where $\Phi(x_i) \in R^{1\times D}$, [i = 1; 2; …; N], and D is the dimension of feature space. The kernel function must fulfill Mercer's theorem. As stated in eqnarray 13.5, the Gram matrix $K \in R^{N\times N}$ is calculated as follows:

$$K = \Phi(X)\Phi(X)^T \tag{13.6}$$

with $\Phi(X) = [\varphi(x_1)^T, \varphi(x_2)^T, \ldots, \varphi(x_N)^T]$.

There are several kernel functions. The two commonly used kernels are the polynomial kernel (POL) and the RBF kernel (Table 13.1) (Wang, 2012). The kernel parameters are fixed using cross-validation approach (Neffati et al., 2019b).

Algorithm

The kernelized PLS approaches that take into consideration the optimization of data matrices are given by Lindgren et al. (Lindgren et al., 1993). First, mean centering

Table 13.1 Two common kernels with their formula and parameters

Name	Formula & Parameter
Polynomial kernel	$k(x, y) = (x^t y)^d$ $k(x, y) = (x^t y + c)^d$ d: order of polynomial kernel c: positive constant
Radial basis function kernel	$k(x, y) = \exp\left(\frac{-\|x - y\|^2}{2\sigma^2}\right)$ $k(x, y) = \exp(-\sigma \|x - y\|^2)$ σ: parameter of RBF kernel

in high-dimensional space is performed. Then, the Gram matrix K is calculated as follows:

$$K \leftarrow \left(I_n - \frac{1}{n} 1_n 1_n^T\right) K \left(I_n - \frac{1}{n} 1_n 1_n^T\right) \tag{13.7}$$

with 1_n is a vector of ones with the length of N and I_n an identity matrix with N dimensions. At this stage, we consider an adjusted form of the PLS algorithm (Zhang and Ma, 2011). Instead of scaling the weight vectors W and C, score vectors T and U are scaled to unit norm (Zhou et al., 2010). Furthermore, we assure deflation phase based on the rank-one reduction of the K and Y matrices using a new extracted score vector T. K and Y are calculated as follows:

$$K \leftarrow \left(I_n - tt^T\right) K \left(I_n - tt^T\right) = K - tt^T - Ktt^T + tt^T Ktt^T \tag{13.8}$$

$$Y \leftarrow Y - tt^T Y \tag{13.9}$$

with I_n an identity matrix with N dimensions.

Once the kernel is computed, we calculate the prediction outputs on training and testing samples. The prediction outputs that corresponds on training samples are written as:

$$\hat{Y} = KU \left(T^T KU\right)^{-1} T^T Y \tag{13.10}$$

The prediction outputs that corresponds on testing samples are written as:

$$\hat{Y}_t = K_t U \left(T^T KU\right)^{-1} T^T Y \tag{13.11}$$

where K_t denotes the kernel matrix of the test samples. The standard phases of KPLS can be summarized in ten steps presented in Algorithm 2 and Fig. 13.3.

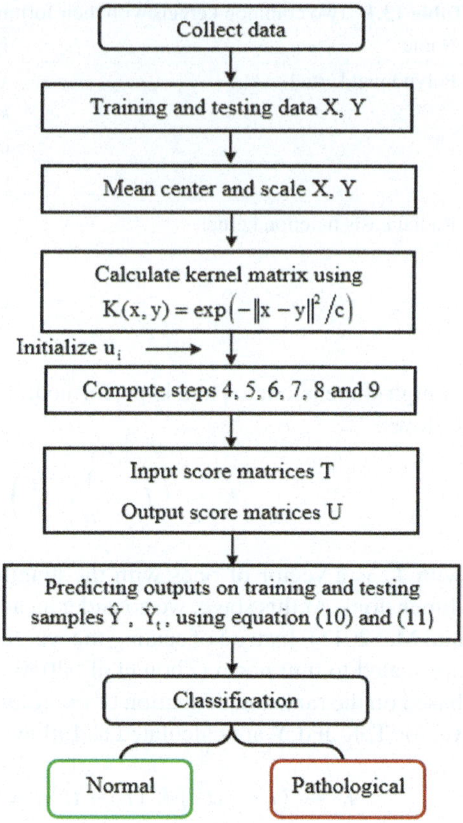

Fig. 13.3 Flowchart of KPLS

Algorithm 2:

1. Compute the kernel matrix and then center.
2. Set $i = 1$, $K_1 = K$, $Y_1 = Y$.
3. Randomly initialize u_i as any column of Y_i.
4. $t_i = K_i^T u_i$, $t_i = \dfrac{t_i}{\| t_i \|}$.
5. $c_i = Y_i^T t_i$.
6. $u_i = Y_i c_i$, $c_i = \dfrac{c_i}{\| c_i \|}$.
7. If t_i converges go to (7) else return to (3).
8. Deflate K and Y.
9. Repeat (3) to (6) to extract further latent variables.
10. Obtain the cumulative matrices T and U.

13.4 Proposed Downsized KPLS Method (DPLS)

In this work, we propose a Downsized KPLS (DKPLS) model. In this approach, a downsized number of observations among the measurement variables of the information matrix is selected. The retained observations can be used as a new data matrix.

Principle

The main aim of the suggested work is to reduce computation time and to reduce complexity of the MRI by selecting only the relevant features. In this approach, a downsized number of information is selected from the feature matrix. The dimension of the resulted matrix is equal to the number of retained LCs. As a result, in this work, each image is represented by a reduced KPLS matrix that satisfies two conditions at once. From one hand, the matrix is chosen to be the most reduced KPLS matrix, to decrease the execution time, and from another hand, It is chosen to be the richest one in information to improve the classification results. The selected matrix is called DKPLS matrix. This latter is the one that will affect directly the results of the DKPLS-ANN algorithm. The proposed DKPLS approach resides in approaching each LC $\{w_i\}_{i=1...P}$ by a modified input data $\phi(x^{(i)}_{Latent}) \in \phi\{x^j\}_{j=1...M}$ that have the highest projections in w_i direction (Taouali et al., 2015).

The projection of $\phi(x^{(i)}_{Latent})$ can be written as in eqnarray 13.12 with L is the number of the LC:

$$\phi(x^{(i)}_{Latent}) = \alpha_i \times k_i(x), i = 1 \ldots L \tag{13.12}$$

Then, all the vectors of the modified input data $\phi\{x^j\}_{j=1...M}$ are projected on the LC w_i and only $x^{(i)}_{Latent} \in \phi\{x^j\}_{j=1...M}$ that satisfy eqnarray 13.13 are retained.

$$\begin{cases} \phi(x^i_{Latent})_i = \max_{j\in[1,M]} \phi(x^j)_i \\ \phi(x^i_{Latent})_{i\neq j} < \zeta \end{cases} \tag{13.13}$$

where ζ is a given threshold.

When the downsized dataset is fixed, the downsized matrix is determined as follows:

$$X_r = [x^1_{Latent}; x^2_{Latent}; \ldots; x^L_{Latent}]^T \tag{13.14}$$

Moreover, a downsized kernel matrix K_r associated to a kernel function k is defined as indicated in the following eqnarray:

$$K_r = \begin{bmatrix} k(x_1, x_1) & \ldots & k(x_1, x_L) \\ \vdots & \ddots & \vdots \\ k(x_L, x_1) & \ldots & k(x_L, x_L) \end{bmatrix} \in R^{L\times L} \tag{13.15}$$

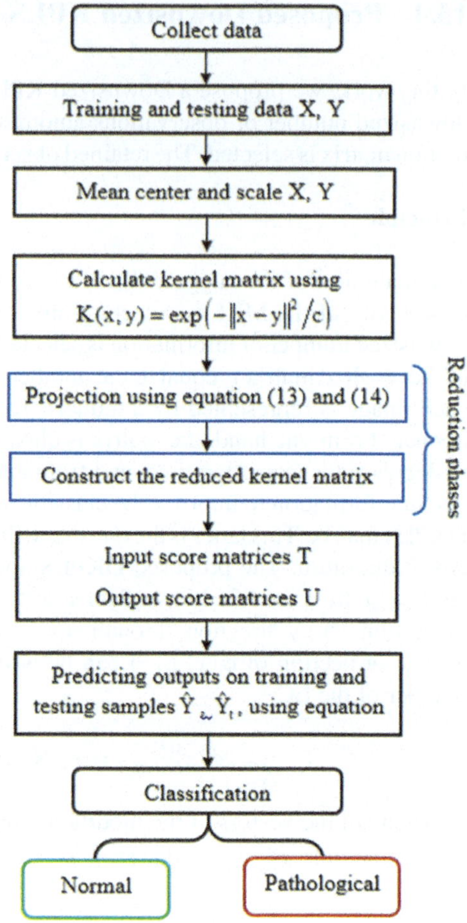

Fig. 13.4 Flowchart of DKPLS

Algorithm

The main steps of the proposed DKPLS algorithm are presented in Algorithm 3 and the associated organigram is presented by Fig. 13.4.

Algorithm 3:

1. Collect and an initial standardized blocks of data $\{x_i\}_{i=1..N}$ to be trained and scale them.
2. Compute and scale the kernel matrix K.
3. Project $\{x_i\}_{i=1..N}$ on the LC $\{w_i\}$ and select $x_{latent}^{(i)}$ that that satisfy eqnarray 13.13
4. Construct the downsized kernel matrix $K_r \in R^{L \times L}$ as in eqnarray 13.15
5. Evaluate the DKPLS model.
6. Predict with ANN.

13.5 Optimization with Multi-objective Optimization Algorithm

13.5.1 Principle

MOO, also known as multi-objective programming, multi-criteria optimization or Pareto optimization, deals with optimization problems that include multiple parameters or functions to be optimized synchronously. MOO issues happen to be faced in different fields, like engineering, logistics and economics, when ideal setting of parameters need to be made with two or more of these parameters, which leads to conflicting objectives (Mirjalili et al., 2017). For instance, to develop a new component, it might be needed to minimize the weight and at the same time to maximize strength. Another example, to choose a portfolio, it might be needed to maximize the expected return should be maximized and the risk should be simultaneously minimized (Mirjalili et al., 2017; Riquelme et al., 2015; Donoso and Fabregat, 2016).

Usually there is no unique solution considered as the optimal solution for all objectives. Alternatively, there is a set of optimal solutions or Pareto optimal solutions (Donoso and Fabregat, 2016). If no one of the objective functions is ameliorated without lowering one or more of the other objective values, then this solution is named non-dominated or Pareto optimal. Without supplementary information about subjective preferences, all Pareto optimal solutions are seen as identically acceptable (Riquelme et al., 2015). In the following section, we focus on selecting the optimal parameter of the RBF kernel used in the DKPLS method only.

13.5.2 Selection of Kernel Parameter with Multi-Objective Optimization Algorithm

To optimize DKPLS, we have to estimate the optimized parameter of the method, which is in this case the scaling factor. We used MOO to estimate the optimal value of $\widehat{\sigma}$. This method was built upon two objective functions:

* Reduced size (R)
* Classification Error Rate (CER)

Let σ be the optimization vector or the decision vector as $\sigma = OV$. In order to compute $\widehat{\sigma}$, which is the optimal parameter of DKPLS, the MOO algorithm was tested using two objective functions (R and CER). The non-dominated optimization trick was to find the optimal solutions through a minimization trick. The MOO mainly tries to find the vector of decision variables $\widehat{D_E}$ that minimizes the objective functions R and CER at the same time.

A non-dominated solution $D_{i,E}$ is turned out to be dominate if for the kth objective function f_k is as follows:

$$f_k(D_{i,E}) \leq f_k(D_{j,E}) \forall k \in [1, N] \tag{13.16}$$

The non-dominated solutions D_{ND} from the set of the solutions D_E are not dominated by any member of the set D_E. Thus, this technique allows determining the optimal $\widehat{D_E}$ that minimizes R and CER as follows:

$$\widehat{D_E} = \arg[\min(R(D_E)_{OV \in R^N}, CER(D_E))] \tag{13.17}$$

Once the optimal vector $\widehat{D_E}$ with the optimal values of R and CER are determined, then $\widehat{\sigma}$ is selected. Optimized DKPLS (ODKPLS) with optimized $\sigma = OV$ is then calculated. The ODKPLS algorithm is presented in Algorithm 4.

Algorithm 4:

---

1. Set $\sigma = 1, ..., 50$
2. Calculate the DKPLS model for all σ
3. Determine the reduced size R and CER for all σ
4. Generate and plot a Pareto front D_E
5. Extract the non-dominated solution D_{ND}
6. Determine the corresponding $\widehat{\sigma}$
7. Calculate the optimized DKPLS model

13.6 Classification Using Neural Networks

Classification is the last phase of pattern recognition, during which a class is assigned to a given test sample on the basis of the knowledge gained by the classifier during the training phase (Hau, 2015). Therefore, the main task of classification is assigning an input pattern represented by a vector to one of many pre-specified classes (Hau, 2015). In this work, the ANN (Neffati et al., 2019c) classifier was utilized for the classification of brain MRI into healthy or diseased subjects. ANN is a flexible system that learns relationships through recurrent presentation of information (Fathabadi, 2016).

In pattern recognition, the goal is to assign the input pattern to one of several classes or categories. ANNs produce excellent features; like fault tolerance, noise immunity, capability and generalization. Thus, decisions made by ANNs cannot be affected by any variation in the system conditions since they are built upon relaying algorithms (Kahouli et al., 2019).

In ANN, two layers of neurons interact by means of a weight connection network. The weighted connections utilized in used networks of our work is the feedforward neural networks. This latter is composed of:

* an input layer
* one or more hidden layers
* an output layer.

The neuron is modeled as follows:

$$y = f\left(\sum_{i=1}^{n} w_i x_i - b\right) \quad (13.18)$$

with:

* x_i / $i \in [1, n]$ are the inputs for the neurons (of the layers).
* w_i / $i \in [1, n]$ are the corresponding weights.
* f is a nonlinear function
* b is the bias of the neuron

In this work we use a Multi Layer Perceptron (MLP); three layer feedforward neural networks in Matlab environment. The classification of the NN starts with the sum of multiplication of weights and inputs, plus bias at the neuron. Only output elements fire if the sum is positive. Otherwise, they do not fire. ANN is an adaptive system; it adapts itself as it changes the weights of the system during the operation (updated during the on-line operations) (Fathabadi, 2016; Drira and Derbel, 2011).

13.7 Experiments

In this section, we present the results of the proposed method and the datasets of the used AD MR images. We start by introducing the utilized dataset and then we introduce the results according to the used workflow. Simulation was carried out using MATLAB 2016a, on a personal computer with a 3.10 GHz Core-i5 processor and 6 GB RAM, running under a Windows 10 operating system. The performances of the proposed scheme are compared with other algorithms in terms of accuracy, specificity and sensitivity. These parameters are defined as follows:

* **Accuracy**: The total number of images that are correctly classified:

$$Accuracy = \frac{TP + TN}{TP + TN + FP + FN} \quad (13.19)$$

* **Sensitivity**: The number of abnormal images correctly classified out of the total number of pathological images:

$$Sensitivity = \frac{TP}{TP + FN} \quad (13.20)$$

* **Specificity**: The number of normal images correctly classified out of the total number of normal images:

$$Specificity = \frac{TN}{TN + FP} \tag{13.21}$$

where *TP* is the true positive (correctly classified positive images), *TN* is the true negative (correctly classified negative images), *FN* is the false negative (incorrectly classified positive images) and *FP* is the false positive (incorrectly classified negative images).

13.7.1 Experiments on ADNI Dataset

The MRIs utilized in this chapter were downloaded from ADNI dataset. The dataset contains subjects having ages ranging from 55 to 90 years, having a companion capable of providing an independent assessment of functioning abilities. In this research work, we focused only on the subjects, which have structural MRI data available. We randomly selected 200 subjects including 100 case suffering from AD and 100 healthy subjects. More details about the dataset are presented in Table 13.2.

The good results of the proposed DKPLS-ANN approach tested on ADNI dataset are presented in Tables 13.3 and 13.4.

The classification results given in Table 13.4 report that the accuracy, sensitivity, and specificity of the proposed DKPLS-ANN approach are respectively 91.00, 90.50, and 84.5%. We notice that the proposed DKPLS-based classifier provides better results compared to the classical classifiers based on kernel methods such as KPLS, KPCA, PLS and PCA. From the results, it is obvious that the suggested approach presents high classification accuracy when tested on a ADNI dataset.

Table 13.2 Baseline characteristics of MRI subjects

Characteristics	AD	No dementia
No. of subjects	100	100
Male	64	49
Female	36	51
Average education (years)	12.3	12.5

Table 13.3 Confusion matrix of ANN-DKPLS method

DKPLS-ANN	No dementia (%)	AD (%)	Total (%)
No dementia	88	12	
AD	6	94	
Total			91

Table 13.4 Classification results of ANN-DKPLS method

Method	Accuracy (%)	Specificity (%)	Sensitivity (%)
DKPCA + ANN	91.00	90.50	84.50
KPCA + ANN	88.20	90.00	86.40
PCA + ANN	86.40	87.20	85.6

This work suggests that the proposed kernel method entitled DKPLS is extremely efficient in increasing the accuracy of classifying MRI in general and in detecting AD in particular.

13.7.2 Experiments on OASIS Dataset

This section we present the results of the experiments conducted on the MRIs taken from the "Open Access Series of Imaging Studies (OASIS)" dataset (OASIS, 2019). OASIS was made available by Dr. Randy Buckner at the Howard Hughes Medical Institute (HHMI) at Harvard University, the Neuroinformatics Research Group (NRG) at Washington University School of Medicine, and the Biomedical Informatics Research Network (BIRN). OASIS dataset consists of 416 subjects aged 18 to 96. For each subject, 3 or 4 individual T1-weighted MRI scans obtained in single scan sessions are included. The subjects are all right-handed and include both men and women. 100 of the included subjects over the age of 60 have been clinically diagnosed with very mild to moderate Alzheimer's disease (AD). The images were divided using the CDR scaling as presented in Table 13.5. The OASIS database does not contain any images for severe dementia (CDR = 3).

We have used in this work samples of 198 individuals, 98 individual subjects presented Non-Cognitive Impairment (NCA) and 100 subjects were clinically diagnosed with AD. All the used subjects are over the age of 60. Some images from the OASIS dataset are illustrated in Fig. 13.5. The characteristics of the utilized set of MRIs are presented in Table 13.6 and the general characteristics of OASIS database are given in Table 13.7.

Table 13.5 CDR scaling

CDR scale	Cognitive impairment (CI)
0	No dementia
0.5	Very mild dementia
1	Mild dementia
2	Moderate dementia
3	Severe dementia

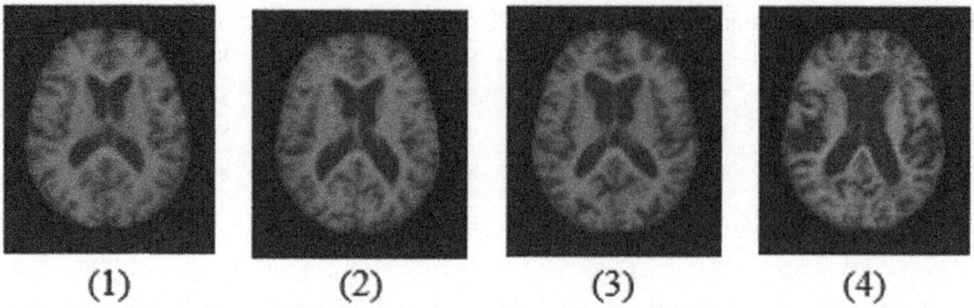

Fig. 13.5 Magnetic resonance images for (1) a patient with no dementia, (2) a patient with very mild AD, (3) a patient with mild AD (4) a patient with moderate AD

Table 13.6 General database characteristics

Group	N	Gender (F/M)	Age	MMSE
NCA	98	72/26	75.9 ± 0.9 (60–94)	29.0 ± 0.1 (25–30)
AD	100	59/41	78.5 ± 2.0 (62–96)	20.7 ± 0.7 (14–30)

Table 13.7 Specificity of the OASIS database

Age group	Total	Without dementia			With dementia		
		n	Male	Female	*n*	Male	Female
18–20	19	19	10	9	0	0	0
20s	119	119	51	68	0	0	0
30s	16	16	11	5	0	0	0
40s	31	31	10	21	0	0	0
50s	33	33	11	22	0	0	0
60s	40	25	7	18	15	6	9
70s	83	35	10	25	48	20	28
80s	62	30	8	22	32	13	19
90s	13	8	1	7	5	2	3
Total	416	316	119	197	100	41	59

To determine the DKPLS model, the RBF kernel, whose expression is shown in Table 13.1, was used. The optimal scaling factor σ for the used kernel was determined using the MOO algorithm as explained in the previous sections. To show the performance of the DKPLS approach, the ANN-classifier was associated to DKLPS in order to classify MRIs. Ten fold cross-validation (10-CV) approach was used in our experiments. During our tests, four separated classes were used for group partition. We used one-against-all method while realizing these experiments and we defined *no dementia* as the reference group. The confusion matrix, obtained during the first

Table 13.8 Confusion matrix of the DKPLS method

DKPLS	No dementia	Very mild AD	Mild AD	Moderate AD
No dementia	96.1%	3.9	0	0
Very mild AD	7.8%	84.7	7.5%	0
Mild AD	0	5.2%	84.9	9.9%
Moderate AD	0	0	2%	98%

Table 13.9 Comparison of the proposed work with other algorithms based on kernel methods on the OASIS database

Method	Sensitivity (%)	Specificity (%)	Accuracy (%)
PCA + ANN	81.2	92.0	87.1
PLS + ANN	87.9	82.6	85.3
KPCA + ANN	86.4	89.9	87.2
KPLS + ANN	88.4	91.3	89.9
DKPLS + ANN	90.5	95.2	90.3

run of CV, is presented in Table 13.8. The findings in Table 13.9 are obtained after applying 10 CV. The results report that the accuracy, sensitivity and specificity of the proposed method are 90.3, 90.5, and 95.2% for DKPLS-ANN.

We conclude from the results in Tables 13.8 and 13.9 that the proposed kernel method entitled DKPLS-ANN present better results for AD classification compared to the other traditional kernel-based approaches. In this section, we presented the results of the DKPLS-ANN classifier applied to OASIS dataset. Results were found to be excellent compared to other well-known kernel based algorithms in terms of classification accuracy.

13.8 Conclusion and Future Work

In this work, we have proposed a new method to select relevant features from MRIs to be used in CAD systems. Firstly, we used the MOO algorithm to select the optimal kernel parameter, in order to build the optimal classifier for MRI classification. Once the parameter is selected, we build the DKPLS model for feature selection. The designed method was coupled with ANN. The experimental results of diversified MRIs from different datasets have shown that the proposed DKPLS-ANN classifier present the best accuracy rate.

In future work, we will try to improve the proposed model by focusing on detecting earlier AD stages (mild cognitive impairment and preclinical AD), and by trying to generalize the model by including more imaging modalities.

Acknowledgements In this study MRIs from OASIS dataset were used. The OASIS project is supported in part by grants P50 AG05681, P01 AG03991, R01 AG021910, P20 MH071616, and U24 RR021382.

References

Baffi, G., Martin, E., & Morris, A. (1999). Non-linear projection to latent structures revisited: The quadratic pls algorithm. *Computers & Chemical Engineering, 23*(3), 395–411.

Cao, J., & Lin, Z. (2015). Extreme learning machines on high dimensional and large data applications: A survey. *Mathematical Problems in Engineering*

Cheng, J.-Z., Chou, Y.-H., Huang, C.-S., Chang, Y.-C., Tiu, C.-M., Chen, K.-W., & Chen, C.-M. (2010). Computer-aided us diagnosis of breast lesions by using cell-based contour grouping. *Radiology, 255*(3), 746–754.

Doi, K. (2007). Computer-aided diagnosis in medical imaging: Historical review, current status and future potential. *Computerized Medical Imaging and Graphics, 31*(4–5), 198–211.

Donoso, Y., & Fabregat, R. (2016). Multi-objective optimization in computer networks using metaheuristics. *Auerbach* Publications.

Drira, A., & Derbel, N. (2011). Classification of rotor fault in induction machine using artificial neural networks. In *Eighth international multi-conference on systems, signals & devices* (pp. 1–6). IEEE.

El-Dahshan, E.-S.A., Hosny, T., & Salem, A.-B.M. (2010). Hybrid intelligent techniques for MRI brain images classification. *Digital Signal Processing, 20*(2), 433–441.

El-Dahshan, E.-S.A., Mohsen, H. M., Revett, K., & Salem, A.-B.M. (2014). Computer-aided diagnosis of human brain tumor through MRI: A survey and a new algorithm. *Expert Systems with Applications, 41*(11), 5526–5545.

Erickson, B. J., & Bartholmai, B. (2002). Computer-aided detection and diagnosis at the start of the third millennium. *Journal of Digital Imaging, 15*(2), 59–68.

Fathabadi, H. (2016). Novel filter based ANN approach for short-circuit faults detection, classification and location in power transmission lines. *International Journal of Electrical Power & Energy Systems, 74*, 374–383.

Hau, C. C. (2015). *Handbook of pattern recognition and computer vision*. World Scientific.

Kahouli, O., Alshammari, B., & Dhouib, B. (2019). Application of ann and anfis techniques for pss tuning in a multimachine power system. In *2019 16th international multi-conference on systems, signals & devices (SSD)* (pp. 435–440). IEEE.

Khachaturian, Z. S. (1985). Diagnosis of alzheimer's disease. *Archives of Neurology, 42*(11), 1097–1105.

Kostic, A. D., Howitt, M. R., & Garrett, W. S. (2013). Exploring host-microbiota interactions in animal models and humans. *Genes & Development, 27*(7), 701–718.

Lindgren, F., Geladi, P., & Wold, S. (1993). The kernel algorithm for PLS. *Journal of Chemometrics, 7*(1), 45–59.

McCarville, M. B., Lederman, H. M., Santana, V. M., Daw, N. C., Shochat, S. J., Li, C.-S., & Kaufman, R. A. (2006). Distinguishing benign from malignant pulmonary nodules with helical chest CT in children with malignant solid tumors. *Radiology, 239*(2), 514–520.

Mirjalili, S., Jangir, P., & Saremi, S. (2017). Multi-objective ant lion optimizer: A multi-objective optimization algorithm for solving engineering problems. *Applied Intelligence, 46*(1), 79–95.

Motai, Y., Siddique, N. A., & Yoshida, H. (2017). Heterogeneous data analysis: Online learning for medical-image-based diagnosis. *Pattern Recognition, 63*, 612–624.

Neffati, S., Abdellafou, K. B., Taouali, O., & Bouzrara, K. (2019a). A new bio-cad system based on the optimized KPCA for relevant feature selection. *The International Journal of Advanced Manufacturing Technology, 102*(1–4), 1023–1034.

Neffati, S., Ben Abdellafou, K., Taouali, O., & Bouzrara, K. (2019b). Enhanced SVM-KPCA method for brain mr image classification. *The Computer Journal*

Neffati, S., Taouali, O., & Bouzrara, K. (2019c). New kernel method for MRI classification. In *2019 16th international multi-conference on systems, signals & devices (SSD)* (pp. 1–6). IEEE.

OASIS. (2019). Retrieved September 18, 2019, from http://www.oasis-brains.org/.

Qin, S. J. (2012). Survey on data-driven industrial process monitoring and diagnosis. *Annual Reviews in Control, 36*(2), 220–234.

Riquelme, N., Von Lücken, C., & Baran, B. (2015). Performance metrics in multi-objective optimization. In *2015 Latin American computing conference (CLEI)* (pp. 1–11). IEEE.

Rosipal, R., & Trejo, L. J. (2001). Kernel partial least squares regression in reproducing kernel hilbert space. *Journal of Machine Learning Research*, *2*(Dec):97–123.

Talukdar, U., Hazarika, S. M., & Gan, J. Q. (2018). A kernel partial least square based feature selection method. *Pattern Recognition, 83*, 91–106.

Taouali, O., Elaissi, I., & Messaoud, H. (2015). Dimensionality reduction of RKHS model parameters. *ISA Transactions, 57*, 205–210.

Wang, Q. (2012). Kernel principal component analysis and its applications in face recognition and active shape models. arXiv:1207.3538.

Zhang, Y., & Ma, C. (2011). Fault diagnosis of nonlinear processes using multiscale KPCA and multiscale KPLS. *Chemical Engineering Science, 66*(1), 64–72.

Zhou, D., Li, G., & Qin, S. J. (2010). Total projection to latent structures for process monitoring. *AIChE Journal, 56*(1), 168–178.

The manufacturer's authorised representative in the EU is Springer Nature Customer Service Centre GmbH, Europaplatz 3, 69115 Heidelberg, Germany. If you have any concerns regarding our products, please contact ProductSafety@springernature.com

Printed and bound by CPI Group (UK) Ltd, Croydon, CR0 4YY
30/09/2026
02225324-0010